Statistics Made Easy

WILLIAM W. MODUGU, Ph.D

Department of Forestry and Wildlife,

Faculty of Agriculture,

University of Benin, Benin City, Nigeria

i

ISBN: 1981799788

ISBN-13: 978-1981799787

DEDICATION

I owe my darling, my one and only wife of my youth and mother of my seven children, my best friend and confidant, and God's precious gift to me, Margaret Oyemhomhe Modugu, my heart-felt appreciation and profound gratitude for her unwavering love, loyalty, support and tender loving care (TLC) through every trial of life for more than four decades. My darling, I am most grateful.

I also acknowledge the "contribution" of my indefatigable grandchildren, Jesse, Gabriella, Emmanuella, and Josiah, who scribbled unreadable things on the manuscript, while in preparation.

I wish to sincerely thank and appreciate Professor U.J. Ikhatua, then Dean of the Faculty of Agriculture, University of Benin, Benin City, who insisted that I should teach "Statistics and Field Experimentation" (AGR311), to adequately lay a solid foundation and prepare the students for higher courses in statistics.

I also wish to deeply appreciate and express my profound gratitude to Professor I.A. Ogboghodo, who became the Dean, Faculty of Agriculture, University of Benin, Benin City, and further insisted that I continue to teach AGR311, until my retirement from the University of Benin, Benin City.

My profound thanks also goes to my dear friend, Dr. Ekeoba Matthew Isikhuemen, one of the foremost Forest Ecologists of our time, for his unfailing assistance and consistent encouragement that brought the completion of this Text to its final conclusion. I'm most grateful to you all.

I owe Dr. S.A. Ogedengbe, Department of Crop Science, Faculty of Agriculture, University of Benin, Benin City, copious gratitude for his persistent encouragement, even when I felt like giving up on the completion of this Text, and for reviewing and correcting portions of the manuscript.

I acknowledge and profoundly appreciate the immense contribution of my first son and Computer Guru, Engineer Peter Ikerebe Modugu, who assisted me in the many intractable computer problems, both hardware and software, which threatened to abruptly end the compilation of the Text. Without his dogged efforts, the work would have been abandoned altogether.

I also deeply appreciate the painstaking work of Engr. A. A. Muhammed who typeset and edited all the statistical formulae and equations. His untiring efforts certainly brought the manuscript to its final completion.

The contribution of Dr. O. Ojogho, Department of Agricultural Economics and Extension, for correcting many of the typographical and statistical errors in the entire manuscript, is acknowledged and appreciated.

Finally, I owe lots of appreciation and many thanks to all my ex-students, both undergraduate and post-graduate, whose active participation in class over a period of more than fifteen years, helped me to fine-tune the statistical methodologies presented in this textbook.

I say to God be all the glory in Jesus name.

DR. WILLIAM WILLOWS MODUGU, Ph.D.

PREFACE

The idea to write this book **"Statistics Made Easy"** birthed from my experience in teaching both undergraduate and post-graduate courses in statistics and experimental designs for fifteen years at the Department of Forestry and Wildlife, Faculty of Agriculture, University of Benin, Benin City.

I discovered that a large number of students have a phobia for any course that deals with figures, especially statistics. Therefore, this text was designed to accommodate all levels of students-from the uninitiated to the advanced. The statistical procedures have been simplified to such a level that any student who can add, divide, multiply, subtract, square and take square-roots of numbers, will successfully take any course in statistics. Most of the examples in the text are from agricultural experiments in forestry, animal and crop science, fisheries, as well as pharmaceutical sciences, sociology, etc.

The text is designed for application of statistical principles and methodologies in experiments rather than the gory mathematical proofs required for engineering and mathematics. The omission of mathematical details sometimes necessitates more lengthy verbal justification of the statistical ideas for the understanding of students that are not mathematically inclined, or endowed.

The general format used in this text is to discuss the statistical methodology followed by practical numerical worked examples to acquaint the student with the step-by-step approach to solving the statistical problems. By working through many examples of analyses, it is possible to fully understand how each analysis provides the estimate of the accuracy of the results of each experiment and simultaneously apply the statistical methods.

Looking at the detailed contents of the book show that chapter 1 presents the basic statistical ideas in the organization and presentation of data, while chapters 2 to 6 discuss probability and probability distributions, as well as estimation procedures. Chapter 7 presents the determination of sample sizes required for experimentation under various conditions like whether the population variance is known or unknown, and the estimation of confidence intervals for sample means and proportions. Chapters 8 and 9 concern the testing of statistical hypotheses on one mean (or variance) and two means (or variances) respectively.

Chapters 10 and 11 examine Analysis of Variance (ANOVA) and various Experimental Designs from the Completely Randomized Design to the more complex Factorial Experiments, as well as, the various methods of multiple comparisons of treatment means.

Chapters 12,13 and 14 deal with the Problem of Missing data, Regression and Correlation Analysis, from Simple, Multiple, to No-Linear (Curvilinear) Models and Polynomial Models.

Finally, chapter 15 discusses Analysis of Covariance.

I have only listed a few of the references here, even though I consulted many more references for the text.

TABLE OF CONTENTS

DEDICATION	iii
PREFACE	iv
TABLE OF CONTENTS	v
1 INTRODUCTION	**1**
1.2 GRAPHICAL METHODS	2
1.3 SUMMARY	4
1.4 NUMERICAL DESCRIPTIVE METHODS	4
1.5 POPULATIONS AND SAMPLES	4
1.6 MEASURES OF CENTRAL TENDENCY	5
1.6.1 THE ARITHMETIC MEAN OR AVERAGE (x)	5
1.7 MEASURES OF VARIABILITY (OR DISPERSION)	11
2 PROBABILITY	**19**
2.1 SAMPLE SPACE	19
3 RANDOM VARIABLES AND PROBABILITY DISTRIBUTIONS	**23**
3.1 RANDOM VARIABLES	23
3.2 PROBABILITY DISTRIBUTIONS FOR DISCRETE RANDOM VARIABLES	23
3.3 SOME DISCRETE PROBABILITY DISTRIBUTIONS	31
3.4 THE MEAN AND VARIANCE OF A PROBABILITY DISTRIBUTION	46
4 CONTINUOUS RANDOM VARIABLES AND PROBABILITY DISTRIBUTIONS	**52**
4.1 THE NORMAL (OR GAUSSIAN) DISTRIBUTION	53
4.2 PROPERTIES OF THE NORMAL CURVE	54
4.4 HOW TO TRANSFORM RANDOM VARIABLES TO STANDARD NORMAL DISTRIBUTION	56
4.5 CONVERSIONS AND APPLICATIONS OF STANDARD NORMAL VARIABLES	56
4.6 NORMAL APPROXIMATION TO THE BINOMIAL	62
4.7 HOW TO USE THE STANDARD NORMAL DISTRIBUTION TABLE TO ESTIMATE PROBABILITIES	65
4.8 THINGS TO NOTE ABOUT THE STANDARD NORMAL TABLE	67
5 SAMPLING AND SAMPLING DISTRIBUTIONS	**72**
5.0 POPULATION	72
5.1 SAMPLES	72
5.2 SAMPLING DISTRIBUTIONS	72
5.3 SAMPLING DISTRIBUTIONS OF THE MEAN AND ITS VARIANCE	73
5.4 EXAMPLES OF SAMPLING WITHOUT REPLACEMENT	76
5.5 CHI-SQUARED DISTRIBUTION	80
5.6 F-DISTRIBUTION	83

6 ESTIMATION THEORY   **85**

6.1 INTRODUCTION   *85*

6.2 A POINT ESTIMATOR OF A POPULATION MEAN   *85*

6.3 A POINT ESTIMATOR OF A POPULATION PROPORTION   *86*

6.4 ESTIMATING A POPULATION VARIANCE   *88*

6.5 ESTIMATING THE DIFFERENCE BETWEEN TWO MEANS $(\bar{X}_1 - \bar{X}_2)$: A CASE WHERE σ_1^2 AND σ_2^2 ARE KNOWN   *88*

6.6 ESTIMATING THE DIFFERENCE BETWEEN TWO MEANS: $(\bar{X}_1 - \bar{X}_2)$A CASE WHERE σ_1^2 AND σ_2^2 ARE UNKNOWN BUT EQUAL   *90*

6.7 ESTIMATING THE DIFFERENCE BETWEEN TWO MEANS: $\bar{X}_1 - \bar{X}_2$ A CASE WHERE σ_1^2 AND σ_2^2 ARE UNKNOWN AND UNEQUAL   *91*

6.8 ESTIMATING THE DIFFERENCE BETWEEN TWO PROPORTIONS   *93*

6.9 ESTIMATING THE DIFFERENCE IN PAIRED OBSERVATIONS   *94*

7 DETERMINING THE NECESSARY SAMPLE SIZE   **97**

7.1 DETERMINING SAMPLE SIZE WHEN σ IS KNOWN, OR ESTIMATED FROM $N \geq 30$ *97*

7.2 DETERMINING SAMPLE SIZE WHEN σ IS UNKNOWN AND A PRELIMINARY SAMPLE OF $N < 30$ *98*

7.3 DETERMINING THE SAMPLE SIZE FOR ESTIMATING A PROPORTION   *99*

7.4 INTERVAL ESTIMATION: CONFIDENCE INTERVALS   *101*

8 TESTING STATISTICAL HYPOTHESES   **110**

8.1 THE TWO TYPES OF POSSIBLE DECISION ERRORS   *110*

8.2 PROCEDURE FOR TESTING HYPOTHESES   *112*

8.3 STATING THE CONCLUSION   *112*

8.4 "P-VALUE"   *113*

8.5 POWER OF THE TEST FOR H_0: $\mu = \mu_0$ *114*

8.6 ONE-SIDED (ONE-TAILED) TESTS OF HYPOTHESES   *115*

8.7 TWO-SIDED (TWO-TAILED) TESTS OF HYPOTHESES   *117*

8.8 THE POWER OF A STATISTICAL TEST   *118*

8.9 POWER AND OPERATING CHARACTERISTIC CURVES FOR ONE-TAILED AND TWO-TAILED TESTS   *122*

9 TESTS OF HYPOTHESES CONCERNING THE MEANS OF TWO POPULATIONS   **134**

9.1 TESTING THE EQUALITY OF MEANS OF TWO NORMAL POPULATIONS WHEN THE TWO VARIANCES, σ_1^2 AND σ_2^2 ARE KNOWN   *135*

9.2 TESTING THE EQUALITY OF MEANS WHEN THE VARIANCES, σ_1^2 AND σ_2^2 ARE UNKNOWN, BUT THE SAMPLE SIZES ARE LARGE $(N_1, N_2 \geq 30)$ *138*

9.3 TESTING THE EQUALITY OF MEANS WHEN SAMPLE SIZES ARE SMALL $(N_1; N_2 < 30)$ AND POPULATION VARIANCES, σ_1^2 AND $\sigma_{2'}^2$ ARE UNKNOWN, BUT EQUAL $(\sigma_1^2 = \sigma_2^2 = \sigma^2)$ *139*

9.4 TESTING THE EQUALITY OF MEANS WHEN SAMPLE SIZES ARE SMALL $(N_1, N_2 < 30)$ AND POPULATION VARIANCES ARE UNEQUAL $\left(\sigma_1^2 \neq \sigma_2^2\right)$ AND UNKNOWN   *143*

9.5 TESTING THE EQUALITY OF TWO PROPORTIONS   *146*

9.6 TESTING THE EQUALITY OF TWO VARIANCES $\left(\boldsymbol{\sigma_1^2 = \sigma_2^2}\right)$ *150*

9.7 ESTIMATION OF VARIANCES USING SEVERAL SAMPLES   *154*

9.8 ENUMERATION STATISTICS   *155*

10 THE ANALYSIS OF VARIANCE (ANOVA   **168**

    10.1 TERMINOLOGIES AND SYMBOLISM OF ANOVA   *168*

11 THE ONE-WAY (SINGLE-FACTOR) CLASSIFICATION: COMPLETELY RANDOMIZED DESIGN (CRD)   **176**

12 PROBLEM DATA   **271**

    12.1 MISSING DATA   *271*

    12.2 COMMON CAUSES OF MISSING DATA   *271*

    12.3 MISSING DATA FORMULA TECHNIQUE   *272*

    12.4. EXPECTED MEAN SQUARES.   *280*

13 REGRESSION ANALYSIS.   **287**

    13.1. SIMPLE LINEAR REGRESSION   *288*

    13.2. MEANING OF REGRESSION PARAMETERS   *291*

    13.3. ESTIMATING THE REGRESSION PARAMETERS   *291*

    13.4 .MULTIPLE LINEAR REGRESSION AND CORRELATION   *314*

    13.5. SIMPLE NON-LINEAR (CURVILINEAR) MODELS.   *320*

    13.6 .SIMPLE POLYNOMIAL REGRESSION MODEL   *337*

14 SIMPLE LINEAR CORRELATION ANALYSIS   **341**

    14.1. COMPUTATION OF THE PRODUCT-MOMENT CORRELATION COEFFICIENT.   *341*

    14.2 WHEN n > 50, MAKE USE OF THE Z – TRANSFORMATION.   *344*

    14.3.NON-PARAMETRIC METHOD (OR RANK CORRELATION), OR SPEARMAN CORRELATION.   *346*

15.ANALYSIS OF COVARIANCE   **348**

    15.1 DEVELOPMENT OF COVARIANCE MODEL   *348*

    15.2   GENERAL PROCEDURE FOR ANALYSIS OF LINEAR COVARIANCE   *350*

REFERENCES   **358**

INDEX   **360**

APPENDIX   **361**

       361

       363

1 INTRODUCTION

Statistics is the science, pure and applied, of creating, developing and applying techniques such that the uncertainty of inductive inferences may be evaluated. The term statistics pertains to a listing of facts, systematic methods of arranging and describing data and inferring generalities from specific observations (Dixon and Massey, 1969). Therefore, the laws of physical, biological, and social sciences have their proof in statistical facts.

There is a general theory of statistics, according to Dixon and Massey (1969), which is applicable to any field of study in which observations, or measurements are made. Statistical procedures now form an important part of all fields of science, medicine, economics, education, or agriculture, and procedures which have been developed for use in this field and have almost invariably found important applications in many other fields of study.

Sokal and Rohlf (1969) defined the application of statistical methods to the solution of biological problems "or, the scientific study of numerical data based on natural phenomenon" as Biometry, or Biological Statistics, or simply as Biometrics or Biostatistics.

Biostatistical or biometrical methods have been developed to help in the quantitative study of variation and in the analysis and interpretation of data from experiments which are subject to variation. In biology and agriculture, the amount of variation and the relative importance of the different causes of variation are often of interest in themselves.

In experiments, variation tends to obscure, or mask, the effects of different treatments, thus making comparisons difficult and can lead to mistakes in interpreting the results of experiments and mistaken judgments.

The concept of variation is fundamental to scientific experimentation. For example, tree seedlings planted the same day will not grow at the same rates: a number of pigs given exactly the same diet, in the same environment, will not increase their body weight at the same rate; similarly, the yield of maize, or okra, on UNIBEN plots will vary, no matter the care in selecting the seed. The amount of variation is generally much greater with biological material than it is with inanimate material.

Therefore, statistics, or biometrics, or biostatistics, is an area of science concerned with the extraction of information from numerical data and its use in making inferences, or decisions, about a population from which the data are obtained (Mendenhall, 1975).

1.1 ORGANIZATION AND PRESENTATION OF DATA

Statistics is the science of classifying and manipulating data in order to draw inferences. An important aspect of classifying data is the efficient and effective organization and presentation of data in order to derive meaning from numerical data (Byrkit, 1975).

The term data (singular=datum) refers to the set of observations, experimental units, values, elements, or objects, under consideration. Such an individual observation, value, or object is called a **"datum"**, or **"data point"**, or an experimental unit, or piece of data.

Example 1: Suppose the University of Benin (UNIBEN) farm raised 100 pullets, or chickens, of different colors:

8 speckled chickens

25 black   "

32 brown   "

35 white   "

Here, the chicken color is used to organize the 100 chickens into meaningful categories or groups to make the data easy to understand or interpret. One way of organizing data is known as the **"frequency distribution"**. Using the chicken color in the example above, one can organize the data as the color distribution of one hundred chickens raised in UNIBEN Farm:

Table 1: Color Distribution of 100 Chickens Raised in UNIBEN FARM.

Chicken Color	Frequency (Number of chickens) f_i	Relative Frequency (f_i/n) (Proportion)	Cumulative Frequency
Speckled	8	$8/100 = 0.08$	8
Black	25	$25/100 = 0.25$	33
Brown	32	$32/100 = 0.32$	65
White	35	$35/100 = 0.35$	100
Total	100	$100/100 = 1.0$	

We used the color of the chickens as an attribute to classify the 100 chickens. Color as an attribute cannot be measured, but can only be described to categorize the chickens. This type of data is called **"Attribute"**, or **"Categorical"** data.

The 100 chickens have been organized and presented in Table 1 as frequency, or as frequency distribution. The data would be more visually appealing if they are presented in pictorial or graphical form.

1.2 GRAPHICAL METHODS

Graphical methods of presenting a set of data are more visually appealing because they draw attention to the outstanding or important features of the distribution at a glance. There are different forms of graphical methods which are generally referred to as **Histograms**, which include:

i.   Bar graphs or Bar charts.
ii.   Relative frequency histogram.
iii. Line graph or frequency polygon.
iv.   Cumulative frequency polygon or Ogive.
v.   Pie Charts or Pictograms.

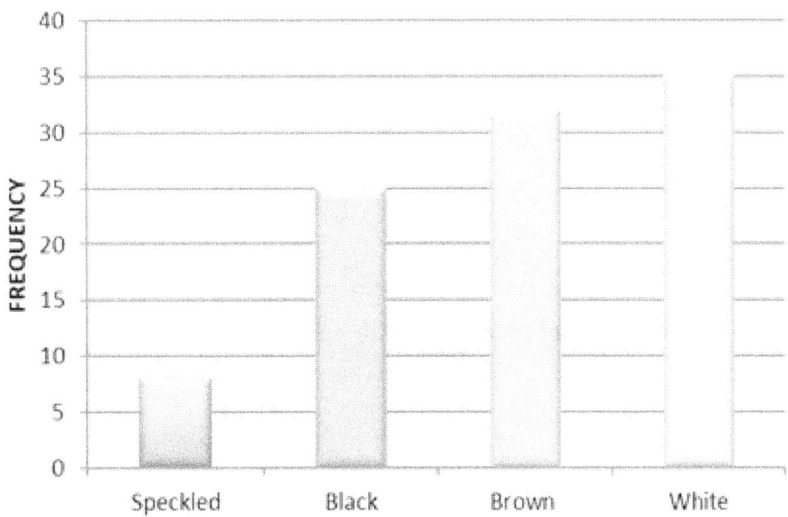

Figure. 1.0: Chicken Color Relative Frequency Histogram

2

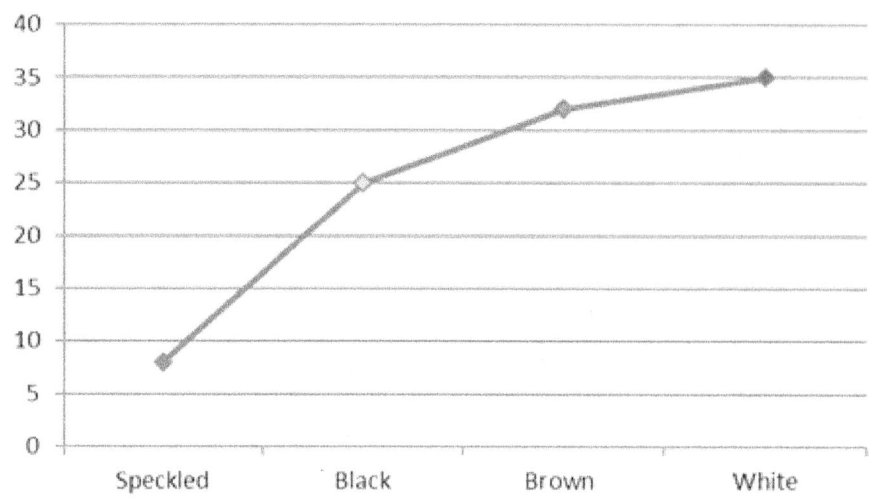

Figure 1.1: Chicken Color Relative Frequency Polygon (Line Graph)

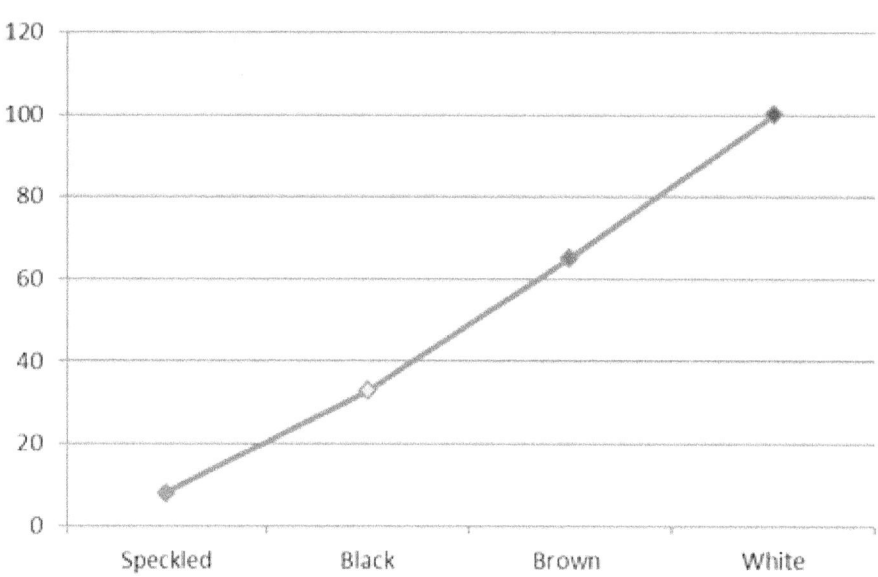

Figure 1.2: An Ogive or Cumulative Frequency Polygon of Chicken Color

The color distribution among the 100 chickens is also presented in a pie chart as shown below:

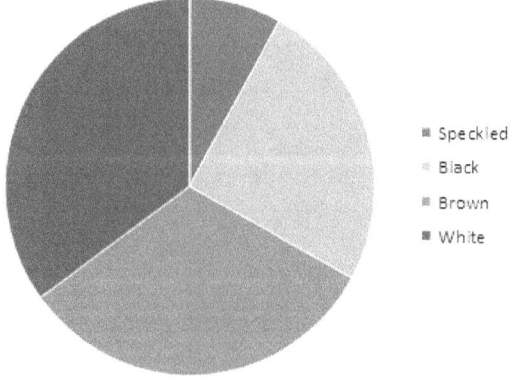

Figure 1.3: Pie Chart of Chicken Color

1.3 SUMMARY

For both large and small amounts of data to be meaningful and understandable, they must be organized and presented through descriptive statistics.

One of the most commonly used methods of organizing data is the use of frequency distributions ...viz: relative frequency, cumulative frequency, or relative cumulative frequency distributions.

To present data in a more visually appealing manner, bar charts, frequency polygons, histograms, ogives and pie charts are frequently used. But histograms are statistically more preferable.

1.4 NUMERICAL DESCRIPTIVE METHODS

As useful and visually appealing as graphical methods of organizing and presenting data are, they are still limited for data description and analysis. This leads us to the use of Numerical Descriptive Methods which can be used for statistical inference.

To illustrate the use of numerical descriptive methods, let us use the germination and height growth data of **Garcinia cola** (bitter kola) randomly selected from an experiment of 100 seeds planted in the Forest Nursery, University of Benin, Benin-City:

Table 1.1: Height (cm) Growth of 30 Seedlings of *Garcinia cola*

1.7	1.6	2.6	2.2	2.0
2.0	2.4	2.1	2.1	2.0
1.9	2.9	2.5	2.2	2.2
2.2	1.9	2.0	2.4	2.2
2.6	1.6	2.3	1.5	2.5
1.11	1.4	2.3	1.8	2.0

11.8	11.8	13.8	12.2	12.9

$\Sigma x_i = 62.5$: $\bar{x} = \dfrac{62.5}{30} = 2.083 = 2.1$

If each data point is designated as X_i, then the sum of all the data points is designated by the symbol:

$\sum Xi = 1.7+2.0+1.9+ \ldots +2.2+2.5+2.0 = 62.5$. The average of these values is designated as $\overline{X}$ which is :

$\sum Xi/n$, where n is the number of data points and i is a counter from 1 to n, the last data point.

Before using numerical descriptive methods, we need to define certain statistical terms.

1.5 POPULATIONS AND SAMPLES

A Population, or Universe, is the totality of all members or observations of a defined population, having some discernible characteristics. In this case, the 100 Garcinia cola seedlings can be treated as a population. Every defined population has its unique properties – that is, its mean (or average, designated as μ and its variance, designated as σ^2, are constant or fixed, but they are usually known. As long as we remain within a defined population, its mean and variance are constant, but unknown. These constant measurable population characteristics of mean, µ, and variance, σ^2 are called **parameters**.

A Sample is a subset, or a representative part, of a population from which the sample, used to make inferences or decisions about the population, is drawn. Because individuals are rarely uniform, the principle of **"randomness"** must be employed to obtain a representative sample - - it ensures that individual/personal biases, either known or unknown in nature, do not influence the selection of sample observations or experimental units.

The mean, or average, of a sample, designated as $\overline{X}$or $\overline{Y}$or $\overline{T}$or , $\overline{A}$tc.) and its variance, designated as S^2, change or vary from one sample to the other and are therefore called **"Statistics"**. The 30 seedlings in Table 1.1 constitute a random sample obtained from the population of 100 seedlings.

1.6 MEASURES OF CENTRAL TENDENCY

Frequently, it is desirable to describe a set of data, like the 30 seedlings of *Garcinia cola*, by using one number called the **"middle"** as a measure of the center of the distribution. The most useful numbers, or descriptive measures, are measures of **Central Tendency** or averages. Among such measures of central tendency are:-

    i.   Mean or Average.
    ii.   Median.
    iii.   Mode.

Table 1.2: Grouped Data of 30 Seedlings of Garcinia cola

Class Frequency (C)	Class Boundaries	Class Marks (Mid-Points)	Frequency (f_i)	Relative Frequency (f_i/n)
1	1.4 – 1.6	1.5	3	$3/30 = 0.10$
2	1.6 – 1.8	1.7	3	$3/30 = 0.10$
3	1.8 – 2.0	1.9	3	$3/30 = 0.10$
4	2.0 – 2.2	2.1	7	$7/30 = 0.23$
5	2.2 – 2.4	2.3	7	$7/30 = 0.23$
6	2.4 – 2.6	2.5	4	$4/30 = 0.13$
7	2.6 – 2.8	2.7	2	$2/30 = 0.07$
8	2.8 – 3.0	2.9	1	$1/30 = 0.03$
			n = 30	0.99 = 1

Let us use the data set in Table 1.1 above to illustrate how to estimate these measures of central tendency for grouped and ungrouped data.

1.6.1 The Arithmetic Mean or Average ($\overline{x}$)

The **"Arithmetic Mean"**, or Average, symbolized as $\overline{X}$ (or any letter of the alphabet with a dash (-) on top of it) is the best known measure of central tendency. If we designate each data point (or observation, or experimental unit) as Xi, the Arithmetic Mean of the 30 (n) Garcinia cola data set (Table 1.1) X1, X2, X3…Xn, is equal to the sum of the measurements divided by n as:

$$\overline{X} = \frac{\sum x_i}{n} = \frac{1.7+2.0+1.9+...+2.2+2.5+2.0}{30} = \frac{62.5}{30} = 2.083 = 2.1 \text{ is the sample mean.}$$

For grouped data in Table 1.2, the Arithmetic Mean ($\overline{X}$) is estimated as:

$$\overline{X} = \frac{\sum_{i=1}^{n} x_i f_i}{n} = \frac{1.5 \times 3 + 1.7 \times 3 + 1.9 \times 3 + 2.1 \times 7 + 2.3 \times 7 + 2.5 \times 4 + 2.7 \times 2 + 2.9 \times 1}{30}$$

i.e $\dfrac{\text{sum of class mark x Frequency}}{n} = \dfrac{63.7}{30} = 2.1$

Table 1.2a: Arithmetic Mean of Grouped Data

Class Mark (X_i)	Frequency (f_i)	$X_i \times f_i$
1.5	3	1.5 x 3
1.7	3	1.7 x 3
1.9	3	1.9 x3
2.1	7	2.1 x 7
2.3	7	2.5 x 4
2.5	4	2.3 x 7
2.7	2	2.7 x 5
2.9	1	2.9 x 1
	n = 30	$\sum X_i f_i = 63.7$

$$\overline{X} = \frac{\sum_{i=1}^{n} X_i f_I}{n} = \frac{63.7}{30} = 2.1$$

The Arithmetic Mean ($\overline{X}$), has certain appealing advantages and some disadvantages:

Advantages of Using Arithmetic Mean ($\overline{X}$):

1. Its computation is an algebraic procedure.
2. A Mean exists for every set of data.
3. It provides certain computational advantages in the calculation of some other statistical measures.
4. The arithmetic mean is an integral part of certain statistical analytical techniques.
5. Every value of the variable in the set of data influences the value of the arithmetic mean.

Disadvantages of Using Arithmetic Mean ($\overline{X}$):

1. The arithmetic mean can be overly influenced by extreme values of the variable.
2. It may be somewhat more difficult or laborious to calculate than other measures of central tendency.
3. It cannot be determined easily from open-ended frequency distributions.

1.6.2 The Geometric Mean (Gm)

The Geometric mean is the nth root of the product of n numbers and is calculated as:

$Gm = \sqrt[n]{X_1.X_2.X_3....X_n}$, which is generally used for determining average percentage change and ratios:

$Gm = $ Anti-log of $\dfrac{1}{n}\sum \log X_i$.

1.6.3 The Harmonic Mean ($\bar{X}_H$)

The Harmonic mean of a set of numbers, X_1, X_2, X_3, $---X_n$, is the reciprocal of the arithmetic mean of the reciprocals of the numbers:

$$X_H = \frac{n}{\frac{1}{x_1}+\frac{1}{x_2}+\frac{1}{x_3}...+} = \therefore \frac{n}{\sum_{i=1}\frac{1}{x_i}}$$

$$\frac{1}{H} = \frac{1}{n}\frac{1}{\sum X_i}$$

The harmonic mean is used in the calculation of relative values such as index numbers and in averaging ratios and rates. Generally, the values obtained for harmonic mean is smaller than geometric and arithmetic means.

1.6.4 The Median (MD)

The second average measure of central tendency is the Median which is the positional middle of the data that is arranged, either in increasing or decreasing order of magnitude. In an ordered array of the data, one-half of the values of the variable precede the median and one-half of the values follow it.

For example, in the ordered array containing an odd number (in decreasing order of magnitude) like: 10, 9, 8, 7, 6 – the middle number, the median, is 8 because the number of items in the data set is five, an odd number. In the ordered even number array: 10, 9, 8, 7, 6, 5, the median is the average of the two middle numbers:

$$\frac{8+7}{2} = \frac{15}{2} = 7.5$$

Advantages:

1. The median is not affected by extreme values of the variable, only by the position of values.
2. The median is also used when one desires to know the positional middle.

Disadvantages:

1. It cannot be used with many statistical analytical procedures.
2. Like the arithmetic mean, the median is sometimes an artificial number.
3. The median can be tedious to locate.

1.6.5 The Mode

The mode, in a set of data, is the observed value that occurs most frequently. For example, in a data set: 2, 4, 5, 7, 7, 9, 10, 12, — 7 occurs twice and is therefore the mode. But in a data set: 2, 4, 5, 7, 9, 10, 12 – there is no mode because all the values occur only once. If there are two or more values in a set of data and the most frequently occurring values are separated by other values, the data set is said to be **bi-modal**. For example, a data set as: 3, 6, 14, 20, 22, 28, 28, 32, 33, 36, 36, 37, 38, 39, 40, 41, 44, 47 – has two modes or bimodal:

Mode 1 = 28

Mode 2 = 36

If the data have been grouped, identification of the modal class is about the best thing to do (Cangelosi et al, 1976).

The mode is not influenced by extreme values, but rather by the frequency of occurrence. Furthermore, the mode can be located at one extreme of the data set. As a result, its use could result in misleading conclusions.

1.6.6 Skewness and the Relationship of the Mean, Median and Mode

According to Cangelosi, Taylor and Rice (1976), skewness refers to the shape of a frequency distribution and can be related to symmetry.

If a frequency distribution is bell-shaped, it is said to be symmetrical i.e. 50% of the values lie to the right of the mode, it is also a unimodal distribution. Thus, the mode, median, and mean are equal.

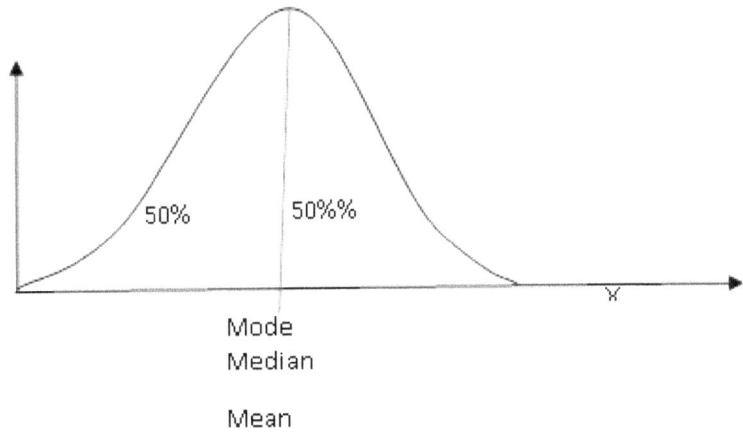

Figure 1.5: Symmetrical or Bell-shaped Frequency Distribution

If the distribution is skewed to the right as in Figure 1.6a, the value of the mode is less than the median which, in turn, is less than the mean, as shown below:

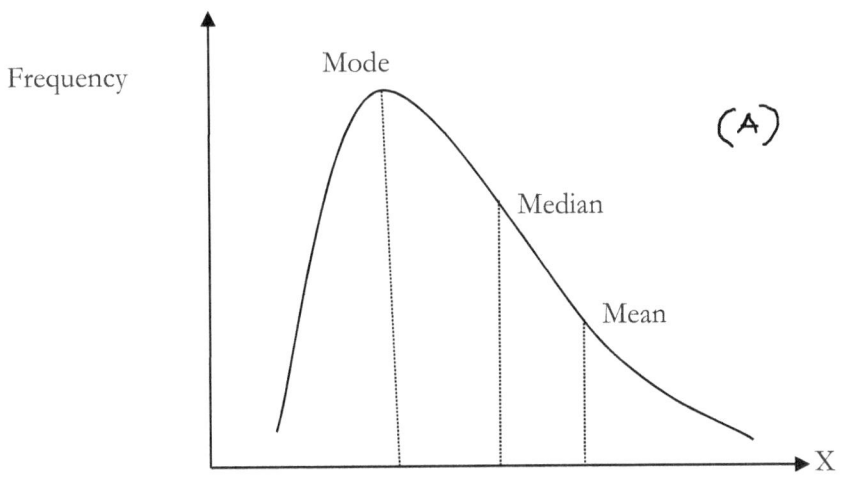

Figure 1.6a: Distribution Skewed right.

8

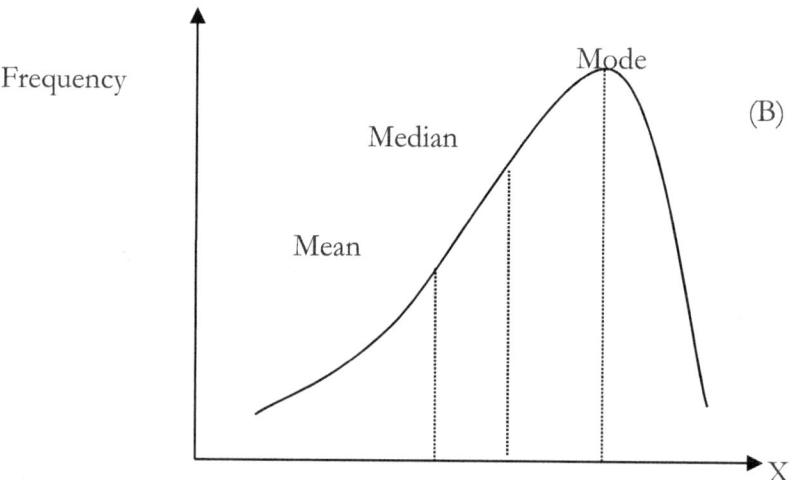

Figure 1.6b: Distribution Skewed Left

But if the distribution is skewed left, the value of the mode is greater than the median which, in turn, is greater than the mean. In skewed distributions, the median value lies between the values of the mean and the mode. The mean is always in the direction of the skewness, relative to the mode.

1.6.7 The Position (or Location) of the Median in an Arrayed Data Set (Median location) = $(\frac{1}{2})$ (n +1)

To find out which number is at the median, Donald R. Byrkit (1975) suggested that we simply take: Median position $(\frac{1}{2})$ (n +1),, where n is the total number of cases. For odd n, this will give a specific number, for even n, the number will not be specific. If n = 30, $(\frac{1}{2})$ (30 +1) = 15.5 i.e. the median will lie midway between the fifteenth and sixteenth values – when the data set is arrayed first, then just count from the bottom or top, until you arrive at the 15th and the 16th values

1.6.8 Estimating a Sample Mean ($\overline{X}$) from an Assumed Mean ($\hat{x}$) in Grouped Data

An **ASSUMED MEAN**, designated as $\hat{X}$ is usually an integer close to the true mean ($\overline{X}$) The sum of the

$$\overline{X} = \frac{\sum(Xi - \hat{x})}{n} + \hat{X}$$

mean difference between $\hat{X}$ and the individual values of X_i,, is divided by n, to obtain the true mean as follows:

To estimate the mean and determine the median from grouped data below:

Table 1.3a: Grouped Data (Byrkit, 1975)

Class Boundaries	Frequency
171 -175	4
166-170	8
161-165	14
156-160	22
151-155	27
146-150	19
141-145	17
136 -140	11
131 -135	3
Total	125

Note: The grouped data in Table 1.3a has nine classes. The middle class is class number 5 (151 -155), if we count from the top or bottom - - there are four classes above and four classes below class (151 – 155). This is the class where the mean is assumed to be at the class mark (153) of this class. The assumed mean $(\hat{X})$ is therefore 153. Using the Class Marks and an Assumed Mean $(\hat{X})$ of 153, which is close to the midpoint of the distribution we have:

Class Marks	Frequency (f_i)	$X_i - 153(X_i - \hat{X})$	$(X_i - 153) \times (f_i)$
173	4	20	80
168	8	15	120
163	14	10	140
158	22	5	110
153	27	0	0
148	19	-5	-95
143	17	-10	-170
138	11	-15	-165
133	3	20	-60
Total	125		-40

Table 1.3b: Estimating Sample Mean ($\overline{X}$) and Class Marks

$$\overline{X} = \frac{\Sigma(Xi-\hat{x})}{n} + \hat{X}$$

$$= \frac{-40}{125} + 153 = -0.32 + 153 = 152.68$$

$$\therefore \overline{X} = 152.68.$$

To determine the median, just find the location of the median in the data set as:

$$\text{Median location} = \left(\frac{1}{2}\right)(n + 1) = \left(\frac{1}{2}\right)(125 + 1) = \frac{126}{2} = 63$$

Thus, counting the data points from the bottom, we see that 50 pieces of data fall in the classes through 146-150, while 77 are contained through 151 – 155. Therefore, the median is located in the interval with class mark 153. Hence, the Median is approximately 153.

Another approach is to assume that the data are distributed evenly over the interval. Thus, data point number 63 in the distribution will be number 13 from the bottom of the interval containing the median in Table 1.3a. The actual class limits of that interval are 150.5 and 155.5, so that 27 pieces of data are distributed evenly over 5 units- - meaning that there are actually 28 intervals. Thus, the 13th data point in the interval lies above 13/28 of the interval. Since interval is 5 units we have:

$$\text{Median} = 150.5 + \left(\frac{13}{28}\right)(5) \doteq 152.82 \approx 152.8$$

1.7 Measures of Variability (or Dispersion)

Measures of central tendency only provide just one description of the distribution of data. The description of the distribution of any data set is incomplete without a measure of the variability, or dispersion, or spread of the data about the mean. A measure of variability tells us whether values in the distribution cluster closely or are spread widely in both directions as in Figure 1-7b. In Figure 1.7a, the data cluster more closely to the mean, and hence the smaller dispersion; the more widely dispersed are the data, the larger the dispersion and vice versa.

There are several measures of viability; but a few important ones will be discussed here:

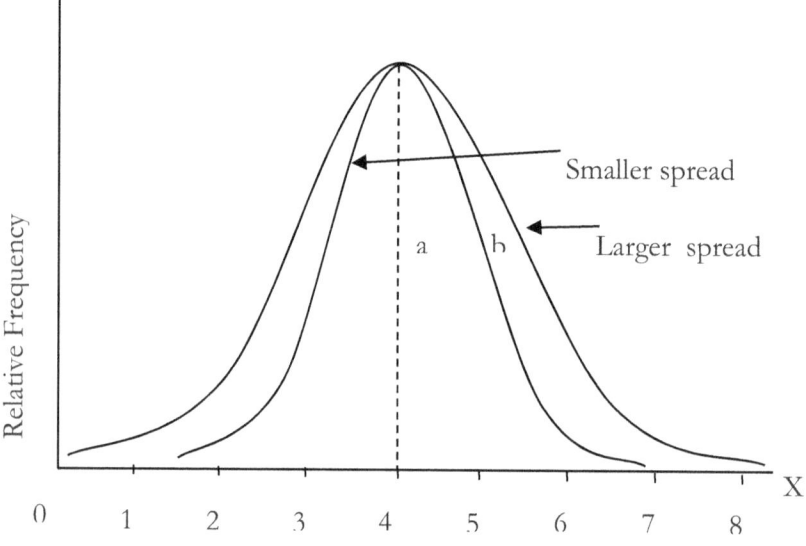

Figure 1.7 (a and b): Two distributions with same mean but different spreads.

1.7.1 The Range (R)

The Range (R) is the simplest measure of dispersion defined as the difference between the largest and the smallest values in a data set. For example, in a data set: 2, 4, 5, 7, 9, 10 and 12, the range is called: R = largest value – Smallest value.

$$R = 12 - 2 = 10.$$

The range is however, not a completely satisfactory measure of dispersion especially when the data set contains extreme values. In such a situation, the range gives a misleading measure (or overestimate dispersion). But when the data set has no extreme values, the range is a very quick and easy measure of dispersion.

The range is related to the standard deviation (S) as:

$$R = 4S$$

Relative Frequency

$$S = \frac{R}{4} \quad \text{i.e} \quad \frac{10}{4} = 2.5$$

Thus, one-fourth of the range $(R/4)$ is the **"rule of thumb"** to estimate the standard deviation(s).

1.7.2 The Quartiles

Byrkit (1975) and Mendenhall (1975) have observed that Quartiles themselves are not measures of dispersion, but their placement gives some indication about the dispersion of the distribution. Quartiles are values of the variable that divide the area of the histogram into quarters:

  i.   First Quartile, Q_1, is the value of the variable that exceeds $\frac{1}{4}$ of the data.
  ii. Second Quartile, Q_2, is the Median, which exceeds $\frac{1}{2}$ of the data.
  iii. Third Quartile, Q_3, is the value that exceeds $\frac{3}{4}$ of the data.

Since Q_2 is the median, we locate Q_1 and Q_3 as:

Q1 is at $\frac{1}{4}(n + 1)$ and Q_3 is at $\frac{3}{4}(n + 1)$ from the bottom of an arrayed distribution.

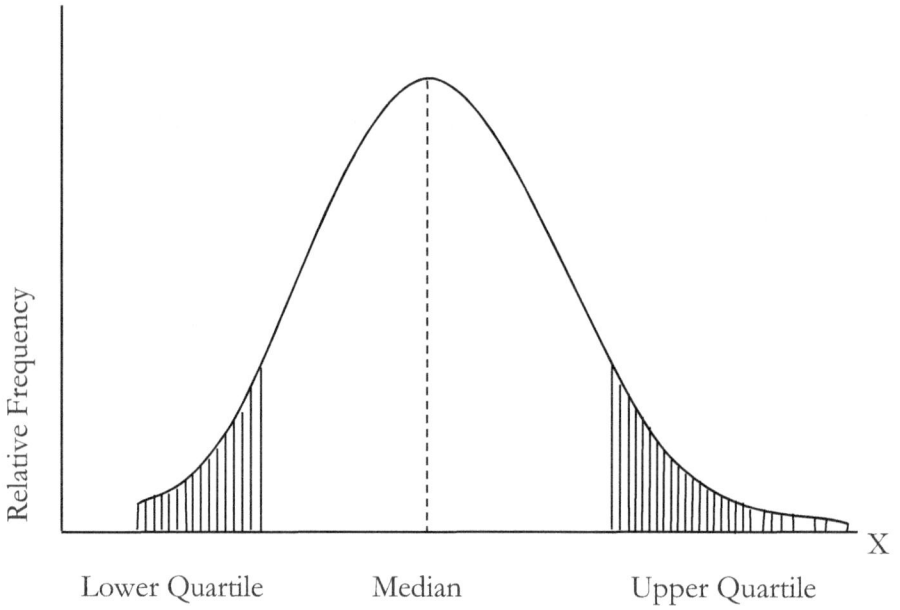

Lower Quartile   Median   Upper Quartile

Figure 1.7c: Location of Quartiles

    i.   Inter-Quartile Range = Distance between Q_3 and Q_1
    ii. A more widely used measure of spread (Byrkit, 1975) is the Semi-Interquartile Range, Q, which is:

Semi-Interquartile Range $(Q) = \frac{1}{2}(Q_3 - Q_1)$

Example 1.7.2: Find the Median, Q1, Q3 and Q (semi-Interquartile Range) of the following data on the ages of 100-level Students of the Faculty of Agriculture, University of Benin, Benin- City.

Table 1.4: Ages of 100-Level Students of Agriculture, UNIBEN

Age	Frequency
23	1
22	13
21	16
20	18
19	6
18	2
Total	56

Solution:

There are 56 pieces of data:

Median location (Md) $= \left(\frac{1}{2}\right)(n+1) = \left(\frac{1}{2}\right)(125+1) = \frac{126}{2} = 63$ i.e. data nos. 28 and 29 in the array.

$Q_1 = \frac{1}{4}(n+1) = \frac{1}{4}(56+1) = \frac{57}{4} = 14.25$ i.e. data piece nos. 14 and 15 in the array.

$Q_3 = \frac{3}{4}(n+1) = \frac{3}{4}(56+1) = \frac{3}{4}(57) = \frac{3\,(57)}{4} = \frac{171}{4} = 42.75$

13

From the frequency minimum age, the data piece nos. 14 and 15 are both located at age 20. Therefore, $Q_1 = 20$.

The Median is located between data piece nos. 28 and 29 in the array which is 21.

$Q3 = 42.75$ i.e. 42nd measure (or piece of data) from the bottom of the array is age 21; 43rd piece of data from the bottom of the array is age 22. So the data piece no. 42.75 is $\frac{3}{4}$ of the distance from 21 to 32. Hence:

$Q3 = 21 + \frac{3}{4}(22 - 21) = 21.75$

Semi-Interquartile Range = $Q = \frac{1}{2}(Q3 - Q1) = \frac{1}{2}(21.75 - 20) = \frac{1}{2}(1.75) = 0.88$

Compared with the median, 21, Q, the Semi-Interquartile Range is only 0.88, hence the spread is very small.

1.7.3 Mean Deviation of a Sample.

The mean deviation is the sum of the absolute values of the differences between each data point and the mean, divided by the number of data points (n):

$$\text{(Mean deviation)} = \frac{\sum x_i - \bar{x}}{n}$$

The mean deviation is a valid measure of spread, but it tends to be unsatisfactory for purposes of statistical inference.

Example 1.7.3: $\bar{X} = \dfrac{\sum X_i}{n} = \dfrac{19}{5} = 3.8$

Table 1.7 Mean Deviation of a Sample

X_i	$X_i - \bar{X}$	$\lvert X_i - \bar{X} \rvert$
5	1.2	1.2
7	3.2	3.2
1	-2.8	2.8
2	-1.8	1.8
4	0.2	0.2
$\sum X_t$ = 19	0	9.2

$$\frac{\sum \lvert Xi - \hat{X} \rvert}{n} = \frac{9.2}{5} = 1.84$$

Note:

$$\sum_{t-1}^{n} X_t - \sum_{t-1}^{n} \overline{X}$$

$$But \quad \sum_{t-1}^{n} \overline{X} = n\overline{X}$$

$$Hence \quad \sum_{t-1}^{n}(X_t - \overline{X}) = \sum_{t=1}^{n} X_t - n\overline{X}$$

$$\sum_{t-1}^{n} X_t - n\frac{(\sum nX_t)}{n}$$

$$= \sum_{i=1}^{n} X_i - \sum_{i=1}^{n} X_i = 0$$

1.7.4 The Sample Variance (S^2) and Standard Deviation (S)

The variance, sometimes referred to as the **MEAN SQUARE DEVIATION**, which is just the average of the squares of the deviations of the values of measurements from their mean, tells us something about the variability of the measurements (Walpole, 1974; Mendenhall, 1975).

If the measurements are all close to the mean, the deviations from the mean will be small and hence S^2 will be small. Alternatively, if the measurements are widely dispersed about their mean, the S^2 will be large.

The positive square root of the variance (S^2) is called the **Standard Deviation, or Root-Mean-Square Deviation (S)**.

   a. Estimation of a Sample Variance (S^2) and its Standard Deviation (S)

Mathematically, an "unbiased" estimate of a sample variance is given by:

i (a) $\quad S^2 = \frac{\sum(X_i - \bar{X})^2}{n-1}; S = \sqrt{\frac{\sum(X_i - \bar{X})^2}{n-1}}$

(b). $S^2 = \frac{n\sum X_i^2 - (\sum X_i)^2}{n(n-1)}: S = \sqrt{\frac{n\sum X_i^2 - (\sum X_i)^2}{n(n-1)}}$

The above formula, though a good working formula, it is not a good computational formula because it is cumbersome to apply. Other computational formulae that are more appealing and easier to compute are:

ii. $\quad S^2 = \frac{\sum X_i^2 - \frac{(\sum X_i)^2}{n}}{n-1}: S = \sqrt{\frac{\sum X_i^2 - \frac{(\sum X_i)^2}{n}}{n-1}}$

iii. $S^2 = \frac{n\sum X_i^2 - (\sum X_i)^2}{n(n-1)}: S = \sqrt{\frac{n\sum X_i^2 - (\sum X_i)^2}{n(n-1)}}$

Using the data in Example 1.7.3 above, the three formulae would produce the same result of the standard deviation (s):

Table: 1.7.4 Standard Deviation, or Root-Mean-Square Deviation (S).

X	X²
5	25
7	49
1	1
2	4
4	16
19	95

Sum $= \sqrt{\dfrac{\sum X_i^2 - n(\bar{X})^2}{n-1}} = \sqrt{\dfrac{300 - 5(3.8)^2}{4}} = \sqrt{\dfrac{300 - 72.2}{4}}$

$= \sqrt{\dfrac{227.8}{4}} = 7.5465 = 7.5$

(i) $\quad S = \sqrt{\dfrac{\sum X_i^2 - \dfrac{(\sum X_i)^2}{n}}{n-1}} = \sqrt{\dfrac{300 - \dfrac{(19)^2}{5}}{4}} = \sqrt{\dfrac{300 - 72.2}{4}}$

$= \sqrt{\dfrac{227.8}{n-1}} = \sqrt{56.95} = 7.5465 = 7.5$

(ii) $\quad S = \sqrt{\dfrac{\sum X_i^2 - (\sum X_i)^2}{n(n-1)}} = \dfrac{5(300) - 361}{5(4)}$

$= \sqrt{\dfrac{1,500 - 361}{20}} = \sqrt{\dfrac{1,139}{20}} = \sqrt{56.95} = 7.5465 = 7.5$

1.7.5 The Percentile

The percentile is a measure of location in a set of data that is arranged in increasing order of magnitude from the lowest to the highest. The pth percentile is the value which exceed p-percent of the scores.

Therefore, the minimum score in an array is the zero percentile, since it does not exceed any other value, while the highest score in the array would be the 99th percentile, since it exceeds 99% of the score except itself.

To determine the percentile of a particular score, Byrkit (1975) suggested that we count the scores below it and divide by the total number of scores. Thus, if r is the rank of a score (counted from the bottom), its

16

percentile is given by:

$$P = \frac{r-1}{n} \times 100,$$ Where n is the number of scores.

1.7.6 The Practical Significance of the Standard Deviation

The practical significance of the standard deviation of a population or a sample can be seen from Tchebysheff's theorem on the distribution of measurements from which the mean and standard deviation were obtained:

Tchebysheff's Theorem states that: "Given a number k greater than or equal to 1, and a set of n measurements X₁, X₂, ----, Xn, at least $(1 - \frac{1}{k^2})$ of the measurements will lie within k-standard deviations of their mean".

Tchebysheff's Theorem is true for any number we wish to choose fork, as long as it is greater than, or equal to 1 (k ≥ 1). (Mendenhall, 1975). Tchebysheff's Theorem enables an interval to be constructed by measuring a distance for kσ (or ks) on the either side of the mean, (or$\bar{x}$):

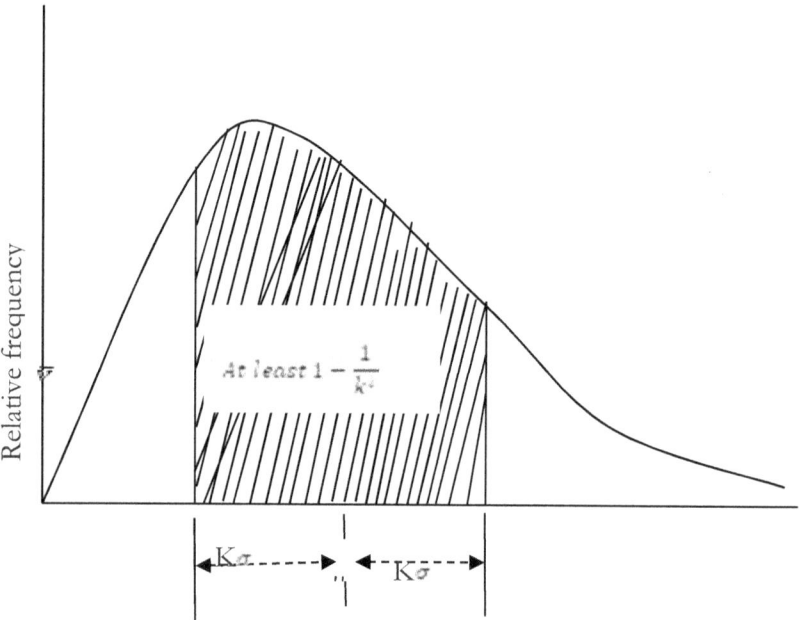

Figure 1.7: Tchebysheff's Theorem illustrated

Thus, in most situations, the fraction of measurements that will fall in the specified interval will exceed

$$1 - \frac{1}{k^2}.$$

Example 1.7.5: : Suppose n = 25; $\bar{X}$ = 75 and S² = 100.

$$S = \sqrt{100} = 10$$

Example 1.7.5: Illustrating Tchebysheff's Theorem

17

K	$(1 - \frac{1}{k^2})$	Interval $(x \pm ks)$
1	0	75 0=±75
2	$\frac{3}{4}$	75±2 (10) = 55 – 95
3	$\frac{8}{9}$	75 ± 3 (10) = 45 –105
4	$\frac{15}{16}$	75 ± 4 (10) = 35 – 115

Tchebysheff's Theorem applies to any set of measurements. But for Bell-shaped and mound-shaped distribution of data, the **Empirical rule** describes accurately the variability.

1.7.7 The Empirical Rule:

The Empirical Rule states that "Given a distribution of measurements that is approximately bell-shaped, the interval:

1. $\mu \pm \sigma (or\bar{x} \pm S)$ will contain approximately 68% of the measurements
2. $\mu \pm 2\sigma (or\bar{x} \pm 2S)$ will contain approximately 95% of the measurements
3. $\mu \pm 3\sigma (or\bar{x} \pm 3S)$ will contain almost all of the measurements.

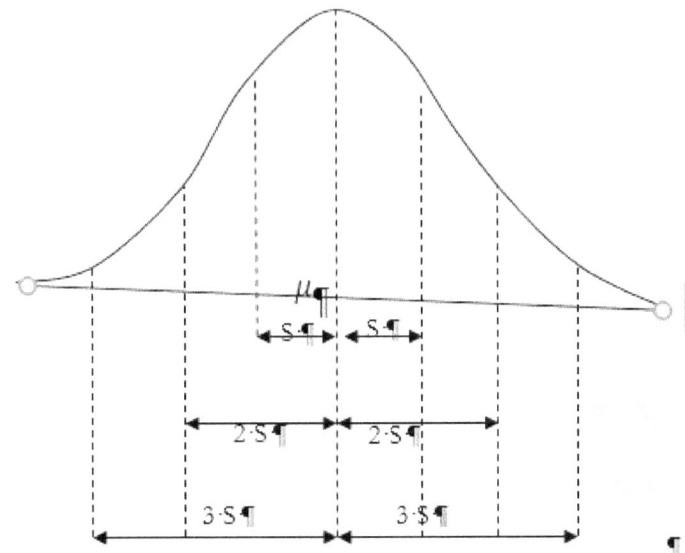

Figure 1.8: Empirical rule illustrated

18

2 PROBABILITY

According to Freund (1971), probability plays an important role, directly or indirectly, in all problems of science, business and everyday life, where decisions and understanding involve an element of uncertainty and risk. A basic knowledge of probability is necessary to be able to make proper use of appropriate statistical techniques and to understand the results of their use.

The most widely held concept of probability is the "frequency concept of probability" which says that: the probability of an event is interpreted as the proportion of time that the events of the same kind will occur in the long run i.e. the chances that a particular outcome will occur in a known population of possible outcomes.

For example, in Table 1.1 where we considered the height growth of thirty seedlings of <u>Garcinia cola,</u> there were nine seedlings whose height growth in six months was less than 2.0cm. Therefore, the probability that a Garcinia seedling will grow less than 2.0cm in six months, is an event (or an outcome) which we will designate as 'A', and is stated as: P (A < 2.0) = Number of seedlings less than 2.0cm divided by the total number of seedlings: $9/30 = 0.3$. That is, the proportion of the thirty seedlings whose height growth is less than 2.0cm. This proportion or probability can also be expressed as a percentage: $9/30 = 0.3$ x 100% is 30%.

2.1 SAMPLE SPACE

Let us use our example in Table 1 on the color distribution of 100 chickens in UNIBEN FARM to illustrate the mathematics of probability:

Table 1: Color Distribution of Chickens in UNIBEN FARM

Chicken Color	Frequency (No. of Chickens) f_i	Relative Frequency (f_i/n) (Proportion)
Speckled	8	$8/100 = 0.08$
Black	25	$25/100 = 0.25$
Brown	32	$32/100 = 0.32$
White	35	$35/100 = 0.35$
Total	100	$100/100 = 1.0$

All the chickens in UNIBEN FARM constitute the set of all possible outcomes (color in this case) called the **sample space**, generally represented by the symbol S. We shall denote the event of selecting a speckled chicken as A, a subset of S. All the events, A, B, C and D which make up the sample space, S, are called **Sample Points or Subsets of S**.

The probability of selecting a speckled chicken is mathematically stated as: P(A) = Number of speckled chickens, divided by the total number of chickens i.e.

$$P(A) = \frac{8}{100} = 0.08$$

Denote the event B as selecting a black chicken. Thus, the probability of selecting a black chicken is stated mathematically as:

$$P(B) = \frac{B}{S} = \frac{25}{100} = 0.25 \text{ i.e. number of black chickens divided by total number of chickens.}$$

Denote event C as selecting a brown chicken; the probability of selecting a brown chicken is stated as:

$$P(C) = \frac{C}{S} = \text{No. of brown chickens divided by the total number of chickens}$$

$$= .\frac{32}{100} = 0.32$$

Denote the event D as selecting a white chicken; the probability of selecting a white chicken is mathematically stated as:

$$\frac{D}{S} =$$

$$P(D) = \text{No. of white chickens divided by the total number of chickens}$$

$$= \frac{35}{100} = 0.35$$

Note:

The probability of an Event or Outcome as seen above, is stated in capital letters e.g. P(A) is the probability of Event A. Similarly, the Sample space is designated as capital S.

In the example above, we have four Events or Outcomes: A, B, C and D. The probability of any of these Events is always a real number which is non-negative - - i.e., $P(A) \geq 0$; $P(B) \geq 0$; $P(C) \geq 0$; $P(D) \geq 0$. The minimum value that a probability can take is zero, while its maximum value is 1. No probability is less than zero (0), or more than 1: $0 \leq P(A) \leq 1$ - - the probability of any event within a given sample space, is always between zero (0) and 1.

The sum of all probabilities of individual outcomes within any given sample space is always approximately equal to 1.

Thus: P (A) + P (B) + P (C) + P (D) = 1.

= 0.08 + 0.25 + 0.32 + 0.35 = 1. i.e.

$$\sum_{i-1}^{s} P(E_i) = 1$$

Where the E_i are the Events or Sample Points.

The probability of an event, or an outcome, is an expression of our confidence that the event/outcome will occur. The closer the probability of an event is to 1, the surer, or more confident we are that that event will occur. For example, the probability that a selected chicken will be white is more certain than that of a speckled chicken because: P (D) = 0.35 > P (A) = 0.08. Therefore, selecting a white chicken is more likely or more probable.

If two or more events within a sample space have equal probabilities, we say the events are equiprobable.

The respective probabilities that an event will occur and that it will not occur always add up to 1.

Notationally: P (A) + (P (A') = 1, where P (A) is the probability that the event will occur; P (A') is the probability that the event will not occur.

2.2 COUNTING SAMPLE POINTS

In many cases, we are able to solve a probability problem by counting the number of points in the sample space. Walpole (1974) stated the fundamental principle of counting in two theorems:

Theorem 1: If an operation can be performed in n_1 ways, and if for each of these, a second operation can be performed in n_2 ways, then the two operations can be performed together in $n_1 n_2$ ways.

Example: If two fair coins are tossed together, the number of sample points are $(2)(2) = 4$, since the first coin can land in any one of two ways, head or tail i.e. H or T. For each of these two ways, the second coin can also land in any one of two ways. Therefore, a toss of two fair coins together has $(2)(2) = 4$ sample points in the sample space.

Theorem 2: If an operation can be performed in n_1 ways, and if for each of these a second operation can be performed in n_2 ways, and for each of the first two, a third operation can be performed in $n3$ ways, etc., then the sequence of the k operations can be performed in $n_1 n_2 \ldots n_k$ ways.

Example: If two dice are thrown together, how many sample points are in the sample space? The first die can land in six (6) different ways and the second die can also land in six (6) different ways. Hence, the two dice can land in $(6)(6) = 36$ ways – that is, the sample points are 36. These different arrangements are called Permutations.

2.3 PERMUTATIONS

Mendenhall (1975) defined Permutation as: "An ordered arrangement of r distinct objects". The number of ways of ordering n distinct (different) objects, taken r at a time, will be designated by the symbol P^n_r, or $_nP_r$.

$$P^n_r = n (n - 1) (n - 2) \ldots (n - r + 1)$$ which is expressed in terms of factorials as:

$$Pn_r = \frac{n!}{(n-r)!}.$$

The number of permutations of n distinct objects is n!

$$n! = n (n - 1) (n - 2) \ldots 3.2.1 \text{ and } 0! = 1.$$

Permutation Rule A: If r objects are to be selected from a set of n different objects in such a way that the order of selection is important, the number of permutations is given by:

$$P^n_r = \quad n.(n-1).(n-2) \ldots (n-r+1) \text{ or } \frac{n!}{(n-r)!} \cdot \cdot$$

Permutation Rule B: If a set contains n elements of which r1, are of one kind, r2 are of a second kind, r3 are of a third kind, and so on through rk, the number of permutations of all n objects is given by:

$$\frac{n!}{r_1!r_2!r_3!...r_k!}.$$

2.4 CIRCULAR PERMUTATIONS

Permutations that occur by arranging objects in a circle are called circular permutations.

The number of permutations of n distinct objects arranged in a circle is (n − 1)!

2.5 COMBINATIONS

If the order is immaterial, or not significant, the formulas (or formulae) for permutations do not apply (Byrkit, 1972), and hence some modifications must be made. The number of ways of selecting r objects from n objects, without regard to order, is called Combinations (Walpole, 1974). A combination is actually a partition into two cells - - one cell containing the r objects selected and the other cell containing the (n − r) objects that are left. The number of such combinations is denoted by the symbol C_r^n or simply $\binom{n}{r}$.

$$C_r^n = \binom{n}{r} = \frac{n\,!}{r!(n-r)!}.$$

Example: How many committees of 4 can be chosen from an organization containing 30 members?

$$C_r^n = C_4^{30} = \frac{30}{\frac{4!}{26!}} = \frac{30.29.28.27}{4.3.2.1} = 27.405 \text{ committees.}$$

3 RANDOM VARIABLES AND PROBABILITY DISTRIBUTIONS

3.1 RANDOM VARIABLES

The term Statistical Experiment is used to describe any process by which several chance observations are obtained. Walpole (1974) and Tsokos (1972) described a **Random Variable** as a function (or rule) that assigns a number to every outcome of an experiment. All possible outcomes of an experiment comprise a set that is called the **Sample Space**. Walpole (1974) defined a **Random Variable as a function whose value is a real number determined by each element in the Sample Space**. A Random Variable is generally denoted by a capital letter, X, and corresponding small letters to represent its specific values, x.

In mathematics and many of the physical sciences, we encounter very basic formulas like the area of a circle given by $A = \pi r^2$. Therefore, to determine the area of a circle, we must specify its radius (r). For each independent variable, r, there is a corresponding value of area (A). The collection of these pairs of numbers is a function, or a relationship between the values of r and A. Other synonyms for a Random Variable are: Chance Variable, Variate and Stochastic Variable.

Example: If a pair of fair dice is rolled together once, the sample space consists of 36 ordered pairs of numbers, namely: S = [(1, 1), (1, 2), (1, 3) …. (6,6)].

Let the Random Variable, X, represent the total number of spots showing on the dice. Hence, X, assigns to each ordered pair in S its sum: X (1, 1) = 2; X (1, 2) = 3, …, X (6, 6) = 12, which values are in Random Variable, X. This has established that X is a real valued function, defined on S, that has a range that is a subset of the Random Variable.

Thus, it can be seen that a random variable can be used to assign numbers to mutually exclusive events which make up a sample space. The rules of probability described earlier will enable us determine the probability of each of the events designated by a particular value of a random variable.

The entire collection of values that the random variable can take on, together with the associated probabilities for each of these values, define a PROBABILITY DISTRIBUTION.

3.2 PROBABILITY DISTRIBUTIONS FOR DISCRETE RANDOM VARIABLES

Random Variables are generally divided into two classes namely:

3.2.1 Discrete Random variables

Discrete Random variables with one common characteristic - - their sample spaces contain a finite, or countable number of sample points, which enable us to assign probabilities to the sample points, so that the sum of their respective probabilities is equal to one. Mendenhall (1975) defined **the probability distribution (or function) for a discrete random variable as "a formula, table, or graph that provides the probability associated with each value of the random variable"**. The two requirements for a probability distribution are:

a) $f(x_i) = P(X = x_i) \geq 0$, for all x_i i.e. $0 \leq P(x_i) \leq 1$ i.e. probabilities are positive values.

b) $\sum P(X = x_i) = 1. \; ; \; \sum f(x_i) = 1.$

Discrete probability distributions can be graphed by various techniques like the histogram, as seen in section 1.2 - - graphical methods of presenting data.

Example 3.2.1: Find a formula for the probability distribution of the number of heads when a fair coin is tossed 4 times.

Solution:

There are 2 possible outcomes for each toss; when tossed 4 times, the sample space contains $2^4 = 16$ equally likely outcomes. Therefore, the formula will have 16 as the denominator. In general, x heads and 4 – x tails can occur in $\binom{4}{x}/16$, x = 0, 1, 2, 3, 4.

Substituting the values of x into the formula, we obtain the probability distribution:

$$\text{For x=0, } f(x) = P(X=0) = \binom{4}{0}/16 = \left(\frac{4!}{0!\,4!}\right)/16 = \frac{1}{16}$$

$$\text{x=1, } f(x) = P(X=1) = \binom{4}{1}/16 = \left(\frac{4!}{1!\,3!}\right)/16 = \frac{4}{16}$$

$$\text{x=2, } f(x) = P(X=2) = \binom{4}{2}/16 = \left(\frac{4!}{2!\,2!}\right)/16 = \frac{6}{16}$$

$$\text{x=3, } f(x) = P(X=3) = \binom{4}{3}/16 = \left(\frac{4!}{3!\,1!}\right)/16 = \frac{4}{16}$$

$$\text{x=4, } f(x) = P(X=4) = \binom{4}{4}/16 = \left(\frac{4!}{4!\,0!}\right)/16 = \frac{1}{16}$$

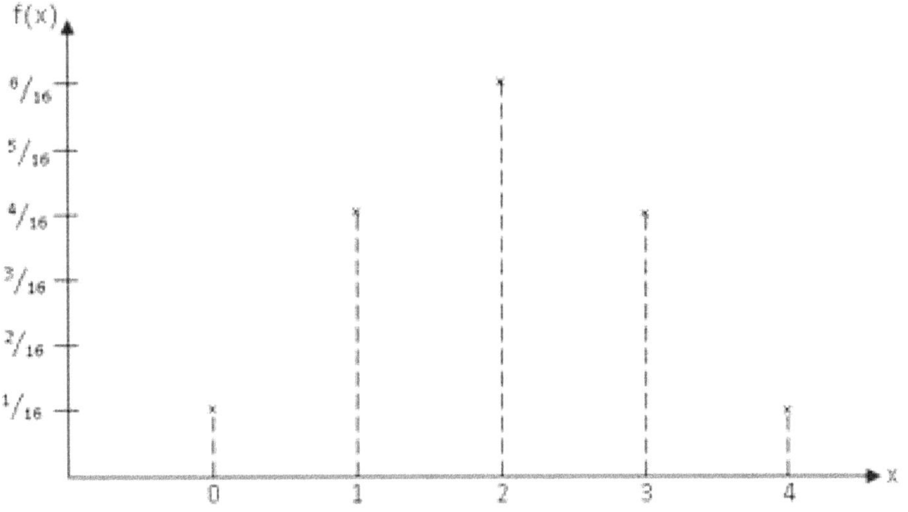

Figure 3.1a: Bar Chart of (x, f (x)) points

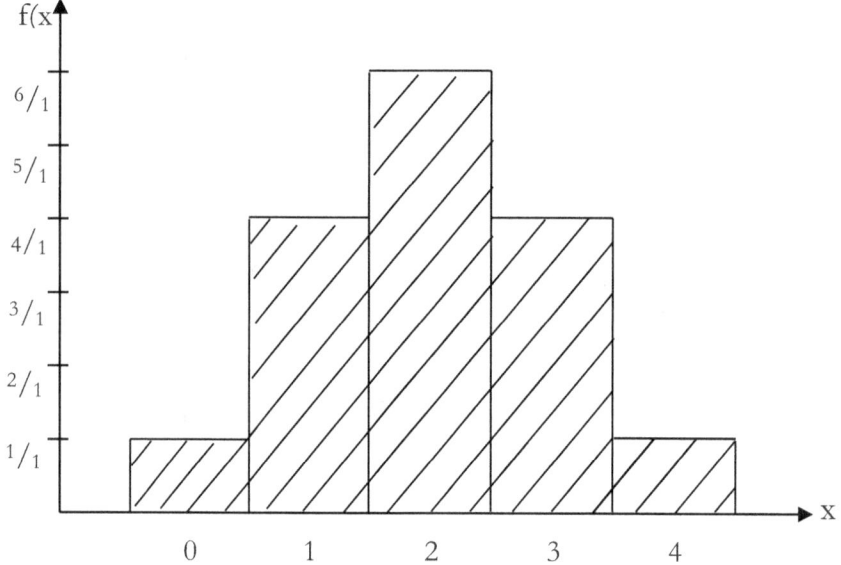

Figure 3.1b: Probability Histogram of (x, f(x)) points

The rectangles of the probability histogram are of equal width and centered at each value of x; their heights are equal to the corresponding probabilities given by f (x).

Example 3.2.2: Five fair coins are tossed together. If the random variable, X, represents the number of heads that show, construct its probability density function.

Solution:

Each toss of a coin has two possible outcomes; hence the sample space of this experiment contains 2^5. The event of getting x heads can occur in $\binom{5}{x}$ ways. Hence,

$$f(x) = P(X = x) = \binom{5}{x}/2^5, x = 0, 1, 2, \ldots 5. = \binom{5}{x}/32$$

$$\text{For } x = 0, f(x) = P(X = 0) = \binom{5}{0}/32 = \left(\frac{5!}{0! \, 5!}\right)/32 = \frac{1}{32}$$

$$x = 1, f(x) = P(X = 1) = \binom{5}{1}/32 = \left(\frac{5!}{1! \, 4!}\right)/32 = \frac{5}{32}$$

$$x = 2, f(x) = P(X = 2) = \binom{5}{2}/32 = \left(\frac{5!}{2! \, 3!}\right)/32 = \frac{10}{32}$$

$$x = 3, f(x) = P(X = 3) = \binom{5}{3}/32 = /32 = \frac{10}{32}$$

$$x = 4, f(x) = P(X = 4) = \binom{5}{4}/32 = \left(\frac{5!}{4! \, 1!}\right)/32 = \frac{5}{32}$$

$$x = 5, f(x) = P(X = 5) = \binom{5}{5}/32 = \left(\frac{5!}{5! \, 0!}\right)/32 = \frac{1}{32}$$

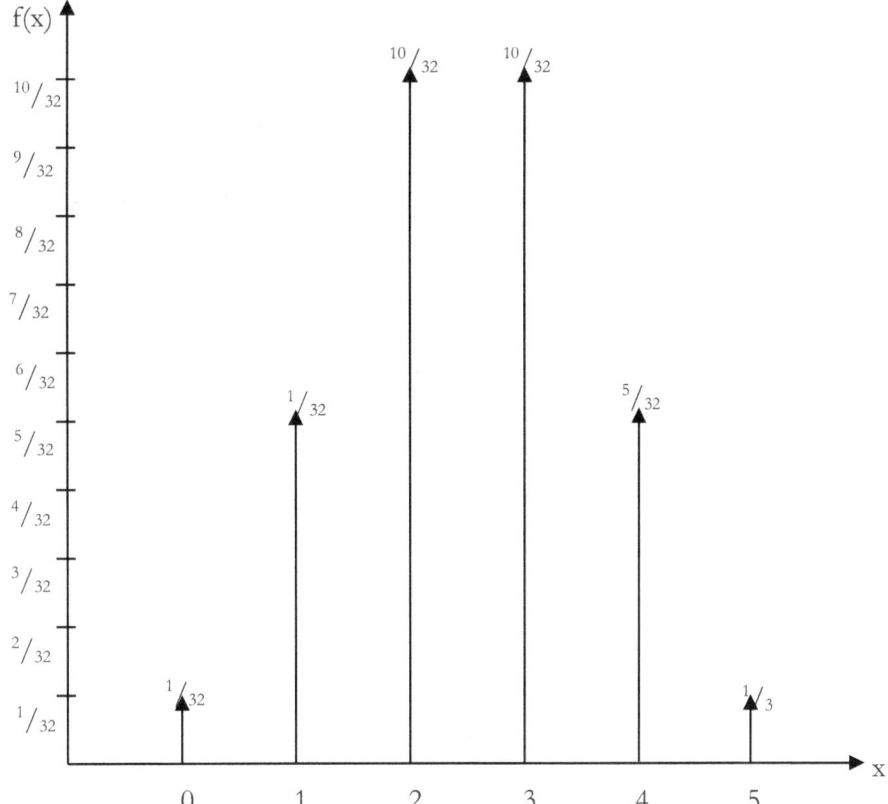

Figure 3.1c: Graph of the Density Function f(x), as vertical lines

Note: The length of each vertical line indicates the probability of the value assumed by the random variable. These probabilities can also be used to produce a probability histogram.

Apart from discrete probability distribution function f (x), other types of probabilities can be estimated from any sample space. For example, we may wish to estimate the following probabilities when five coins are tossed together:

(i) What is the probability of obtaining exactly three heads i.e.:

$$P (X = 3) = f (3) = \binom{5}{3}/2^5 = \left(\frac{5!}{3!\ 2!}\right)/32 = \left(\frac{5x4}{2x1}\right)/32 = \frac{10}{32} = \frac{5}{16}$$

(ii) What is the probability that the random variable, X, is at most equal to four heads? This is mathematically stated as:

$$P (X \leq 4) = \sum_{x=0}^{4} \frac{\binom{5}{x}}{32}$$

This can be re-written as: P (X > 4) = 1 – P (X = 5)

$$= 1 - \binom{5}{5}\frac{}{32} = 1 - 1/32 = \,^{31}/_{32}$$

(iii) What is the probability that the random variable, X, is at least equal to three heads? This is mathematically stated as:

$$P(X \geq 3) = \sum_{x=3}^{5} \binom{5}{x} / 2^5$$

Alternatively, this may also be stated as: $1 - P(X < 3) = 1 - \sum_{x=0}^{3} \binom{5}{x} / 2^5$

Note: The summation, Σ, in (ii) and (iii) implies that the values of the probabilities of x = 0, 1, 2, etc., must be calculated and added to obtain the probability that X is at least equal to three heads.

Thus, $1 - P(X < 3) = 1 - (f(0) + f(1) + (f(2))$ which we estimated in Example 3.2.2 above i.e.

$$1 - \frac{1}{(32} + \frac{5}{32} + \frac{10}{32)} = 1 - \frac{1}{32} = 1 - \frac{1}{2} = \frac{1}{2}.$$

(iv) What is the probability that the random variable, X, assumes a value in the closed interval between 2 and 4?

This is mathematically stated as:

$$P(2 \leq X \leq 4) = \sum_{x=2}^{4} \binom{5}{x} / 2^5$$

$$= f(2) + f(3) + f(4), \text{ as above}$$

$$=$$

$$\frac{10}{32} + \frac{10}{32} + \frac{5}{32} = \frac{25}{32}$$

(v) What is the probability that the random variable is equal to two, given that the number of heads is less than or equal to three?

This is mathematically stated as:

$P(X = 2/X \leq 3)$. This implies that it is already known that less than or equal to three heads occurred.

Thus,

$$P\left(X = \frac{2}{X} \leq 3\right) = \frac{\binom{5}{2}/2^5}{\sum_{x=0}^{3}\binom{5}{x}/2^5} = \binom{5}{2} / \sum_{x=0}^{3}\binom{5}{x} = \left(\frac{5!}{2!\,3!}\right) / \sum_{x=0}^{3}\binom{5}{x}$$

$$\sum_{x=0}^{3}\binom{5}{2} = \frac{10/32}{f(0) + f(1) + f(2) + f(3)} = \frac{10/32}{1/32 + 5/32 + 10/32 + 10/32} = \frac{10/32}{26/32}$$

$$= \frac{10}{32} \times \frac{32}{26} = \frac{10}{26} = \frac{5}{13}.$$

(vi) What is the probability that the number of heads is less than or equal to two, when it is already known that less than four heads occurred? Mathematically stated:

P (X ≤ 2/X < 4) =

$$\left[\sum_{x=0}^{2}\binom{5}{x}/2^5\right]\bigg/\left[\sum_{x=0}^{3}\binom{5}{x}/2^5\right]$$

$$=\left[\frac{f(0)+f(1)+f(2)}{32}\right]\bigg/\left[\frac{f(0)+f(1)+f(2)+f(3)}{32}\right]$$

$$=\frac{\left(\frac{1}{32}+\frac{5}{32}+\frac{10}{32}\right)}{32}\bigg/\frac{\left(\frac{1}{32}+\frac{5}{32}+\frac{10}{32}+\frac{10}{32}\right)}{32}=\frac{\left(\frac{16}{32}\right)}{32}\bigg/\frac{\left(\frac{26}{32}\right)}{32}=\frac{\left(\frac{16}{32}\right)}{32}\bigg/\frac{\left(\frac{26}{32}\right)}{32}$$

$$=\left(\frac{16}{32}\times\frac{32}{1}\right)\bigg/\left(\frac{26}{32}\times\frac{32}{1}\right)=16/26=\underline{8/13}$$

From the foregoing, we see that for a given discrete random variable, X, various probabilities concerning the behavior of X, can be determined.

3.2.2 Cumulative Distribution Function (F (x))

Frequently, Tsokos (1972) observed that we are interested to know that a specified value of a random variable X_i, is less, or equal to some real number x_j. The probability that X assumes a value less than x_j can be written as:

$P(X \le x_j) = \sum_{x_i \le x_j} f(x_i)$, where f ($x_i$) is the probability density function of the random variable X.

According to Tsokos (1972) and Freund (1971), the function, F (x), defined by:

$$F(x) = P\,(X \le x) = \sum_{x_i \le x_j} f(x_i),$$

where the summation is extended over all points, x_i, for which $x_i \le x$, is called a **Cumulative Distribution Function of the discrete random variable, X, having a probability density function, f (x_i).** As the name implies, the cumulative distribution function, F(x), accumulates the values of f(x_i) from $-\infty$ to x. Also, its graph is a **"step function" that is continuous from the right.**

3.2.3 Properties of the Cumulative Distribution Function

If F(x) is the cumulative distribution function of the discrete random variable, X, then F(x) has the following properties:

(i) $F(\infty) = 1$.
(ii) $F(-\infty) = 0$.
(iii) F(x) is a non-decreasing function of x.
(iv) F(x) is continuous from the right.

Example 3.2.3: Find the probability distribution of the sum of the numbers when a pair of dice is tossed.

Solution: : Let X be a random variable whose values, x_i, are the possible totals. Thus, x, can be any integer from 2 to 12. Total number of sample points = (6) (6) = 36, each with a probability of $^1/_{36}$.

The cumulative distribution function of the probability density of the experiment is given by:

$$F(x) = P(X \le x) = \begin{cases} 0, x < 2 \\ \displaystyle\sum_{x_i \le x} \frac{6 - |7 - x_i|}{36}, 2 \le x \le 12 \\ 1, x > 12 \end{cases}$$

By substituting the values of x = 2, 3, 4, ... 12, we have the following probability distribution:

$$F(2) = P(X \le 2) = \frac{6 - |7 - 2|}{36} = \frac{6 - 5}{36} = \frac{1}{36}$$

$$F(3) = P(X \le 3) = \frac{6 - |7 - 3|}{36} = \frac{6 - 4}{36} = \frac{2}{36}$$

Table 3.2.3: Sum of the outcomes of tossing two dice, their Probability Density and Cumulative Probability.

X	f(x)	F(x)	Note:
2	$1/36$	$1/36$	F(0) = 0
3	$2/36$	$3/36$	F(1) = f(0) + f(1) = 0
4	$3/36$	$6/36$	F(2) = f(0) + f(1) + f(2) = $1/36$
5	$4/36$	$10/36$	F(3) = f(2) + f(3) = $3/36$
6	$5/36$	$15/36$	F(4) = f(2) + f(3) + f(4) = $6/36$
7	$6/36$	$21/36$	:
8	$5/36$	$26/36$	:
9	$4/36$	$30/36$	:
10	$3/36$	$33/36$	:
11	$2/36$	$35/36$	:
12	$1/36$	$36/36 = 1$	F(12) ≡ f(2) + f(3) + ... + f(12) = 1

Probability Density and Cumulative Probability

The graph of this distribution functions is a "step-function" shown in Figure 3.2.

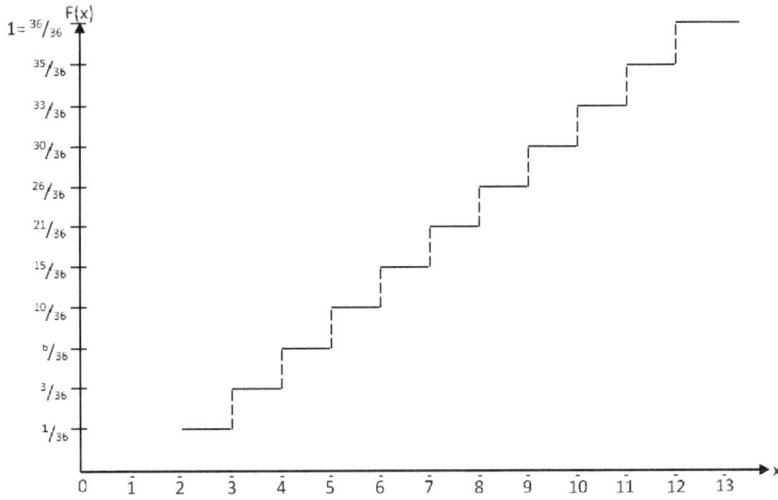

Figure 3.2: Discrete Cumulative Distribution Function of Total Obtained with a Pair of Dice

Example 3.2.4: Find a formula for the probability distribution of the number of heads when a fair coin is tossed four times.

Solution:

The sample space contains $2^4 = 16$ points of equiprobable, or equally likely outcomes. The number of ways of getting x heads in 4 tosses and therefore the probability distribution is given by:

$$P(X = x) = f(x) = \binom{4}{x}\Big/16, = 0,1,2,3,4.$$

For $x = 0$: $\binom{4}{x}\Big/16 = \left(\frac{4!}{0!\,4!}\right)\Big/16 = \frac{1}{16}$

$$x = 1 = \binom{4}{x}\Big/16 = \left(\frac{4!}{1!\,3!}\right)\Big/16 = \frac{4}{16}$$

$$x = 2 = \binom{4}{x}\Big/16 = \left(\frac{4!}{2!\,2!}\right)\Big/16 = \frac{6}{16}$$

$$x = 3 = \binom{4}{x}\Big/16 = \left(\frac{4!}{3!\,1!}\right)\Big/16 = \frac{4}{16}$$

$$x = 4 = \binom{4}{41}\Big/16 = \left(\frac{4!}{4!\,0!}\right)\Big/16 = \frac{1}{16}$$

But the cumulative distribution is given by:

$$F(x) = P(X \leq x) = \sum_{x_i \leq x} x_i$$

$$P(X \leq 0) = F(0) = f(0) = \frac{1}{16}$$

$$F(1) = f(0) + f(1) = \frac{1}{16} + \frac{4}{16} = \frac{5}{16}$$

30

$$F(2) = f(0) + f(1) + f(2) = \frac{1}{16} + \frac{4}{16} + \frac{6}{16} = \frac{11}{16}$$

$$F(3) = f(0) + f(1) + f(2) + f(3) = \frac{1}{16} + \frac{4}{16} + \frac{6}{16} + \frac{4}{16} = \frac{15}{16}$$

$$F(4) = f(0) + f(1) + f(2) + f(3) + f(4) = \frac{1}{16} + \frac{4}{16} + \frac{6}{16} + \frac{4}{16} + \frac{1}{16} = 1.0$$

Table 3.2.4:Probability Density and Cumulative Probability of Tossing a Coin Four Times

Variables	Density Variation	Cummulative Function
x	f(x)	F(x)
0	$^1/_{16}$	$^1/_{16}$
1	$^4/_{16}$	$^5/_{16}$
2	$^6/_{16}$	$^{11}/_{16}$
3	$^4/_{16}$	$^{15}/_{16}$
4	$^1/_{16}$	$^{16}/_{16} = 1.0$

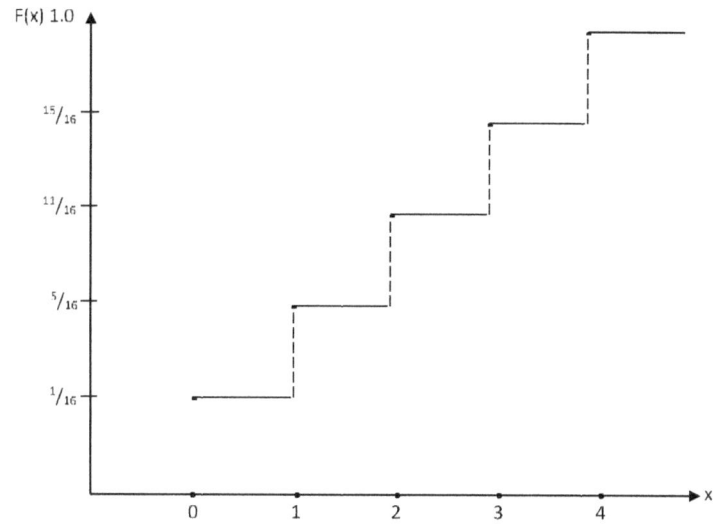

Figure 3.3: Discrete Cumulative Distribution of the Number of Heads in 4 Tosses of a Coin

3.3 SOME DISCRETE PROBABILITY DISTRIBUTIONS

The terms **"probability function"** and **"probability distribution"** in the literature of probability and statistics, are often used interchangeably (Freund, 1971). But some writers have made the distinction that the term **"probability distribution" applies to all the probabilities associated with a random variable, and not only those given directly by its probability function.**

31

According to Walpole and Myers (1972), many random variables associated with statistical experiments, have similar properties and can be described essentially by the same probability distribution. For example, all random variables representing the number of successes in **n** independent trials of an experiment, with a constant probability of a success, have the same general type of behavior and therefore can be represented by a single formula.

The mean, or variance, of a given probability distribution is defined to be the mean of any random variable having that distribution. Of course, not all discrete probability distributions will be discussed here.

3.3.1 The Discrete Uniform Distribution

The simplest of all discrete probability distributions is one where the random variable assumes all its values with equal probability.

Definition: If the random variable, X, assumes the values, $x_1, x_2, x_3, \ldots x_k$, with equal probability, then the discrete uniform distribution is given by:

$$F(x; k) = \frac{1}{k}, x = x_1, x_2, x_3, \ldots, x_k.$$

The notation f(x; k), instead of f(x) indicates that the uniform distribution depends on the parameter k.

Example 3.3.1: : When a balanced die is tossed, each element of the sample space S = (1, 2, 3, 4, 5, 6), occurs with probability $\frac{1}{6}$. Hence, we have a uniform distribution with $f(x, 6) = \frac{1}{6}$, x = 1, 2, 3, 4, 5, 6.

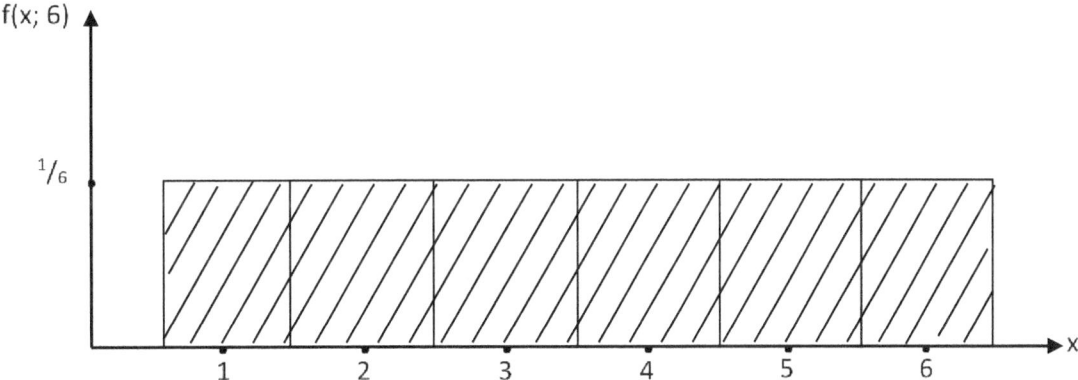

Figure 3.3.1: Probability Histogram for the Tossing of a Die

3.3.2 The Bernoulli Distribution

The Bernoulli probability distribution named after James Bernoulli (1654 – 1705) applies when an experiment has two possible outcomes, often referred to as **"success", with probability p, and "failure" with probability q = 1 – p; where p + q = 1.** We shall denote a failure as 0 and a success as 1.

The probability function of the Bernoulli distribution can be written as:

$$f(x; p) = p^x q^{1-x}, x = 0, 1, 0 < p < 1.$$

The Bernoulli probability distribution is a one-parameter distribution where the parameter is p - - i.e. it depends on the parameter p. The parameter, p, is a constant when we refer to **a specific Bernoulli distribution,** but can and will take on different values in different problems (Freund, 1972).

For example, p = 0.5, when we flip a **"fair"** or an **"honest"** coin, and 1 and 0 represent heads and tails, respectively.

Tsokos (1972) described the Bernoulli-type distribution as "The point Binomial Distribution" as "an experiment entailing a single observation, the result of which will be one of two possible outcomes – success or failure, yes or no".

An experiment to which the Bernoulli distribution applies is referred to as a **"Bernoulli trial", or just a "trial"**, and the repetition of such experiments as **"repeated trials"**.

3.3.3 The Binomial Distribution

Repeated Bernoulli trials play a very important role in probability and statistics, and in fact, it is one of the more widely used discrete probability distributions. It applies to problems that involve independent repeated trials, the outcome of which have been classified into two categories like success and failure, where the probability of success is the same for each trial.

The random variable of interest in a binomial experiment is the total number of times in which the experiment results in a success.

Scheaffer and Mendenhall (1975) described a **binomial experiment** as one that possesses the following properties:

1) The experiment consists of **n** identical trials.
2) Each trial results in one of two outcomes - - one called a success, the other called a failure i.e. it is **dichotomous with only two possible outcomes.**
3) The probability of success on a single trial is equal to p and remains the same from trial to trial. The probability of a failure is equal to q = (1 – p).
4) The trials are independent.
5) The random variable, of interest, Y.

Definition (Tsokos, 1972): A random variable, X, is said to have a binomial or Bernoulli distribution, if its probability density function is given by:

$$P(X = x) = f(x; n, p) = \binom{n}{x} p^x (1-p)^{n-x}, x = 0, 1, 2, \ldots, n; 0 < p < 1$$

Both Freund (1971) and Walpole (1974) used the formula:

$$b(x; n, p) = \binom{n}{x} p^x q^{n-x}, x = 0, 1, 2, \ldots, n; \text{ where} \binom{n}{x} =$$

$$\frac{n!}{x!(k-x)!} \text{ is the binomial coefficient.}$$

Notice that in the binomial distribution functions above, we used **f(x; n, p) or b(x; n, p)** to indicate that the values of the probability function depend on the parameters n and p - - the number of trials and the constant probability of success for each trial.

$$\sum_{x=0}^{n} f(x; n, p) = \sum_{x=0}^{n} \binom{n}{x} p^x (1-p)^{n-x} = [p + (1-p)]^n = 1$$

where:

n = number of trials.

x = number of successes.

p = the probability of a success.

The cumulative distribution is given by:

$$F(x) = P(X \leq x) = \sum_{x=0}^{n} \binom{n}{x} p^x (1-p)^{n-x}, 0 \leq x \leq n$$

Note: The sample points in the sample space are not equiprobable for the binomial experiment (Scheaffer and Mendenhall, 1975).

According to Walpole and Myers (1972) and Walpole (1974), the binomial distribution derives its name from the fact that the $(n + 1)$ terms in the binomial expansion of $(q + p)^n$ correspond to the values of $b(x; n, p)$ for $x = 0, 1, 2, 3, \ldots, n$. Thus,

$$(q + p)^n = \binom{n}{0} q^n + \binom{n}{1} pq^{n-1} + \binom{n}{2} p^2 q^{n-2} + \ldots + \binom{n}{n} p^n.$$

$$= b(0; n, p) + b(1; n, p) + b(2; n, p) + \ldots + b(n; n, p).$$

Since $p + q = 1$, we see that the sum of the value of $b(x; n, p)$, that is, $\sum_{x=0}^{n} b(x; n, p) = 1$, a condition that must hold for any probability distribution.

Fortunately, the binomial distribution has been tabulated in several texts (Tsokos, 1972; Walpole, 1974; Scheaffer, 1975; Freund, 1971) for various values of x, n and p.

To illustrate some of the many applications of binomial distribution, the following examples are presented:

Example 3.3.3i: A newlywed couple expects to have four children. For them, having a male child is a success. What is the probability distribution of the number of male children in the experiment?

Solution: Let X be the number of males in a family of 4 children. Therefore, the probability of having any specified number of males is given by:

$$P(X = k) = \binom{n}{k} p^k q^{n-k}, x = 0, 1, 2, 3, 4;$$ where the probability of a male equals the probability of a female = 0.5.

$P(X = k)$ means the probability of having exactly k number of males out of 4 kids. Substituting the values of $x = 0, 1, 2, 3, 4$, we have:

$$P(X = 0) = \binom{4}{0} (0.5)^0 (0.5)^4 = \frac{4!}{0!4!} (0.5)^0 (0.5)^4 = (0.5)^4 = \underline{0.0625}$$

$$P(X = 1) = \binom{4}{1} (0.5)^1 (0.5)^3 = \frac{4!}{1!3!} (0.5)^1 (0.5)^3 = 4 (0.5)^4 = \underline{0.25}$$

$$P(X = 2) = \binom{4}{2} (0.5)^2 (0.5)^2 = \frac{4!}{2!2!} (0.5)^2 (0.5)^2 = 2 \times 3 \times (0.5)^4 = \underline{0.375}$$

$$P(X = 3) = \binom{4}{3} (0.5)^3 (0.5)^1 = \frac{4!}{3!1!} (0.5)^3 (0.5)^1 = 4 (0.5)^4 = \underline{0.25}$$

$$P(X = 4) = \binom{4}{4} (0.5)^4 (0.5)^4 = \frac{4!}{4!0!} (0.5)^4 = (0.5)^4 = \underline{0.0625}$$

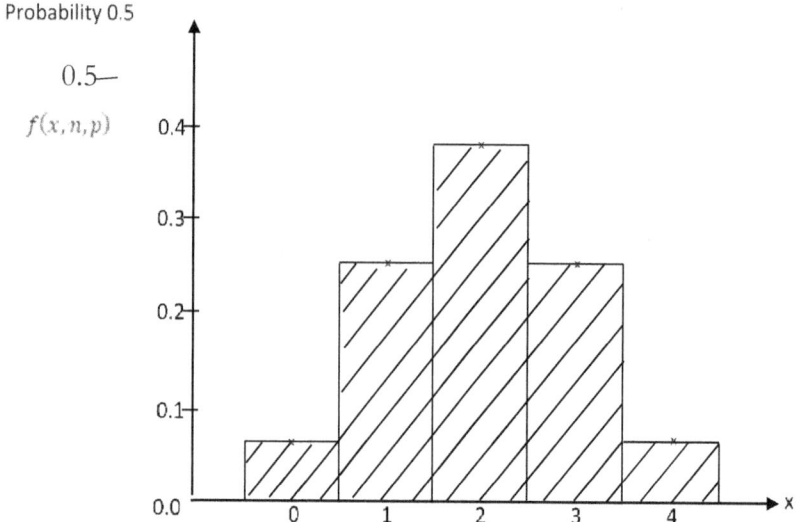

Figure 3.3.3 (i): Probability Distribution of the Number of Males in a Family of Four Kids

Example 3.3.3 (ii): An honest coin is tossed ten times. If heads is a success and tails a failure, find the following probabilities:

a.(i)   The probability of exactly seven successes (heads) i.e.

$$P(X = 7) = f(7; 10, 0.5) = \binom{10}{7} (0.5)^7 (0.5)^3$$

$$= \frac{10 \times 9 \times 8}{3!} = (10 \times 12) (0.5)^{10} = 120 \times 9.765625 \times 10^{-4}$$

$$= 120 \times 0.0009765625 = \underline{0.1171875}$$

3.3.3.1 How to Use the Binomial Probability Table to Estimate Probabilities.

a.(ii)   To apply the **Binomial Tables** to solve this problem, we must remember that the Tables are cumulative. That is, the values shown in the Tables are the probabilities from x = 0, 1, 2, 3 ... up to and including the number of successes. Therefore, P(X = 7) are the summed probabilities from x = 0 to x = 7. Hence, we must subtract the summed probabilities from x = 0 to x = 6 as:

P(X = 7) = P(X = 7) – P(X = 6)

$$\sum_{x=0}^{7} b(x; 10, 0.5) - \sum_{x=0}^{6} b(x; 10, 0.5)$$

From Table A.2. (Walpole, 1974), for n = 10; r = 7; p = 0.5 we have:

$$\sum_{x=0}^{7} b(x; 10, 0.5) - \sum_{x=0}^{6} b(x; 10, 0.5)$$

= 0.9453 – 0.8281 = <u>0.1172</u> same result as in a(i) above!

b.   Find the probability of at least seven successes (heads) i.e.

35

P(X ≥ 7) – This implies that the number of heads is from 7 to 10, written as:

$$\sum_{x=7}^{10} b(x; 10,0.5) = \binom{10}{7}(0.5)^7(0.5)^3 + \binom{10}{8}(0.5)^8(0.5)^2 + \binom{10}{9}(0.5)^9(0.5)^1 + \binom{10}{10}(0.5)^{10}(0.5)^0$$

$$= 0.1171875 + 0.43945312 + 0.009765625 + 0.0009765625$$

$$= 0.171874999$$

Instead of all these gruesome calculations, all we need to do is to re-state the problem as:

1–P(X< 7), since the sum of the probabilities in the sample space is equal to 1. Therefore,

$$1\text{–}P(X < 7) = 1\text{–}P(X \leq 6) = 1 - \sum_{x=0}^{6}(x; 10,0.5)$$

From the Table A.2 for n = 10; r = 6; p = 0.5, we have: $1 - 0.8281 = \underline{0.1719}$.

c. Find the probability of **at most seven successes** i.e. the number of heads is seven or less than seven. This is given as:

$P(X \leq 7) = \sum_{x=0}^{6} b(x; 10,0.5)$. Just read this directly from the Binomial Probability Table with n = 10; r = 6; p = 0.5.

$P(X < 7) = P(X \leq 6)$ i.e. the probability that X is strictly less than 7.

$$= \sum_{x=0}^{6}\binom{10}{x}(0.5)^x(0.5)^{10-x} = \underline{0.828125}$$

From the Binomial Distribution Table:

d.   Find the probability that the number of successes is from 3 to 8, i.e.

$$P(3 \leq X \leq 8) = \sum_{x=3}^{8} b(x; 10,0.5) = \sum_{x=0}^{8} b(x; 10,0.5) - \sum_{x=0}^{2} b(x; 10,0.5)$$

From the Binomial Distribution Table $= 0.9050 - 0.0271 = \underline{0.8779}$

e.   Find the probability that the number of successes is more than 7 i.e.: P(X > 7).

Note: Before solving this problem, we need to understand that P(X > 7) means that the random variable, X, takes on values that are strictly greater than 7. But we do not know how much greater than 7 it is. Is it 8, 9 or 10? Therefore, to estimate the probability of **"greater than"** 7, we must estimate the probability of P(X = 8) + P(X = 9) + P(X = 10) and subtract this from 1.

$$P(X > 7) = 1 - P(X \leq 7) = 1 - \sum_{x=0}^{7} b(x; 10,0.5)$$

From the Binomial Distribution Table =

$$1 - \sum_{x=0}^{7} b(x; 10,0.5) = 1 - 0.9453 = \underline{0.0547}$$

The **Binomial Probability Distribution Table** is designed for the number of trials that are fewer than, or equal to twenty i.e. when n ≤ 20. When the sample size, n, is greater than 20 (n > 20), the Binomial Distribution Table is no longer useful for the estimation of probabilities. For larger sample sizes, we will adopt other methods like the **"Normal Approximation to the Binomial"** in Section 4.6.

3.3.4 The Poisson Distribution

The **Poisson distribution is a discrete frequency distribution of the number of times a rare event occurs** (Sokal and Rohlf, 1969). But in contrast to the binomial distribution, the number of times that an event does not occur is infinitely large.

The purpose of fitting a Poisson distribution to numbers of rare events in nature is to test whether the events occur independently with respect to each other. If the occurrence of one event enhances the probability of a second such event, we obtain a **Clumped or Contagious Distribution.**

The Poisson distribution is named after the French Mathematician, Poisson, who described it in 1837. It is an infinite series whose terms add to one. The series can be represented as:

$$\frac{1}{e^\mu}, \frac{\mu}{1!e^\mu}, \frac{\mu^2}{2!e^\mu}, \frac{\mu^3}{3!e^\mu}, \dots, \frac{\mu^r}{r!e^\mu}.$$

The mean of the rare event is the only quantity that we need to know in order to calculate the relative expected frequencies of a Poisson distribution.

Experiments yielding numerical values of a random variable, X, the number of successes occurring in a given time interval, or in a specified region, are often called Poisson Experiments. For example, the given time interval may be of any length – a minute, a day, a month, or a year; the specified region could be a line segment, an area, a volume, or a piece of material. For example, X may represent the number of field mice per hectare; the number of bacteria in a given culture; or the number of typing errors per page, etc.

A Poisson experiment is one that possesses the following properties:

1. The average number of successes, μ, (or λ) occurring in the given time interval, or specified region is known.
2. The probability that a single success will occur during a very short time interval, or in a small region, is proportional to the length of the time interval, or the size of the region and does not depend on the number of successes occurring outside this time interval, or region.
3. The probability that more than one success will occur in such a short time interval or falling in such a small region is negligible.

The probability distribution of the Poisson variable, X, is called the Poisson distribution and is denoted by p(x;μ), since its values depend only on μ, the average number of successes occurring in the given time interval, or specified region.

The probability distribution of the Poisson random variable, X, is given by the formula:

$$p(x; \mu) = \frac{e^{-\mu}\mu^x}{x!}, \ x = 0, 1, 2, \dots,$$

where:

μ = the average number of successes occurring in the given interval or specified region.

e = 2.71828

The unique property of the Poisson distribution is that the mean, μ, equals the variance and are both equal to the parameter, λ (Lambda) i.e:

$$\mu = \sigma^2 = \lambda.$$

Therefore, $P(X = k) = \dfrac{e^{-\lambda}\lambda^k}{k!}$, $x = 0, 1, 2 ....$, where $\lambda > 0$ is the mean. λ is a constant that is specified for the particular circumstances.

Example 3.3.4.1: A random variable, X, has a Poisson distribution with mean, $\mu = 2$. Find the variance and compute $P(1 \le X)$.

Solution: The density function for the random variable, X, with a Poisson distribution is:

$$P(X = k) = \dfrac{e^{-\lambda}\lambda^k}{k!}, \quad x = 0, 1, 2 ....$$

But, recall that Poisson distribution has the unique property that the mean is equal to the variance. $\mu = \lambda = \sigma^2$. Hence: $\sigma^2 = \mu = \lambda = 2$.

$$P(1 \le X) = 1 - P(X = 0)$$

$$= 1 - \dfrac{e^{-\lambda}\lambda^0}{0!} = 1 - \dfrac{e^{-2}.2^0}{0!} = 1 - e^{-2} = 1 - \dfrac{1}{e^2}$$

where $e = 2.71828$

$$\dfrac{1}{(2.71828)^2} = 1 - 0.135335465 = 1 - 0.135 = \underline{0.865}$$

Example 3.3.4.2: The average number of days a Forest Nursery Site is inaccessible during the rainy season, in Sakponba Forest Reserve (Benin-City), is 4 days. What is the probability that the Forest Nurseries in Sakponba area will be inaccessible for 6 days, during the rainy season?

Solution: We will solve this problem in two ways: (a) Using the Poisson distribution formula and (b) Using Poisson Cumulative Distribution Table (Walpole, 1974: Table A.3).

a) Using the Poisson distribution formula, with $x = 6$ and $\mu (\lambda) = 4$, we find that:

$$P(X = 6) = p(x; \mu) = \dfrac{e^{-\mu}\mu^x}{x!}, \quad x = 0, 1, 2, ... \text{ Substitute for } x = 6; \mu = 4.$$

$$= \dfrac{e^{-4}4^6}{6!} = \dfrac{e^{-4}x4{,}096}{720} = e^{-4}(5.6888889)$$

$e = 2.71828$

$$= \dfrac{5.6888889}{(2.71828)^4} = \dfrac{5.6888889}{54.59800313} = 0.104195914$$

$0 = 0.1042.$

b) Using the Poisson Distribution Table: that Forest Nurseries will be inaccessible for exactly six days, during the rainy season:

$$P(X = 6) = p(x; \mu) = \frac{e^{-\mu}\mu^x}{x!}, x = 0, 1, 2, 3 \ldots$$

$$= \frac{e^{-4}4^6}{6!} = \sum_{x=0}^{6} p(x; 4) - \sum_{x=0}^{5} p(x; 4) = \sum_{x=0}^{6} p(6; 4) - \sum_{x=0}^{5} p(6; 4)$$

$$= 0.8893 - 0.7851 = \underline{0.1042} - \text{same as (a) above.}$$

Example 3.3.4.3: The average number of field rabbits per hectare in a ten-hectare plantation of <u>Millicia excelsa</u> (Iroko) is estimated to be 10. Find the probability that a given hectare contains more than 15 rabbits.

Solution: Let X represent the number of rabbits per hectare of plantation.

a) Using the Poisson probability function, where $\mu = 10$ and $x = 15$, we have:

$$P(X > 15) = p(x; \mu) = p(15; 10) = \frac{e^{-\mu}.\mu^x}{x!} = \frac{e^{-10}.10^{15}}{15!}$$

$$= \frac{e^{-10}.10^{15}}{1.307674368 \times 10^{12}}$$

$$= \frac{(2.71828)^{-10}.10^{15}}{1.307674368 \times 10^{12}} = \frac{10^{15}}{(2.71828)^{10} \times 1.307674368 \times 10^{12}}$$

$$= \frac{10^{15}}{2.880325099 \times 10^{16}} = \underline{0.034718303}$$

b) Using the Poisson probability table, we have:

$$P(X > 15) = 1 - P(X \leq 15)$$

$$= 1 - \sum_{x=0}^{15} p(x; 10) = \sum_{x=0}^{15} \frac{e^{-\mu}.e^x}{x!} = \sum_{x=0}^{15} \frac{e^{-10}.e^{15}}{15!}$$

$$= 1 - 0.9513 = \underline{0.0487}*$$

*We see that the Table estimation is only an approximation and sometimes not as precise as the formula estimate, as we see in this example.

Example 3.3.4.4: Medical Pathologists say that the white blood cell count of a healthy individual can average as low as 6000 per cubic millimeter of blood. To detect a white cell deficiency, a 0.001 millimeter drop of blood is taken and the number of white cells, X, is found. How many white cells are expected in a healthy individual? If at most two are found, is there any evidence of a white cell deficiency?

Solution: First, estimate the average number of white blood cells in $6000m^3$ millimeter of blood: $\therefore$ n = 6000.

p = 0.001

μ = n.p = 6000 x 0.001 = 6

At most 2, is $P(X \leq 2) = p(x; \mu) = \frac{e^{-\mu}\mu^x}{x!} = \sum_{x=0}^{2} \frac{e^{-6}6^x}{x!}$, x= 0,1,2.

Using the **Poisson Probability Table** under $\mu = 6$ and $x = 2$;

$$P(X \leq 2) = \sum_{x=0}^{2} \frac{e^{-6}6^x}{x!} = \frac{e^{-6}6^0}{0!} + \frac{e^{-6}6^1}{1!} + \frac{e^{-6}.6^2}{2!} = 0.062$$

Example 3.3.4.5: In Nigeria, one out of a thousand persons, is a medical doctor. If a random sample of 8000 Nigerians is taken, find the probability that fewer than 7 medical doctors will be found among them.

Solution:

a) Since the sample size, n, is very large i.e. n = 8000 and p = 0.001 is very small, it is basically a Poisson problem.

First, estimate the average number of doctors among 8000 Nigerians as:

n = 8000

p = 0.001 i.e. one Nigerian out of 1000 Nigerians

μ = n x p = 8000 x 0.001 = 8

Let X represent the number of doctors. We want to estimate:

P(X < 7) i.e. the number of doctors is less than 7

$$P(X < 7) = p\ (x;\ \mu) = \sum_{x=0}^{6} \frac{e^{-\mu}\mu^x}{x!}, \quad x = 0,1,\ 2\ldots$$

$$\sum_{x=0}^{6} \frac{e^{-\mu}\mu^x}{x!} \text{ implies: } \frac{e^{-8}.8^0}{0!} + \frac{e^{-8}.8^1}{1!} + \frac{e^{-8}.8^2}{2!} + \frac{e^{-8}.8^3}{3!} + \cdots \frac{e^{-8}.8^6}{6!}$$

But the Poisson Table under $\mu = 8$ and $x = 6$ will give us this sum.

Thus, $\sum_{x=0}^{6} \frac{e^{-8}.8^6}{6!} = 0.3134$

b) Suppose we are now interested in estimating the probability of at most 7 medical doctors, what is the probability?

Solution:

n = 8000

p = 0.001

μ = n.p = 8000 x 0.001 = 8. The probability of at most 7 is given by:

$$P(X \leq 7) = \sum_{x=0}^{7} \frac{e^{-7}.8^x}{x!} = 4530 \text{ (From Poisson Probability Table)}.$$

3.3.4.1 How to Use the Poisson Probability Tables to Estimate Probabilities

Using the Poisson Probability Tables is very similar to using the **Binomial Probability Tables.** For the Binomial Tables, we need to know the number of trials, n (5 to 20); the number of successes, k (or r), and the probability of a success, p. The numbers of trials (n) and successes (k or r) are listed on the left-hand margins and the various probabilities of (0.1 to 0.9) successes are listed on the right-hand columns.

The Tables are not useful, if the number of trials (n > 20) are larger than 20. To estimate any probability, locate the n and k (or r) and trace the k (or r) to the listed probability of successes and just read off the estimated probability. However, we must know whether:

$$P(X < k) = \sum_{x=0}^{k-1} b(x; n, p).$$

$$P(X \leq k) = \sum_{x=0}^{k} b(x; n, p).$$

$$P(X = k) = \sum_{x=0}^{k} b(x; n, p) - \sum_{x=0}^{k-1} b(x; n, p).$$

$$P(X > k) = 1 - \sum_{x=0}^{k} b(x; n, p)$$

$$P(X \geq k) = \sum_{x=0}^{k} b(x; n, p) = 1 - P(X < k) = 1 - \sum_{x=0}^{k-1} b(x; n, p)$$

For the **Poisson Probability Distribution Tables**, the left-hand margin only lists the number of successes, r, since the Poisson distribution is an infinite series. The number of successes, r's, comes in a series of 0 to 6, with a series of means, μ's from 0.1 to 0.9; r's 0 to 16 with a series of μ's 1.0 to 5.0; r's 0 to 24 with μ's 5.5 to 9.5; and r's 0 to 37 with μ's 10.0 to 18.0.

Once the mean is known, to estimate the probability of any number of successes, trace the value of r to the column of the estimated mean. Then read off the probability value of the intersection of r and μ. Of course, we need to exercise caution concerning the direction of the required probability as to whether the probability is:

$$P(X < r) = \sum_{x=0}^{r-1} \frac{e^{-\mu}.\mu^{x}}{x!}.$$

$$P(X \leq r) = \sum_{x=0}^{r} \frac{e^{-\mu}.\mu^{x}}{x!}.$$

$$P(X = r) = \sum_{x=0}^{r} \frac{e^{-\mu}.\mu^{x}}{x!} - \sum_{x=0}^{r-1} \frac{e^{-\mu}.\mu^{x}}{x!}.$$

$$P(X > r) = 1 - \sum_{x=0}^{r-1} \frac{e^{-\mu}.\mu^{x}}{x!}.$$

$$= 1 - P(X < r) = 1$$

$$P(X \geq r) = \sum_{x=0}^{r} \frac{e^{-\mu}.\mu^{x}}{x!} - \sum_{x=0}^{r-1} \frac{e^{-\mu}.\mu^{x}}{x!}.$$

3.3.4.1 The Geometric Distribution

The Geometric distribution is very similar to the binomial experiment where the number of trials are identical and independent, each of which can result in one of two outcomes of success or failure, and the probability of success is equal to p and remains constant from trial to trail (Scheaffer and Mendenhall, 1975).

However, instead of the number of successes that occur in n trials, **the Geometric Random Variable, X, is the number of trials on which the first success occurs. Thus, the number of trials in a geometric experiment is not fixed but can assume an infinite number of positive integral values, 1, 2, 3, …** (Tsokos, 1972).

The trial x can yield the first success only if we have observed an unbroken run of x – 1 failures during the first x – 1 trials. **The probability that this run of x – 1 failures occurs is $(1 – p)^{x-1}$ and this probability followed by a success is: $(1 – p)^{x-1}p$.**

Thus, a random variable, X, has the geometric distribution, if its probability density function is:

$$f(x; p) = p(1 – p)^{x-1}, x = 1, 2, 3, …$$

The geometric distribution is a one-parameter distribution and its successive terms constitute **a Geometric Progression**.

The cumulative distribution of the geometric density is given by:

$$P(X = x) = \sum_{x=1}^{\infty} p(1 – p)^{x-1} = 1$$

Example 3.3.5.1: In a game of billiards, a player continues to play until he misses a shot. If a particular player misses any of his shots with probability, p = 1/4, what is the probability that this player's turn will last (a) exactly six shots? (b) at most five shots? (c) at least four shots?

Solution:

Assuming that the shots are independent and p = ¼ remains the same for every shot, we can apply the geometric distribution.

a) $P(X = 6) = p(1 – p)^{x-1}, x \geq 1$ i.e., probability of exactly six shots.

Substitute for p = 1/4 and x = 6

$$P(X = 6) = \left(\frac{1}{4}\right)\left(\frac{3}{4}\right)^5 = (0.25)(0.75)^5 \cong (0.25)(0.237304687)$$

$$= 0.059326171 \approx \underline{0.05933}$$

b) What is the probability of at most five shots?

$$P(X \leq 5) = \sum_{x=1}^{5} p(1 – p)^{x-1} = \sum_{x=1}^{5} \left(\frac{1}{4}\right)\left(\frac{3}{4}\right)^{x-1} = \sum_{x=1}^{5}(0.25)(0.75)^{x-1}$$

Substitute the values for x = 1, 2, 3, 4; since geometric distribution has no cumulative Tables.

$\sum_{x=1}^{5}(0.25)(0.75)^{x-1} = (0.25)(0.75)^0 + (0.25)(0.75) + (0.25)(0.75)^2 + (0.25)(0.75)^3 + (0.25)(0.75)^4$

$=0.25 + 0.1875 + 0.140625 + 0.10546875 + 0.079101562$

$$= 0.728320312 \approx \underline{0.72832}$$

c) What is the probability of at least four shots?

$$P(X \geq 4) = \sum_{x=4}^{\infty} \left(\frac{1}{4}\right)\left(\frac{3}{4}\right)^{x-1} = \sum_{x=4}^{\infty}(0.25)(0.75)^{x-1} = 1 - P(X < 4)$$

$= 1 - P(X < 4) = \sum_{x=1}^{3}(0.25)(0.75)^{x-1}$. Substituting for x = 1, 2, 3, we have:

From (b) above, we have:

$1 - [(0.25)(0.75)^0 + (0.25)(0.75)^1 + (0.25)(0.75)^2 + (0.25)(0.75)^3 + (0.66)(0.34)^4 + (0.66)(0.34)^5]$

$- [0.66 + 0.2244 + 0.076296 + 0.2594064 + 0.0088108176 + 0.002998737984] =$

$1 - 0.998455195 = 0.001544804416 \approx \underline{0.001545}$

$P(Y = y) = p(1 - p)^{y-1}$, y = 1, 2, 3, 4, 5 . . .: p = 0.5

y	$p(1 - p)^{y-1}$
1	0.5
2	0.25
3	0.125
4	0.0625
5	0.03125
6	0.015625
7	0.0078125
8	0.0030625

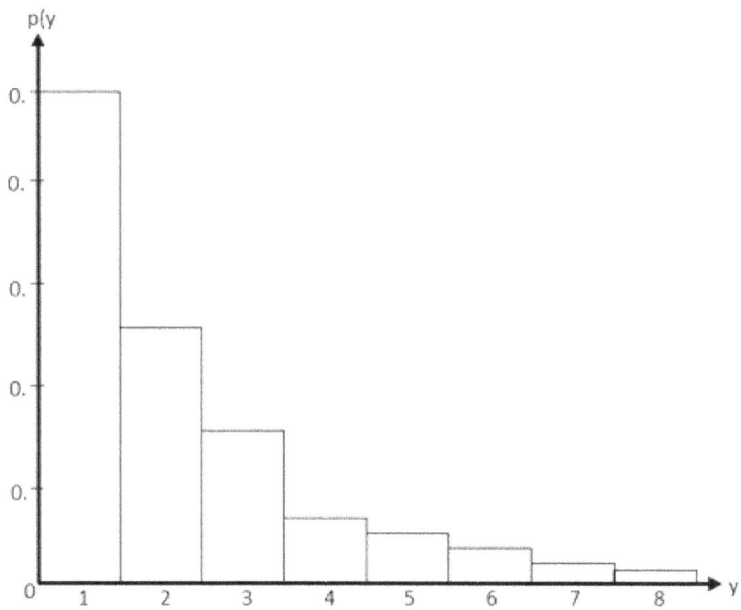

Figure 3.3.5: Geometric Probability Distribution with p = 0.5

3.3.5 The Hypergeometric Probability Distribution

The probability distribution which arises in a binomial population in which sampling is done without replacement is called the Hypergeometric Distribution (Byrkit, 1975). In general, according to Walpole (1974) and Walpole and Myers (1972), in a Hypergeometric Distribution, we are interested in the probability of selecting X successes from the k items labelled "success" and n-x failures from the N-k items labelled "failures", when a random sample of size n is selected from a finite population of size N.

Walpole (1974) stated that a Hypergeometric Distribution possesses the following two properties:

1) A random sample of size n is selected from a population of N items.
2) k of the N items may be classified as successes and N-k classified as failures.

The probability distribution of the hypergeometric variable X is denoted by h (x; N, n, k) and its values depend on the number of successes, k, in the set N, from which we select n items. **Thus, the probability distribution of the hypergeometric random variable X, the number of successes in a random sample of size n is given by:**

$$h\ (x;\ N,\ n,\ k) = \frac{\binom{k}{x}\binom{N-K}{n-k}}{\binom{N}{n}},\ x = 0,\ 1,\ 2,\ \ldots,\ n.$$

The hypergeometric distribution is useful in industrial engineering problems of quality inspection and other physical problems encountered by personnel directors in the selection of the "best" in a finite set of elements.

Some applications of the hypergeometric distribution are shown below:

Example 3.3.6.1: A group of 20 Engineers applied for employment in UNIBEN; 10 of them are selected for employment. What is the probability that the 10 selected include all the 5 best engineers in the group of 20?

Solution:

N = 20, n = 10, k =5, x = 5

$$P(X = 5) = \binom{k}{x}\binom{N-K}{n-k}\Big/\binom{N}{n}$$

$$= \binom{5}{5}\binom{15}{5}\Big/\binom{20}{10} = \left(\frac{15!}{5!10!}\right)\left(\frac{5!}{0!5!}\right)\Big/\left(\frac{20!}{10!10!}\right)$$

$$= \frac{15!}{5!10!}\ \frac{10!10!}{20!} = \frac{10.9.8.7.6}{20.19.18.17.16} = \frac{30,240}{1,868,488}$$

$$= 0.016253869 \approx \underline{0.01625}$$

Like other discrete probabilities, the Hypergeometric distribution also has a cumulative distribution:

$$F(x) = P(X = x) = h(x;\ N,\ n,\ k) = \sum_{x=0}^{n} \binom{k}{x}\binom{N-K}{n-k}\Big/\binom{N}{n},$$ i.e., the probability of exactly x successes and n-k failures in a sample of n items.

$$\frac{1}{\binom{N}{n}}\sum_{x=0}^{n}\binom{k}{x}\binom{N-K}{n-k} = \frac{\sum_{x=0}^{n}\binom{k}{x}\binom{N-K}{n-k}}{\binom{N}{n}}$$

Tsokos (1972) has pointed out that calculating probabilities using the Hypergeometric distribution becomes laborious when n is large. Tsokos (1972) has presented probability and cumulative probability Tables (Table 5).

Example 3.3.6.2: A committee of size 5 is to be selected at random from 5 men and 3 women. Find the probability distribution for the number of women on the committee.

Solution:

Let X be the random variable of the number of women on the committee.

N = 8; n = 5; k = 3

$$H(x; N, n, k) = P(X = x) = \binom{k}{x}\binom{N-K}{n-k}/\binom{N}{n}, x = 0, 1, 2, 3$$

$$= \binom{3}{x}\binom{5}{5-x}/\binom{8}{5}$$

The probability distribution for the number of women in the committee is given by substituting the various values of x into the formula:

$$P(X = 0) = h(0; 8, 5, 3) = \binom{3}{0}\binom{5}{5}/\binom{8}{5} = \frac{3!}{0!3!}\cdot\frac{\frac{5!}{5!}}{\frac{5!3!}{8!}} = \frac{3!}{8.7.6} = \frac{1}{56}$$

$$= 0.017857$$

$$P(X = 1) = h(1; 8, 5, 3) = \binom{3}{1}\binom{5}{4}/\binom{8}{5} = \frac{3!}{2!1!}\cdot\frac{\frac{5!}{1!4!}}{\frac{5!3!}{8!}} = \frac{3.5.3!}{8.7.6} = \frac{15}{56}$$

$$= 0.267857$$

$$P(X = 2) = h(2; 8, 5, 3) = \binom{3}{2}\binom{5}{3}/\binom{8}{5} = \frac{3!}{1!2!}\cdot\frac{\frac{5!}{3!2!}}{\frac{5!3!}{8!}} = \frac{\frac{3.5.4^2}{2!}}{\frac{3!}{8.7.6}} = \frac{(3.5.2)}{56} = \frac{30}{56}$$

$$= 0.535714285$$

$$P(X = 3) = h(3; 8, 5, 3) = \binom{3}{3}\binom{5}{2}/\binom{8}{5} = \frac{5!}{2!3!}\cdot\frac{\frac{5!}{3!2!}}{\frac{5!3!}{8!}} = 1.\frac{\frac{5.4^2}{2!}}{\frac{5!3!}{8.7.6}} = \frac{5.2}{8.7} = \frac{10}{56}$$

$$= 0.178571428$$

Thus, the Hypergeometric distribution of X is:

x	P(X = x)
0	0.17857
1	0.267857
2	0.535714285
3	0.178571428

Example 3.3.6.3: A manufacturer of light bulbs ships them in lots of 20. The manufacturer randomly selects a sample of 5 light bulbs from each lot and rejects the lot, if more than one defective bulb is observed. If a lot is rejected, each bulb in the lot is tested. If a lot contains four defectives, what is the probability that it will be rejected?

Solution: Let X be the number of defectives in the sample (n = 5).

The lot, N = 20; n = 5; k = 4. Lot is rejected if X = 2, 3 or 4.

$h(x; N, n, k) = \binom{k}{x}\binom{N-K}{n-k}/\binom{N}{n}$. P(Rejecting the lot) = P(X≥2) = P(X=2) + P(X=3) + P(X=4).

$$= 1-P(X<2)=1-P(0)-P(1) \quad = 1-\frac{\binom{4}{0}\binom{16}{4}}{\binom{30}{5}} - \frac{\binom{4}{1}\binom{16}{4}}{\binom{30}{5}}$$

$$= 1 - \left(\frac{4!}{0!4!}\cdot\frac{16!}{5!11!}\right)/\left(\frac{20!}{5!15!}\right) - \left(\frac{4!}{1!3!}\cdot\frac{16!}{4!12!}\right)/\left(\frac{20!}{5!15!}\right)$$

$$= 1 - \frac{4,368}{15,504} - \frac{7,280}{15,504} = 1 - 0.281733746 - 0.469556243$$

$$= 1 - 0.751289989 = 0.24871001.$$

3.4 The Mean and Variance of a Probability Distribution

3.4.1 The Mean of a Probability Distribution

The arithmetic average, (or mean) of a set of values is equal to their sum divided by the number of values, (n) – in fact, this is the **EXPECTED VALUE (E(X))** of the set, since each of the n values has a probability of $1/n$. The mean of a probability distribution is usually symbolized by the Greek letter μ (mu) (Byrkit, 1975).

According to Byrkit (1975), "For any discrete probability distribution, μ = E(x); that is, $\mu = \Sigma x_i.P(x_i)$, for each value of the random variable X.

Example 3.4.1: Find the mean value of one toss of a single die and its variance.

Solution (a): Each face on the die has equal probability, that is, they are equiprobable with

$$p = 1/6; \quad \mu = \Sigma x_i.P(x_i).$$

Table 3.4.1: Calculating the Mean of a Random Variable

x_i	$P(x_i)$	$x_i.P(x_i)$
1	$1/6$	$1 \times 1/6 = 1/6$
2	$1/6$	$2 \times 1/6 = 2/6$
3	$1/6$	$3 \times 1/6 = 3/6$
4	$1/6$	$4 \times 1/6 = 4/6$
5	$1/6$	$5 \times 1/6 = 5/6$
6	$1/6$	$6 \times 1/6 = 6/6$
$\mu = \Sigma x_i.P(x_i) = 21/6 = 7/2 = 3.5$		

Tsokos (1972) observed that **when we speak of "expected value", or "mean" of a random variable X, we refer to an ideal, or theoretical average.** If a given experiment is repeated many times, we would expect the average value of the random variable involved to be near its expected value.

3.4.2 The Variance of a Probability Distribution

The variance, symbolized by σ^2 (sigma squared), is a measure that will differentiate between distributions clustered closely about the mean and those more widely scattered. If the measurements are all close to the mean, the deviations from the mean will be small and hence σ^2 will be small; but if the measurements are widely dispersed about the mean, the σ^2 will be large.

For any discrete probability distribution, $\sigma^2 = \Sigma(x_i-\mu)^2$ i.e. $\sigma^2 = \Sigma(x_i - \mu)^2 P(x_i)$, for each value of the random variable X. To estimate the variance:

- we must estimate the mean as: $\mu = \Sigma x_i . P(x_i)$;
- then find $(X_i - \mu)$ i.e., the difference between each value of X_i and the mean;
- then square $(X_i - \mu)2$ and multiply it by the probability, $P(x_i)$; and
- sum it as: $\Sigma(x_i - \mu)2.P(x_i)$.

Thus: $\sigma^2 = \Sigma(x_i-\mu)^2.P(x_i) = (\Sigma x_i^2.P(x_i)) - \mu^2$

Table 3.4.2: Calculating the Variance of a Random Variable.

X	P(x)	x.P(x)	$x - \mu$	$(x - \mu)^2$	$(x - \mu)^2.P(x)$
1	$1/6$	$1/6$	$-5/2$	$25/4$	$25/24$
2	$1/6$	$2/6$	$-3/2$	$9/4$	$9/24$
3	$1/6$	$3/6$	$-1/2$	$\frac{1}{4}$	$1/24$
4	$1/6$	$4/6$	$\frac{1}{2}$	$1/4$	$1/24$
5	$1/6$	$5/6$	$3/2$	$9/4$	$9/24$
6	$1/6$	$6/6$	$5/2$	$25/4$	$25/24$
		$\mu = 21/6 = 7/2$ $= 3.5$	0	70/4	$\sigma^2 = 70/24$ $= 2.9166667 \approx$ 2.92

47

The Standard Deviation of a Probability Distribution

The standard deviation, symbolized by σ, is the positive square root of the variance i.e., the square root of the mean squared deviation:

$\sigma = \sqrt{\sigma^2} = \sqrt{\sum_{i=1}^{n}(x_i - \mu)^2 P(x_i)}$, for each xi in the distribution.

From Table 3.4.2. above,

$\sigma2 = 2.92.$

$\sigma = \sqrt{\sigma^2} = \sqrt{2.92} = 1.71.$

3.4.3 The Algebra of Expectations

According to Winkler and Hays (1975), whether the random variable is discrete or continuous, the expectation of a random variable is a kind of weighted sum of values, and thus the rules for the algebraic treatment of expectations are basically extensions and applications of the rules of summation.

However, we will state the rules of mathematical expectations and variances of the discrete random variables without going through all the gory algebraic proofs:

Rule 1:

If g(x) is some function of X, then: $E(g(x)) = \sum_x g(x)P(x)$.

Rule 2:

If 'a' is some constant number, then: $E(a) = a = \Sigma aP(x) = a\Sigma P(x) = a$

Rule 3:

If 'a' is some constant real number and X is a random variable with expectation E(X), then: $E(aX) = aE(x) = \Sigma ax_iP(x_i) = a\Sigma x_iP(x_i) = aE(X)$.

Rule 4:

If 'a' is a constant real number and X is a random variable, then:

$E(X + a) = E(X) + a$

$= \Sigma(x_i + a).P(x_i)$

$= \Sigma x_iP(x_i) + a\Sigma P(x_i)$

$= E(X) + a$

Rule 5:

If X is a random variable with expectation E(X) and Y is another random variable with expectation E(Y), then,

$E(X + Y) = E(X) + E(Y)$.

3.4.4 The Algebra of Variances

For any distribution, the index V(X), or σ^2, is equal to the expectation of the squared deviations from the mean:

$V(X) = \sigma^2 = E(X - \mu)^2$, is called the variance of the distribution.

Rule 1:

The variance of a random variable is equal to the expectation of the square of the random variable minus the square of the expectation of the random variable: $V(X) = \sigma^2 = E(X^2) - [E(x)]^2$.

Note that: $V(X) = \sigma^2 = E(X - \mu)^2$

$= E(X^2 - 2\mu X + \mu^2)$

$= E(X^2 - E(2\mu X) + E(\mu^2)$

μ is a constant hence $V(X) = E(X^2) - 2\mu E(X) + \mu^2$

Since $\mu = E(X)$,

$V(X) = E(X^2) - 2[E(X)]^2 + [E(X)]^2$

$= E(X^2) - [E(X)]^2 = \sigma^2$

Rule 2:

If 'a' is some constant real number, and if X is a random variable with expectation E(X) and variance $\sigma 2$, then the random variable (X + a) has variance σ^2:

$V(X + 1) = V(X) = \sigma^2$.

We can show that $V(X + a) = E[(X + a) - E(X + a)]^2$

But $E(X + a) = E(X) + a$

$E(X + a) = E[(x + a) - (E(X) + a]^2$

$= E[(X) - E(X)]^2 = \sigma^2$.

Rule 3:

If 'a' is some constant real number, and if X is a random variable with variance $\sigma 2$, the variance of the random variable aX is:

$V(aX) = a^2 V(X) = a^2 \sigma^2$.

Note that $V(aX) = E[(aX)^2] - [E(aX)]^2$

$= a^2 E(X^2) - a^2[E(X)]^2$

$= a^2(E(X) - [E(X)]^2$

$= a^2 \sigma^2$

3.4.5 Means and Variances of Some Common Probability Distributions

In the preceding section, we used the concept of mathematical expectation to develop Rules for finding the expected values and variances of random variables or functions of random variables. Using these rules, we will only state without proofs, the expected values and variances of some common discrete probability distributions.

1. **The Discrete Uniform Distribution (Walpole and Myers, 1972).**

   The mean and variance of the discrete uniform distribution are:

   $$\text{Mean} = \mu = \frac{\sum\limits_{i=1}^{k} x_i}{k}$$

   $$\text{Variance} = \sigma^2 = \frac{\sum\limits_{i=1}^{k} (x_i - \mu)^2}{k}$$

2. **The Bernoulli Distribution** (Tsokos, 1972; Freund, 1971):

   For the Bernoulli Distribution, the mean and variance are:

   $$\text{Mean} = \mu = p$$

   $$\text{Variance} = \sigma^2 = p(1 - p) = pq$$

3. **The Binomial Distribution** (Byrkit, 1975; Sheaffer and Mendenhall, 1975):

   The mean and variance of the binomial distribution are given by:

   $$\text{Mean} = \mu = np$$

   $$\text{Variance} = \sigma^2 = np(1 - p) = npq$$

4. **The Poisson Distribution** (Sheaffer and Mendenhall, 1975; Walpole and Myers, 1972; Tsokos, 1972):

   The mean and variance of the Poisson distribution both have the same value μ, or λ.

   $$\text{Mean} = \mu \text{ (or } \lambda)$$

   $$\text{Variance} = \sigma^2 = \mu \text{ (or } \lambda), \text{ i.e. } \mu = \sigma^2.$$

   Note: It is only the Poisson distribution that has this unique property where the mean and the variance are the same:

   $$\text{Mean} = \mu = \text{Variance} = \sigma^2$$

   When the sample size, n, is very large (or approaches infinity) and the probability of success, p, is very small (or approaches zero), the Poisson distribution can be used to approximate binomial probabilities.

5. **The Geometric Distribution** (Tsokos, 1972).

   The mean and the variance of the Geometric Distribution are given by:

Mean $= \mu = \dfrac{1}{p}$

Variance $= \sigma^2 = \dfrac{q}{p^2}$

6. **The Hypergeometric Distribution** (Tsokos, 1972; Walpole and Myers, 1972).

The mean and the variance of the Hypergeometric Distribution are given by:

Mean $= \mu = \dfrac{nk}{N}$

Variance $= \sigma^2 = \dfrac{N-n}{N-1} . n . \dfrac{k}{N}\left(1 - \dfrac{k}{N}\right)$

where: k = number of successes in the sample.

n = random sample size.

N = number of items in the population.

Note: When n is small, relative to N, the correction factor $(N - n)/(N - 1)$, is negligible and can be approximated to 1.

This reduces the variance to:

$\sigma^2 = n . \dfrac{k}{N}\left(1 - \dfrac{k}{N}\right)$

The means and the variances of these common discrete probability distributions have merely been stated, without the tedious mathematical proofs, because this text is intended for students in agriculture, social sciences, medicine, pharmacy and other non-mathematical sciences. In these areas, statistics is only applied as a tool to solve problems, without the rigorous mathematical proofs that may be required in mathematics or engineering.

4 CONTINUOUS RANDOM VARIABLES AND PROBABILITY DISTRIBUTIONS

Tsokos (1972) stated that a continuous random variable may assume a non-denumerable or non-countable number of values; and Walpole and Myers (1972) explained that a continuous random variable has a probability of zero of assuming exactly any of its values and therefore, its probability distribution cannot be given in a tabular form – but it does have a formula. The probability distribution of a continuous random variable, X, is designated f(x) and usually called **probability density function.**

For example, if X is a random variable whose values are the heights of all 100-level University of Benin undergraduates of 20 years old, between 5 feet and 6 feet, there is an infinite number of heights – from 5.5 to 6.5 feet. For any continuous random variable, the probability of randomly selecting an undergraduate that is exactly 5 feet tall is zero.

Therefore, the probability distribution of a continuous random variable cannot be presented in a tabular form. But a continuous random variable is defined over a continuous sample space, the graph of f(x) will be continuous and may take several forms (Walpole, 1974; Walpole and Myers, 1972) shown here:

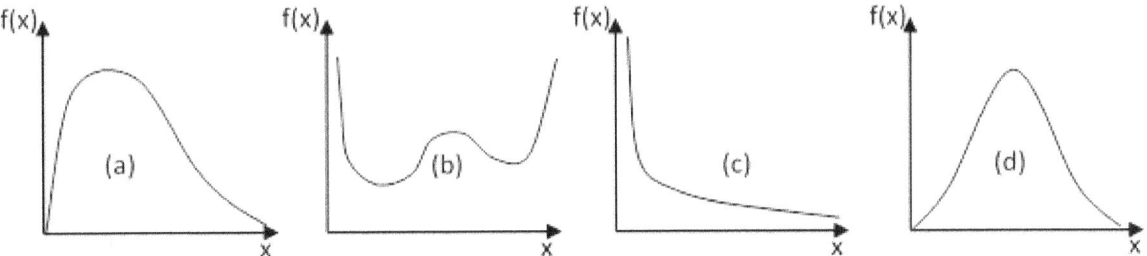

Figure 4.1: Typical density functions

A probability density function is constructed so that the area under its curve bounded by the x-axis is equal to 1, when computed over the range of X for which f(x) is defined.

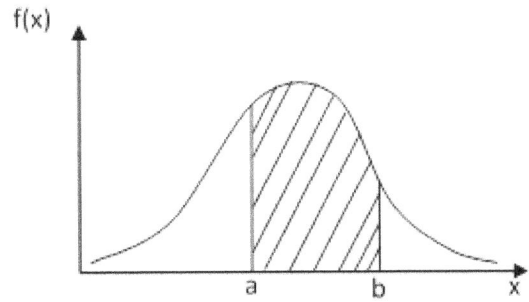

Figure 4.2: The probability that x assumes a value between a and b is equal to the shaded area

Definition: The function f(x) is called a probability density function for the continuous random variable, X, if the total area under its curve bounded by the x-axis is equal to 1, and if the area under the curve between any two ordinates, x = a and x = b, gives the probability that X lies between a and b (Walpole, 1974).

Although we have an estimate of the function of f(x) in Figure 4.2 above, we still have no knowledge of its formula or equation, and therefore cannot find the area that has been shaded. Hence, we resort to some elementary analytic geometry for the shapes of parabolas, hyperbolas, circles, ellipses, etc., which have well-known forms of equation.

From integral calculus, the density function between the ordinates at x = a and x = b is given by:

$$P(a < X < b) = \int_a^b f(x)dx$$

Generally, Walpole and Myers (1972) defined the function f(x) as a probability density function for the continuous random variable, X, defined over the set of real numbers R, if:

1. f(x) ≥ 0 for all x ∈R
2. $\int_{-\infty}^{\infty} f(x)dx = 1$.
3. $P(a < X < b) = \int_a^b f(x)dx$.

In addition, the authors also defined the cumulative distribution, F(x), of a continuous random variable X with density function, f(x), as:

$$F(x) = P(X \leq x) = \int_{-\infty}^{x} f(t)dt.$$

Therefore, P(a < X < b) = F(b) – F(a)

Walpole and Myers (1972) also stressed that many continuous distributions can be represented graphically by the characteristic bell-shaped curve. Its equation of the probability density function, f(x), is as well-known as that of a parabola, or a circle and depends on only the determination of two parameters (μ and σ2).

From the estimated parameters, we can write the estimated equation and using appropriate tables, we may find any probabilities.

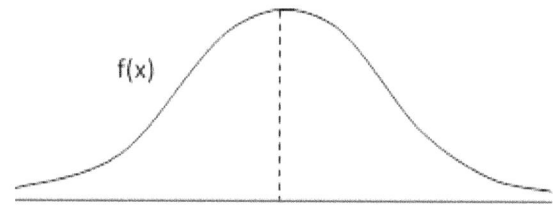

Figure 4.3: The Bell-shaped curve

There are many continuous probability density fractions. However, we shall only discuss the normal distribution, without mathematical derivation and proof - - but only its application to solve everyday problems and hypothesis testing.

4.1 THE NORMAL (OR GAUSSIAN) DISTRIBUTION

The normal density function (Tsokos, 1972; Walpole and Myers, 1972) which is the most important continuous probability distribution in the entire field of statistics is the **Normal Distribution** because it has unique mathematical properties that can be applied to practically any physical problem in nature, industry and research.

Its graph, called the Normal curve, is the symmetrical bell-shaped curve that describes so many sets of data that occur in nature, industry and research. The Normal Distribution, also called the Gaussian Distribution was discovered by De Moivre in 1733 as a limiting form of the binomial distribution but named it in honour of Gauss (1777-1855) who also derived its equation from a study of errors in repeated measurements of the same quantity.

A random variable, X, having the bell-shaped distribution below is called a normal random variable (Walpole and Myers, 1972):

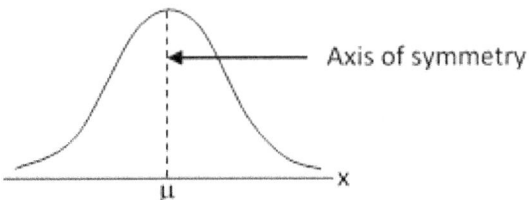

Figure 4.1: The normal curve

The mathematical equation for the probability distribution of the continuous normal variable depends on the two parameters, μ and σ, its mean and standard deviation. The density function of X is denoted by n(x; μ, σ).

Definition (Walpole, 1974): If X is a normal random variable with mean, μ, and variance, σ2, then the equation of the normal curve is:

$$n(x; \mu, \sigma) = \frac{1}{\sqrt{2\Pi\sigma^2}} e^{-\frac{1}{2}\left(\frac{x-\mu}{\sigma}\right)^2}, -\infty < x < \infty$$

where: e = 2.71828 and Π = 3.14159.

Once μ and σ are specified, the normal curve is completely determined. The density function, n(x; μ, σ) or f(x; μ, σ) is symmetrical around the mean, μ, and therefore, the mean, median and mode of the normal distribution are all at the same point.

Thus, **the parameter, μ, is a "location parameter"**, which shifts the mode of the function on the x-axis; **σ is a "shape parameter"**, which alters the shape of the density with respect to a fixed scale - - **as σ increases, the density deviates farther from μ, as shown in Figure 4.2:**

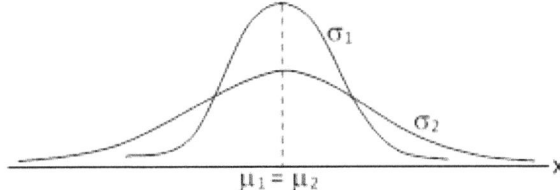

Figure 4.2: Normal Curves with μ1=μ2 and σ2>σ1

Note: The more variable the set of observations are, the lower and wider the corresponding curve will be.

4.2 PROPERTIES OF THE NORMAL CURVE

The normal curve has the following properties (Walpole, 1974; Byrkit, 1975):

1. The mode, which is the point on the horizontal axis where the curve is a maximum, occurs at x = μ.
2. The curve is symmetric about a vertical axis through the mean, μ.
3. The curve has its points of inflection at x = μ ± σ; and is concave downward, if μ − σ < X < μ + σ, and is concave upward otherwise.
4. The normal curve approaches the horizontal axis asymptotically (i.e., the curve never touches the x-axis) as we proceed in either direction away from the mean.
5. The total area under the curve and above the horizontal axis is equal to 1.

4.3 AREAS UNDER THE NORMAL CURVE

The normal curve is dependent on the mean and the standard deviation of the distribution under investigation (Walpole and Myers, 1972). The area under the curve between any two ordinates must also depend on the values μ and σ.

The curve of any continuous probability distribution, or density function, is

constructed so that the area under the curve, bounded by the two ordinates, $x = x_1$, and $x = x_2$ equals the probability that the random variable, X, assumes a value between $x = x_1$ and $x = x_2$. Thus, $P(x_1 < X < X_2)$ is represented by the shaded area in Figure 4.3 below:

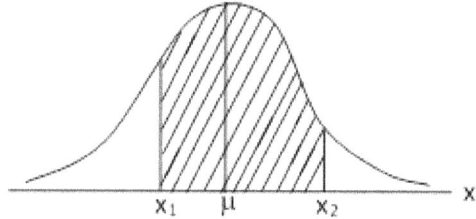

Figure 4.3: $P(x_1 < X < x_2)$ = Area of the shaded portion

The difficulty encountered in solving integrals of the normal density functions (Walpole and Myers, 1972) necessitates the tabulation of normal curve areas for quick reference. **Fortunately, we are able to transform all the observations of any random variable, X, to a new set of observations of a normal random variable, Z, with mean zero and variance 1 as:**

$$Z = \frac{X-\mu}{\sigma} \sim N(0,1).$$

Z is called the standard normal distribution, or a standard deviate.

Whenever X assumes a value x, the corresponding value of Z is given by $z = (x - \mu)/\sigma$.

Therefore, $P(x_1 < X < x_2) = P(z1 < Z < z_2)$.

The transformation to convert observations of a normal random variable into a standard deviate, z, is called **"standardization"** which indicates the number of standard deviation units by which the observation is greater than the mean. We can then look up the Table to find the area under the curve that corresponds to that number of standard deviation units from the mean.

The standard normal distribution is also called the standard deviate, or z-score, or z-value and it is unitless.

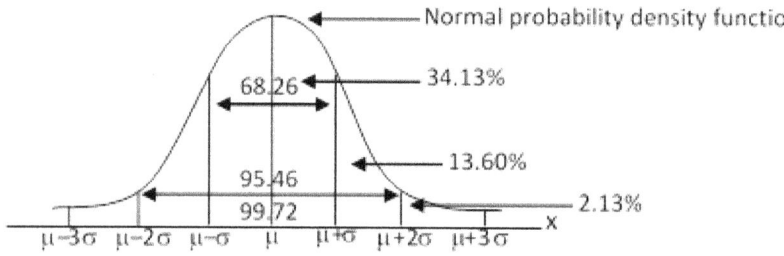

Figure 4.4: Areas under the Normal Curve

The relationship of any value of x to μ can be expressed in terms of the number of standard deviations distant from the mean, μ.

The following percentages of items in a normal frequency distribution lie within the indicated limits:

  (i) $\mu \pm \sigma$ contains 68.26% of the items (or observations).
  (ii) $\pm 2\sigma$ contains 95.46% of the items (or observations).
  (iii) $\pm 3\sigma$ contains 99.73% of the items (or observations

Conversely:

  50% of the items fall between $\mu \pm 0.674\sigma$.

  95% of the items fall between $\mu \pm 1.960\sigma$.

  99% of the items fall between $\mu \pm 2.576\sigma$.

Definition (Walpole, 1974): The distribution of a normal random variable with mean zero and standard deviation equal to 1 is called a standard normal distribution.

4.4 How to Transform Random Variables to Standard Normal Distribution

Any random variable can be easily transformed from its original distribution to standard normal distribution by **"standardizing", or "converting" an x-score to a z-score** by the following equation:

$$z = \frac{x-\mu}{\sigma}.$$

Such standardizations make the calculation of probabilities easier using a **Table of the Standard Normal Distribution** listed at the back of many textbooks on statistics.

Suppose we have original values of the random variable, X, as $x = x_1$ and $x = x_2$, the standard deviate, z, will be:

$$z_1 = \frac{x_1-\mu}{\sigma}$$

$$z_2 = \frac{x_2-\mu}{\sigma}, \text{ As Shown in Figure 4.5:}$$

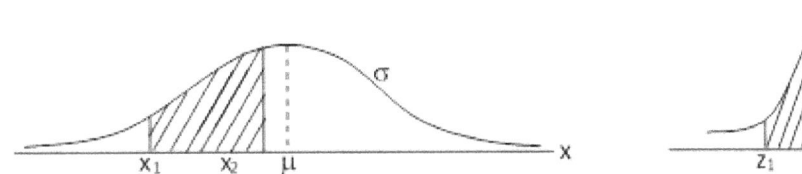

Figure 4.5: The Original (a) and Transformed (b) Normal Populations.

The area under the X curve between the ordinates $x = x_1$ and $x = x_2$ in Figure 4.5a is equal to the area under the z-curve between the transformed ordinates $x = z_1$ and $x = z_2$. Hence $(P(x_1 < X < x_2) = P(z_1 < Z < z_2)$.

4.5 Conversions and Applications Of Standard Normal Variables

Various examples of how to convert x-scores to z-scores will be discussed here to show how to use Normal Distribution to solve several physical problems:

Example 4.5.1:

Given that x has a normal distribution with mean 10 and standard deviation 4, find $P(x < 15)$.

Solution:

$\mu = 10; \sigma = 4$

(i)   It is always advisable to sketch a graph of the density function from the provided data of the problem that we want to solve.

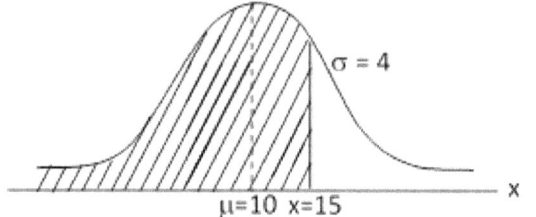

P(x < 15) is the shaded portion

Figure 4.5.1a: Estimating the P(x < 15)

(ii)   Convert the x-score to a z-score as:

$z = \frac{x - \mu}{\sigma}$ by substituting the data given:

$$= \frac{x - 10}{4}$$

$$P(x < 15) = P\left(\frac{x-10}{4} < \frac{15-10}{4}\right)$$

$$= P(z < \frac{5}{4}) = P(z < 1.25)$$

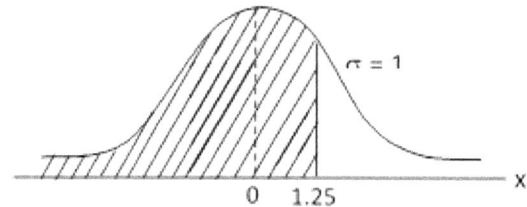

Figure 4.5.1b: P(z < 1.25)

(iii)   Go to the Table A.4 (Walpole, 1974) where areas (i.e. probabilities) under the normal curve are provided.

(iv)   Obtain the probability of the z-value which you calculated i.e. $z = 1.25$. The $P(z < 1.25) = 0.8944$.

Note: How to use z-scores to estimate probabilities will be presented in the next section.

Examples 4.5.2:

Teak seedlings in UNIBEN Nursery grew at a mean height of 50cm in six months, with a standard deviation of 10cm. Find the probability that the growth of the seedlings will be between 45cm and 62cm, assuming the growth is normally distributed.

Solution:

$P(45 < X < 62)$, where $\mu = 50; \sigma = 10$.

(i)   Sketch the graph of the density function from the data provided:

57

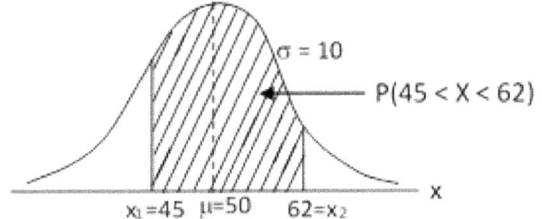

Figure 4.5.2: Estimating P(45 < x < 62)

(ii)   Convert the x-scores to z-scores as:

$$z_1 = \frac{x_1 - \mu}{\sigma} = \frac{45 - 50}{10} = \frac{-5}{10} = -0.5$$

$$z_2 = \frac{x_2 - \mu}{\sigma} = \frac{62 - 50}{10} = \frac{12}{10} = 1.2$$

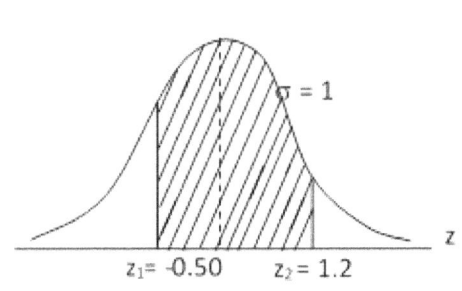

$$P(45 < X < 62) = P\left(\frac{45 - 50}{10} < \frac{X - 50}{10} < \frac{62 - 50}{10}\right)$$

$$= P(-0.5 < Z < 1.2).$$

$$= P(z < 1.2) - P(z < -0.5).$$

(iii)   Go to Table A.4 for areas under the normal curve:

$$P(z < 1.2) - P(z < -0.5)$$

$$= 0.8849 - 0.3085 = \underline{0.5764}$$

Example 4.5.3:

In a normal distribution, a value of 42.1 is 1.3 standard deviations above the mean of 31.7. What is the standard deviation of the distribution?

Solution:

(i)   Sketch the graph of the density function from the data given:

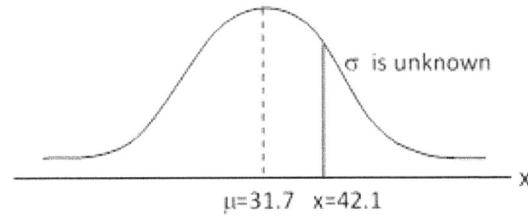

Figure 4.5.3: Estimating the standard deviation where σ is unknown

(ii)   Conversion equation is $z = \frac{x - \mu}{\sigma}$. but σ is unknown.

X = 42.1; μ = 31.7; z = 1.3

Substitute these values in the equation: $z = \frac{x - \mu}{\sigma}$.

$$1.3 = \frac{42.1 - 31.7}{\sigma} = 1.3\sigma = 42.1 - 31.7.$$

$$= 1.3\sigma = 10.4 \quad = \frac{10.4}{1.3} = 8.$$

$$\sigma = \quad 10.4/1.3 = 8.0.$$

Example 4.5.4:

If a normal distribution has a standard deviation of 1.7, and a value of 11.3 lies 2.1 standard deviations below the mean, determine the mean of the distribution.

Solution:

(i)   Sketch the graph of the density function from the data given:

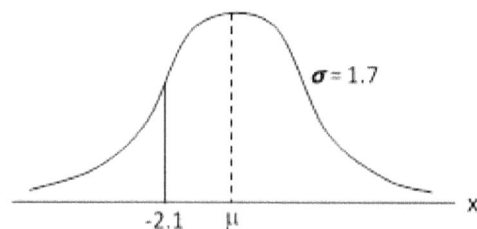

Figure 4.5.4: Estimating the Mean of the Distribution

(ii)   Transformation equation:$Z = \frac{x-\mu}{\sigma}$; hence, $z\sigma = x - \mu$; $x = 11.3$; $\sigma = 1.3$.

z = -2.1 since it is below the mean.

Substitute these values in the transformation equation: $-2.1 = \frac{11.3-\mu}{1.7}$

$= (-2.1) (1.7) = 11.3 - \mu$

$= -3.57 = 11.3 - \mu = -3.57 - 11.3 = -\mu$. Multiply each term by -1

$= 3.57 + 11.3 = \mu$

$\mu = 14.87.$

Example 4.5.5:

The grade-point averages (GPA) of 300 students in the Faculty of Agric, UNIBEN, are approximately normally distributed with a mean of 2.1 and a standard deviation of 1.2. How many of these students of Agric do you expect to have a score between 2.5 and 3.5 inclusive if the GPAs are computed to the nearest tenth?

Solution:

Since the GPAs are recorded to the nearest tenth, we require the area between $x_1 = 2.45$ and $x_2 = 3.55$, i.e. $P(2.45 < X < 3.55) = ?$

(i)   Sketch the graph of the density function according to the data provided.

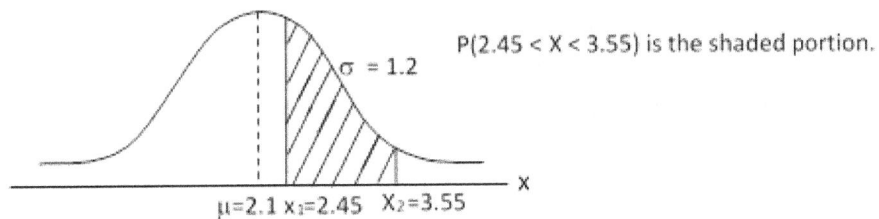

P(2.45 < X < 3.55) is the shaded portion.

$\sigma = 1.2$

$\mu=2.1$ $x_1=2.45$ $X_2=3.55$

Figure 4.5.5: Estimating the P(2.45 < X < 3.55)

(ii)   Convert $X_1 = 2.45$ and $X_2 = 3.55$ to standard normal distribution as:

$$= \frac{X_1-\mu}{\sigma}; Z_2 = \frac{X_2-\mu}{\sigma}$$

Substitute the values of X_1 and X_2:

$$Z_1 = \frac{P(2.45 - 2.1)}{1.2} = \frac{0.35}{1.2} = 0.2917$$

$$Z_2 = \frac{P(3.55 - 2.1)}{1.2} = \frac{1.45}{1.2} = 1.2083$$

$\therefore$ P(2.45 < X < 3.55) = P(0.29 < Z < 1.21)

= P(Z < 1.21) – P(Z < 0.29

= 0.8869 – 0.6141

= 0.2728.

$\therefore$ 0.2728 x 100% = 27.28% of the 300 students would have GPA between 2.5 and 3.6.

Thus, 27.28% of 300 = 0.2728 x 300 = 81.84 students.

= 82 students have GPAs between 2.5 and 3.6.

Example 4.5.6:

The heights of soldiers are normally distributed. If 13.57% of them are taller than 72.2 inches and 8.08% of the soldiers are shorter than 67.2 inches, what are the mean and standard deviation of heights of these soldiers?

Solution:

(i)   Note that 13.57% of the soldiers are taller than 72.2 inches, while 8.08% are shorter than 67.2 inches.

$\therefore$ P(X > 72.2) = 0.1357

i.e., the areas under the normal curves have been given to us.

P(X < 67.2) = 0.0808

(ii)   Convert the areas to z-scores:

$$P\left(\frac{X-\mu}{\sigma} > \frac{72.2-\mu}{\sigma}\right) = 0.1357 \text{ i.e. } P\left(Z > \frac{72.2-\mu}{\sigma}\right) = 0.1357$$

$$P\left(\frac{X-\mu}{\sigma} < \frac{67.2-\mu}{\sigma}\right) = 0.0808 \text{ i.e. } P\left(Z < \frac{67.2-\mu}{\sigma}\right) = 0.0808$$

Because the normal distribution is symmetrical,

$$P\left(Z < \frac{67.2-\mu}{\sigma}\right) = P\left(Z > \frac{-67.2-\mu}{\sigma}\right)$$

(iii)   Using these probabilities (or areas) of 0.1357 and 0.0808, go to the standard normal Table (Table A.4) to obtain the z-scores:

$$P\left(Z > \frac{72.2-\mu}{\sigma}\right) = 0.1357 = P\left(\frac{72.2-\mu}{\sigma}\right) = 1.1 \dotfill (1)$$

$$P\left(Z < \frac{67.3-\mu}{\sigma}\right) = 0.0808 = P\left(\frac{-67.2+\mu}{\sigma}\right) = 1.4 \dotfill (2)$$

(iv)   Thus, multiplying equations (1) and (2) by σ and adding them:

$$72.2 - \mu = (1.1)\ \sigma.$$

$$\frac{-67.2+\mu = (1.4)\sigma}{5 \qquad = 2.5\sigma}$$

$$\therefore\ \sigma = \frac{5}{2.5} = 2.0.$$

(v)   Substituting the value of $\sigma = 2.0$ to either equation (1) or (2), we have:

$$72.2 - \mu = (1.1)\ (2.0) = 72.2 - \mu = 2.2$$

$$\therefore\ \mu = 72.2 - 2.2 = 70.2$$

Thus, the mean and standard deviation of the distribution are:

$$\mu = 70$$

$$\sigma = 2.0.$$

Example 4.5.7:

If the average height of *Gmelina arborea* seedlings in the nursery is 12 inches, with a standard deviation of 1.8 inches, what percentage of the seedlings exceeds 14 inches in height, assuming that the heights follow a normal distribution and can be measured to any desired degree of accuracy?

Solution:

The relative frequency for an interval is equal to the probability of falling in the internal. If we multiply the relative frequency or the probability by 100%, we have the percentage.

$$\mu = 12$$

$$\sigma = 1.8$$

$$x > 14$$

$$P(x > 14) = ?$$

(i)   Sketch the graph of the density function as:

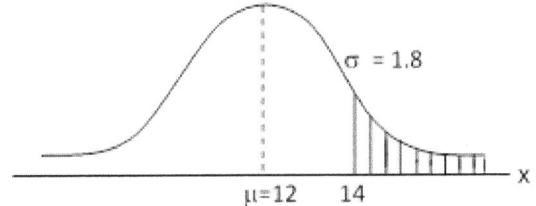

Figure 4.5.7: Probability that x > 14

(ii)   Convert x-score to z-score as:

$$Z = \frac{x - \mu}{\sigma}$$

Substitute the values given above: $Z = \frac{14-12}{1.8} = \frac{2}{1.8} = 1.11$

(iii)   Go to Table A.4 for the area of z = 1.11 = 0.8665.

Hence, P(X > 14) = 1 – P(X < 14) = 1 – P(X ≤ 13)

= 1 – P(Z < 1.11)

= 1 – 0.8665

= 0.1335

∴ 0.1335 x 100% = 13.35% of the seedlings will exceed 14 inches.

4.6 NORMAL APPROXIMATION TO THE BINOMIAL

Probabilities associated with Binomial Experiments are readily obtainable from the formula b(x; n, p) of the binomial distribution Table A.2 (Walpole, 1974). But when n is larger than 20 for the binomial and 40, for the Poisson distributions, Tables A.2 and A.3, can no longer be used to obtain probabilities. Thus, we must compute binomial probabilities by approximation procedures.

Recall that both the Binomial and Poisson distributions are discrete, while the Normal Distribution is continuous. A Theorem that allows us to use areas under the normal curve to approximate binomial probabilities, when n is sufficiently large, states that:

If X is a binomial random variable with parameters n and p, and n is large, X is approximately normal with mean, μ = np and variance σ² = npq, then the limiting form of the distribution of:

$$Z = \frac{x - np}{\sqrt{npq}}$$

as n → ∞, is the standardized normal distribution, N(Z; 0, 1).

Therefore, to use the Normal Approximation for the discrete binomial random variable, we must apply a half-unit (0.5) correction for continuity.

Example 4.6.1:

If X is binomial with n = 20; p = 0.4; Find P(X ≥ 12)?

Solution:

(a) Binomial: $P(X \geq 12) = \sum_{x=12}^{n} b(x; n, p) = \sum_{x=12}^{20} b(x; 20, 0.4)$

$= 1 - P(X < 12) = \sum_{x=0}^{11} b(x; 20, 0.4)$

From the Binomial Probability Table (A.2):

$1 - P(X < 12) = 1 - P(X \leq 11)$

$= 1 - 0.9435$

$= 0.0565$

(b) Using the Normal Approximation to the Binomial: Find $P(X \geq 12)$ by applying the half-unit correction:

$P(X \geq 12) \approx P(X \geq 11.5)$: $\mu = np = (20)\ (0.4) = 8$

$\sigma^2 = npq = (20)\ (0.4)\ (0.6) = 4.8$

Sketch the graph of the density function according to the data provided as shown in Figure 4.6.1

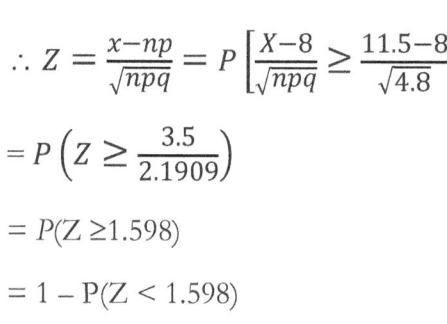

$\therefore Z = \dfrac{x-np}{\sqrt{npq}} = P\left[\dfrac{X-8}{\sqrt{npq}} \geq \dfrac{11.5-8}{\sqrt{4.8}}\right]$

$= P\left(Z \geq \dfrac{3.5}{2.1909}\right)$

$= P(Z \geq 1.598)$

$= 1 - P(Z < 1.598)$

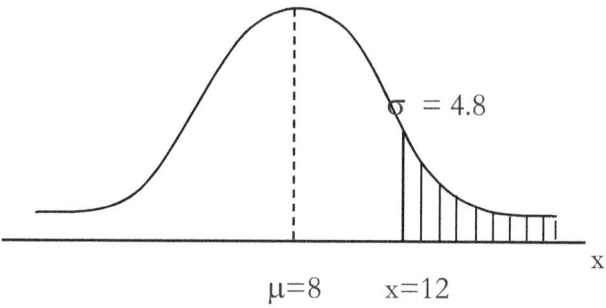

Figure 4.6.1: Probability of $X \geq 12$

Go to the standard normal Table to obtain the probability for $z < 1.598$:

$= 1 - 0.9441 = \underline{0.0559}$

Example 4.6.2:

If X is a binomial random variable with n = 15, p = 0.4, find $P(X = 4)$.

Solution:

(a) Using the binomial distribution to solve the problem, use Binomial Probability Table A.2 (Walpole, 1974):

$P(X = 4) = \sum_{x=0}^{4} b(x; 15, 0.4) - \sum_{x=0}^{3} b(x; 15, 0.4)$

$= 0.2173 - 0.0905 = \underline{0.1263}$

(b) Using the Normal approximation:

$\mu = np = (15)\ (0.4) = 6$

$\sigma^2 = npq = (15)\ (0.4)\ (0.6) = 3.6$

(i)   We must use the half-unit correction on X = 4 in order to transform it from the discrete binomial to the continuous normal distribution as:

$P(X = 4) = P(3.5 < X < 4.5)$ i.e. $x_1 = 3.5$ and $x_2 = 4.5$.

(ii)   Convert these x-scores to z-scores as: $Z = \frac{x - np}{\sqrt{npq}}$

$$Z_1 = \frac{3.5 - 6}{\sqrt{3.6}} = \frac{-2.5}{1.897} \approx \frac{-2.5}{1.90} = -1.3157$$

$$Z_2 = \frac{4.5 - 6}{\sqrt{3.6}} = \frac{-1.5}{1.90} = -0.7894$$

(iii)   Go to the standard normal Table A.4 to obtain the areas (or probabilities) of these z-scores:

∴ $P(3.5 < X < 4.5) = P(-1.3157 < Z < -0.7894) = P(Z < -0.789) - P(Z < -1.316)$

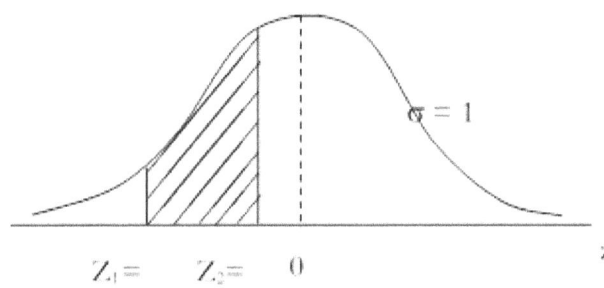

Figure 4.6.2: P(-1.136 < z < -0.789)

$$= 0.2148 - 0.0934 = \underline{0.1214.}$$

Example 4.6.3:

A multipurpose quiz has 200 questions, with 4 possible answers of which only 1 is correct. What is the probability that a guess work yields from 25 to 30 correct answers for 80 of the 200 questions about which the student has no knowledge?

Solution:

The problem gives the importance of normal approximation to the Binomial because the Binomial Probability Table cannot be used to solve this particular problem because n = 25 and 30.

The probability for a correct answer for each of the 80 questions is p =1/4. If X represents the number of correct answers due to guess work, then:

$$P(25 \le X \le 30) = \sum_{x=25}^{30} b\left(x; 80, \frac{1}{4}\right)$$

(i)   Using the Normal approximation, we need to estimate μ = np and $\sigma2$ = npq:

μ = np = (80) (1/4) = 20.

$\sigma2$ = npq = (80) (1/4)(3/4) = 15.

$\sigma = \sqrt{15}$ = 3.8729.

(ii)   Use the half-unit correction on X1 = 25 and X2 = 30 to transform them from the discrete binomial to the continuous normal distribution before we can use the standard normal Table.

(iii)   Sketch the graph of this density function as:

$x_1 = 24.5$ and $x_2 = 30.5$

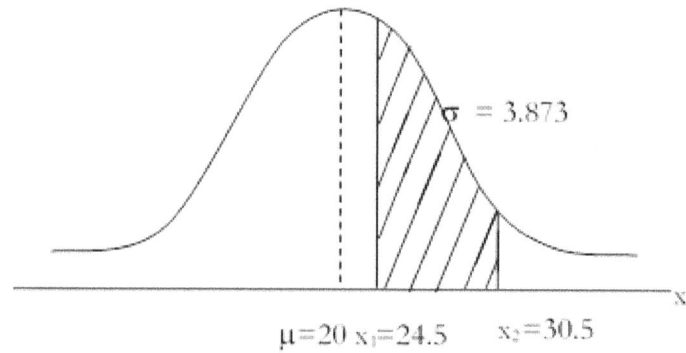

Figure 4.6.3: P(24.5 < X < 30.5)

(iv)   Convert these x-scores to z-scores as:

$$z_1 = \frac{x_1 - np}{\sigma} = \frac{24.5 - 20}{3.8729} = \frac{4.5}{3.87} = 1.16279$$

$$z_2 = \frac{x_2 - np}{\sigma} = \frac{30.5 - 20}{3.8729} = \frac{10.5}{3.87} = 2.71317$$

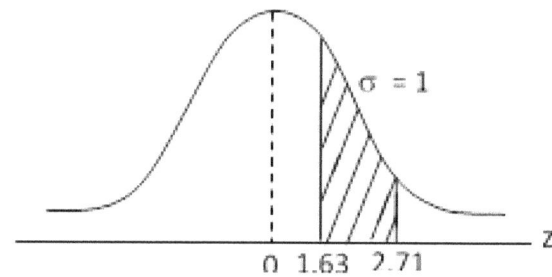

Figure 4.6.4: P(1.163 < z < 2.713)

(v)   Go to the standard normal Table to obtain the probabilities of these z-values.

$P(24.5 \leq X \leq 30.5) = P(1.163 < Z < 2.713)$

$= P(Z < 2.713) - P(Z < 1.163)$

$= 0.9967 - 0.8776$

$= 0.1191$

4.7 HOW TO USE THE STANDARD NORMAL DISTRIBUTION TABLE TO ESTIMATE PROBABILITIES

The mathematical formula for a normal density function is as follows:

$$f(x) = \frac{1}{\sqrt{2\pi\sigma^2}} e^{-z^2/2}$$

Winkler and Hays (1975) have observed that since π, e and 2 in the equation are mathematical constants, the working part of the function is the exponent:

65

$$\frac{-(x-\mu)^2}{2\sigma^2}$$

The more x differs from μ, the larger the quantity in the numerator of this exponent will be. Since the normal distribution is really a family of distributions, statisticians constructing tables, or probabilities, found by the normal rule, find it convenient to think of the variable in terms of a standardized random variable.

The density function of a standardized normal random variable, z, is of the form:

$$f(z) = \frac{1}{\sqrt{2\pi}} e^{-z^2/20}$$

For standardized normal variables, the density depends only on the absolute value of z written as $|Z|$; since both z and $-z$ give the same value of z2, they both have the same density. **The higher the z is in absolute value, the less the associated density is.**

The standardized form of the distribution makes it possible to use one Table of Cumulative Probabilities for any normal distribution, regardless of its particular parameters (Winkler and Hays, 1975).

4.7.1 Computing Probabilities for the Normal Distribution

All probability statements that are made, when using a normal distribution, are in terms, either of cumulative probabilities, or the probabilities of intervals.

The cumulative probability: $F(x) = P(X \le x) = \int_{-\infty}^{x} f(x)dx$ is the area under the normal curve in the interval bounded by $-\infty$ and x, and can be used to find the probability of any interval since $P(a \le x \le b) = F(b) - F(a)$.

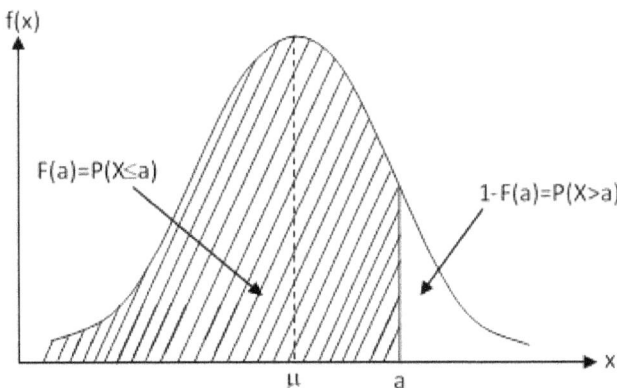

Figure 4.7.1: Probabilities as Areas Under the Normal Density Function

Most Tables of the normal distribution give the cumulative probabilities for various **standardized values**. For a given z, the Table provides the cumulative probability **up to and including that standardized value in a normal distribution.** Due to the symmetry of the normal distribution, it is possible to use the Table to determine cumulative probabilities for negative values of z, by using the relation:

$$F(-z) = 1 - F(z)$$

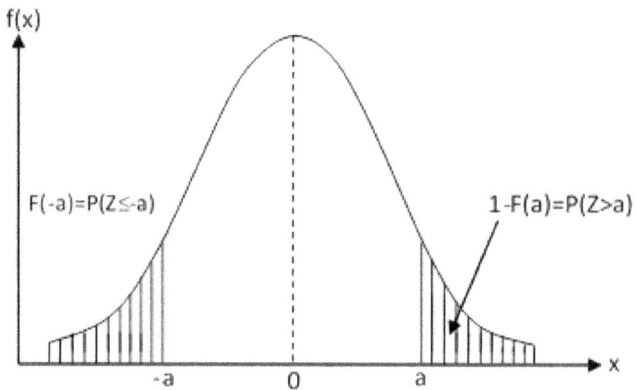

Figure 4.7.2: The Relationship between F(a) and F(-a) for a Standardized Normal Distribution

The shaded area on the right, which is 1-F(a), is the same as the shaded area on the left, F(-a).

4.8 THINGS TO NOTE ABOUT THE STANDARD NORMAL TABLE

1. The equation for the standard normal distribution is given as:

$Z = \frac{x-\mu}{\sigma} \sim N(0, 1)$, i.e. z is normally distributed with mean, $\mu = 0$, and standard deviation, $\sigma = 1$.

Thus, $z\sigma = x-\mu$, where z is equal to the number of standard deviations. Hence, the value of the random variable, $X = z\sigma+\mu$.

The area under the symmetric standard normal curve is shown in Figure 4.8.1:

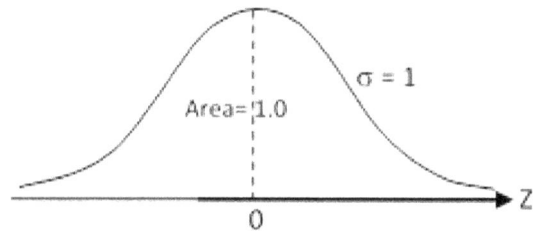

Figure 4.8.1: Area under the Symmetric Standard Normal Curve.

2. The total area, or probability, under the standard normal curve is equal to one. The normal curve is symmetric with its center at $\mu = 0$. From $\mu = 0$ to the end of the curve, on the right-hand side (A2) and on the left-hand side (A1), the area is equal to 0.5 each. See Figure 4.8.2:

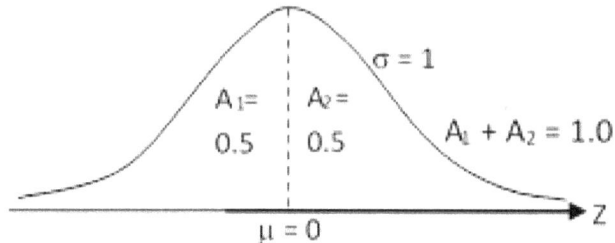

Figure 4.8.2: Symmetric Standard Normal Curve

Note: All z-tabled values are estimated for probabilities that a calculated z-value is less than, or equal to the tabled value: i.e. $P(Z \le z1)$.

3. Before using the Table to compute probabilities, or areas under the standard normal curve, observe the type of the diagram that has been used to compile the Table - - either diagram Type I or Type II as shown in Figure 4.8.3:

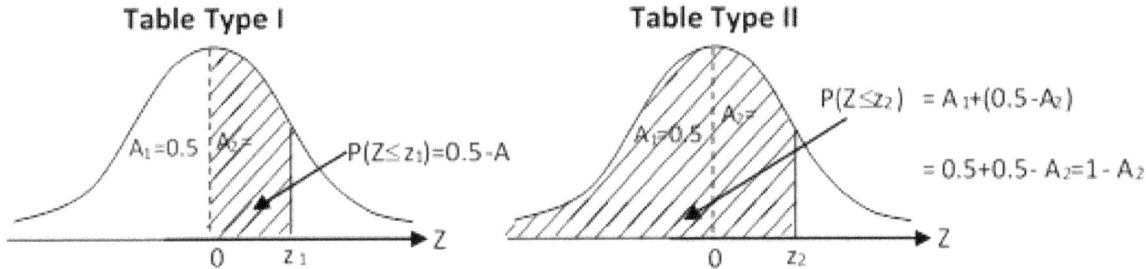

Figure 4.8.3: The twoTypes of Tabled z-values

(i)   Tables based on Type I diagram (Mendenhall, 1975) only present areas, or probabilities between 0 and $z1$ i.e. $0.5 - A2$. The Tabled z-values are only positive from 0.0 to 3.5 - - $P(Z \leq z1) = 0.5 - A2$.

(ii)   Tables based on Type II diagram (Walpole, 1974) present areas, or probabilities from $z2$ to the end of the curve on the left-hand side. The Tabled z-values are both negative and positive from -3.4 to +3.4. Hence the value of $z2$ above includes two areas: $A1 = 0.5$ and $(0.5 - A2)$.

Therefore, to obtain the area, or probability between 0 and $z2$, we must subtract the area of $A1 = 0.5$ from the tabled value.

4. The top row of the two Types (I and II) of the Table are the same i.e. z-values range from .00 to .09 (correct to the nearest hundredths), while on the left-hand side, z-values are from 0.0 to 3.5 (correct to the nearest tenths (Snedecor and Cochran, 1967; Mendenhall, 1975), or from -3.4 to +3.4 (Walpole and Myers, 1972; Walpole, 1974).

5(a) Therefore, to obtain the probability of a z-value of 2.0 from **Table Type I** above (Mendenhall, 1975, Table 3), go to the left-hand side column 1 and locate the value of z = 2.0 and trace it to the right-hand side column 1 whose top-row is listed as .00. Read the value at the intersection of z = 2.0 and .00 which is 0.4772. This value (0.4772) is the probability (or area) of the z-value of 2.0.

(b)If we use **Table Type II** above from Walpole and Myers (1972) Table IV, to obtain the probability of a z-value of 2.0; locate the z-value = 2.0 from the left-hand side column 1 and trace it to the right-hand side column 1 whose top row is listed 0.00 and read the value as 0.9772. But the probability which we want to estimate is from 0 to $z2$. Therefore, we must subtract 0.5 from 0.9772: $0.9772 - 0.5 = 0.4772$, same as in 5(a) above.

6 If two z-values are calculated as $z1 = 1.2083$ and $z2 = 0.2917$ as in Figure 4.8.3, follow this procedure:

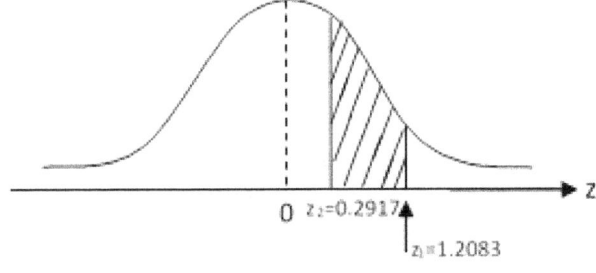

Figure 4.8.3: Estimating the area between two z-values

(a)   Locate the z_1-value of 1.2 in the left-hand side column and trace it to the right-hand side column 2 whose top row is listed as .01 and obtain the area of this intersection as **0.3869**, if you are using **Type I Table** above; but if you are using **Type II Table**, the area of this intersection is **0.8869**.

68

(b) Then, locate z_2-value of 0.2 on the left-hand column and trace it to the right-hand side column 10 whose top row is listed as .09 and obtain the area (probability) at this intersection as **0.1141**, if you are using **Type I Table** above; or 0.6141, if you use **Type II Table**.

(c) The area (probability) between the two z-values, $P(0.2917 \leq Z \leq 1.2083)$ is the difference between the areas in (a) and (b):

$$P(0.2917 \leq Z \leq 1.2083) = P(Z \leq 1.2083) - P(Z \leq 0.2917)$$

$$= 0.3869 - 0.1141 \text{ - - -for Type I Table.}$$

$$= \mathbf{0.2728}.$$

OR

$$P(0.2917 \leq Z \leq 1.2083) = P(Z \leq 1.2083) - P(Z \leq 0.2917)$$

$$= \mathbf{0.8869 - 0.6141} \text{ - -for Type II Table.}$$

$$= \underline{\mathbf{0.2728}} \text{- - -same result.}$$

Note:

(i)   Areas, or probabilities, tabulated are calculated to the **nearest four decimal places** as 0.3869 or 4 digits. Therefore, these areas must be calculated to 5 places of decimal (or 5 digits after the decimal point). Then, if the 5th digit is 5 or higher, it is approximated to 1 and added to the 4th digit.

(ii)   The z-values on the left-hand column are calculated to one place of decimal. However, the z-values must be estimated to 4 places of decimal. But only 3 places of decimal are generally used because the 4th place of decimal is approximated to 1 and added to the 3rd place of decimal.

(iii)   For example, $P(0.2917 \leq Z \leq 1.2083)$ to check the probability of $z_1 = 0.2917$, locate the z-value of 0.2 on the left-hand column and trace it to the right-hand column 10 whose top row is listed as .09 and obtain the probability at their intersection. But for $z_2 = 1.2083$, locate the z-value of 1.2 on the left-hand column; since we are only going to use 2 decimal places, we must approximate the 3rd digit to 1 and add it to the 2nd digit i.e. 1.2083 becomes 1.21. Thus, we locate z = 1.2 and trace it to the top row listed as .01 to obtain the probability as before.

7 To estimate the probability that an estimated z-value is greater than a tabled value as: $P(Z > z_1)$, this cannot be estimated directly from the standard normal table, unless we switch the inequality from: $P(Z > z_1)$ to $1 - P(Z \leq z_1)$. See Figure 4.8.4 shaded portion.

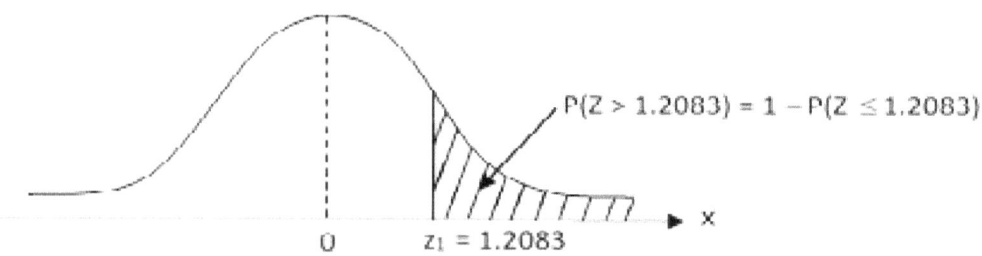

Figure 4.8.4: Estimating the probability $P(Z > z_1)$

8(a) If two z-values are estimated as $-z_1 = -0.50$ and $z_2 = 1.00$ respectively, and we want to estimate the area between the two: $P(-z_1 \leq Z \leq z_2)$, we add the two areas as: $P(-z_1 \leq Z) + P(z_2 \leq Z)$ – see Figure 4.8.5 – shaded portion. Notice that $-z$ (minus z-values) are on the left-hand side of zero in the standard normal (unit normal) figure as in Figure4.8.5.

$$P(-z_1 \leq Z \leq z_2) = P(-z_1 \leq Z) + P(Z \leq z_2)$$

$$P(-0.50 \leq Z \leq 1.00) = P(-0.50 \leq Z) + P(z \leq 1.00)$$

$$= 0.1915 + 0.3413$$

$$= \underline{0.5328}$$

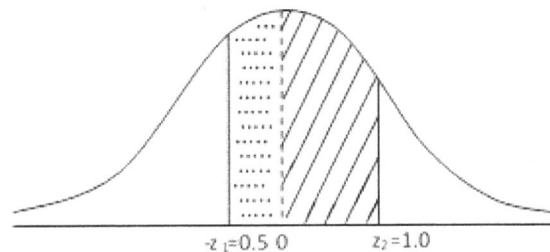

Figure 4.8.5: Estimating the Probability $P(-z_1 \leq Z \leq z_2)$ of the shaded portion

8(b)If the two z-values estimated are both negative (Figure 4.8.5) as $-z_1 = -0.32$ and $-z_2 = -2.22$ and we want to estimate the area between the two negative z-values, i.e. $P(-z_1 \leq Z \leq -z_2)$, we estimate the area between them as if both are positive as:

$$P(-0.32 \leq Z \leq -2.22 = P(z \leq -2.22) - P(z \leq -0.32)$$

$$= 0.4868 - 0.1255$$

$$= \underline{0.3613}$$

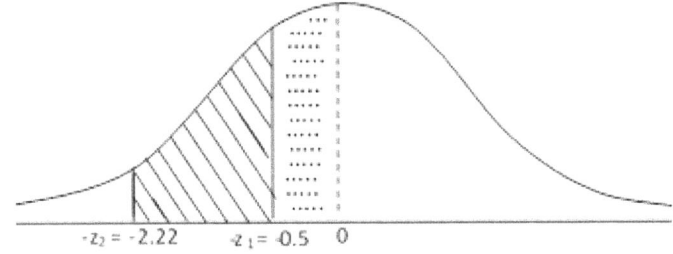

Figure 4.8.6: Estimating the probability of shaded portion: $P(-z_1 \leq Z \leq z_2)$

$$= P(-z_2 \leq Z) - P(-z_1 \leq Z).$$

9. If the area of the standard normal is given as 0.4750, we can use it to estimate the z-scores from the Table:

(i) Go into the body of the Table and locate the area that is close to 0.4750, for Table Type I (Figure 4.8.3) but not exceeding it. When you locate such a value, trace it to the left-hand margin to obtain the z-score and trace the value (0.4750) to the top row that contains the value.

   The value, 0.475, traced to the left-hand margin, is 1.9 and traced to the top row is .06. Thus, the z-score whose area is 0.4750 is 1.960.

(ii) For Table Type II (Figure 4.8.3), the area of the standard normal given as 0.4750 in Table Type I and, in fact, all other areas in Table Type I have 0.5000 added to obtain the values of Table Type II.

Thus, the value, or area (or probability) listed as 0.4750 in Table Type I will become 0.4750+0.5000 = 0.9750 in Table Type II. Hence, we must locate 0.9750 in Table Type II, trace it to the left-hand margin which is 1.9 and traced to the top row which is .06. Therefore, the area of 0.9750 will have a z-value of 1.960 as in 9 (i) above.

However, we should note that only Table Type II (Figure 4.8.3) has negative z-values. Whether the z-values are negative or positive, areas or probabilities are always positive. For negative z-values, the probabilities are read directly from the Table as before. All negative z-values have probabilities less than 0.5000. Negative z-values range from -0.0 to -3.4 and as the z-values decrease from -0.0 to -3.4, their probabilities also decrease. See Walpole (1974) Table A.4.

5 SAMPLING AND SAMPLING DISTRIBUTIONS

5.0 POPULATION

In our definition of statistical terms in section 1.5, we defined a **"population"** or **"universe"** as the totality of all members (or individuals, or objects, or observations, or experimental units) which have some common observable or discernible characteristics. Every defined population has its unique properties – that is, its mean (or average) designated as μ, and its variance, designated as σ2, are constant or fixed, but usually unknown. These constant unknown measurable population characteristics of mean, μ, and variance, σ2, are called **parameters.**

5.1 SAMPLES

Any subset of a population is a **sample.** We know the characteristics of a sample, but not the population from which it was drawn. The extent to which we can apply our knowledge of this sample to estimate the characteristics of the whole population constitutes, in a broad sense, the field of **statistical inference** (Byrkit, 1975).

The characteristics or measures which describe a **sample** are called **statistics**. These sample characteristics vary from one sample to the other. One of the important uses of statistical inference is the determination of how accurately sample statistics estimate the population parameters. A statistic is a random variable that depends only on the observed random sample (Walpole and Myers, 1972).

A **ready-made sample**, such as the set of students in a class, is called an **incidental sample**. The value of a sample depends on the degree to which it represents the original population. Therefore, an incidental sample should not be used, if better and more representative samples are available.

What constitutes a good sample varies among statisticians. Current research tends to indicate that samples obtained at **random** from a population may be more representative of the population. A sample of a population is a **random sample**, if every sample of a given size is equally likely or equi-probable i.e., each sample has an equal chance of being selected.

According to Dixon and Massey, Jr. (1969), a random sample of individuals is chosen, in turn, from the population, with each remaining individual having an equal chance of being chosen on the next draw. If the individuals selected are returned to the population before the next sample member is selected, the process is called **sampling with replacement**; if the next sample member is selected from a population that does not include previously selected members, the sampling is called **sampling without replacement.**

As a general rule, a random sample can be obtained from any population by a random selection procedure, and from a randomly arranged population, by a systematic selection procedure (Byrkit, 1975). A good procedure is to use a **Table of Random Numbers** whose digits are 0, 1, 2, 3, 4, 5, 6, 7, 8, 9, arranged in a sequence completely at random. These numbers can be used singly (0, 1, 2, … 9), or as two-digit numbers (00, 01, 02, 03, … 99).

There are several other methods of sample selection like stratified, cluster, systematic, or even sub-sampling. Those interested in detailed methods of sample selection are advised to consult texts on **"Sampling Techniques"**.

5.2 SAMPLING DISTRIBUTIONS

Dixon and Massey (1969) observed that any statistic, or measurement, on a sample has a **sampling distribution**; Walpole and Myers (1972) stated that **"the probability distribution of a statistic is called a sampling distribution, while the standard deviation of the sampling distribution of a statistic is called the standard error of the statistic"**.

Winkler and Hays (1975) defined a Sampling Distribution as "a theoretical probability distribution that shows the functional relationship between the possible values of some sample statistic and the probability (density) associated with each value of the sample statistic over all possible samples of a particular size from a particular population".

Therefore, the sampling distribution of the statistic in question will always depend, in some specific way, upon the size of the population, the size of the samples, and the method of choosing the random samples.

The sampling distribution of any summary characteristic (mean, mode, median, range, etc.) of a sample may be found – an important fact for the theory of statistics, as it provides ways to make inferences about population parameters, on the basis of random samples, in terms of the probability of a sample statistic's value arising by chance from a certain population.

If the size of the population is large or infinite, the statistic has the same distribution, whether we sample with or without replacement. But sampling with replacement from a small finite population (Walpole and Myers, 1972) gives a slightly different distribution from the statistic than if we sample without replacement. Sampling with replacement from a finite population is equivalent to sampling from an infinite population.

A statistic whose sampling distribution has its mean equal to a population parameter is called an **unbiased estimate** of that parameter; thus, for random samples with replacement, $\bar{x}$ is an unbiased estimate of μ and s^2 is an unbiased estimate of σ^2.

It is common practice to refer to the standard deviation of a statistic as the **standard error** of that statistic. Thus, $\sigma_{\bar{x}}$ is the standard error of $\bar{X}$ and for random samples with replacement, is numerically equal to $\frac{\sigma}{\sqrt{n}}$.

We shall study several important sampling distributions of frequently used statistics.

5.3 SAMPLING DISTRIBUTIONS OF THE MEAN AND ITS VARIANCE

The most frequently used statistics for measuring the center of a set of data is the mean. The sample mean is commonly represented by the statistic $\bar{X}$. If a random sample has a total of n pieces of data (sample size), which need not be different, the mean is given by:

$$\bar{X} = \frac{\sum x_i}{n} = \frac{X_1 + X_2 + X_3 + \cdots + X_n}{n} = \sum \bar{X} f(\bar{X})$$

Every mean, whether of a population, or of a sample, has its variance. The population variance is given by:

$$\sigma^2 = \Sigma(x - \mu)^2 = \Sigma(x - \mu)^2.P(x). \text{ See Section 3.4.0 above.}$$

Each of the n observations, X_i, from a normal population with mean, μ, and variance, σ^2, estimates the population parameters:

$$\mu_{\bar{X}} = \frac{\mu + \mu + \mu + \dots + \mu}{n} = \frac{n.\mu}{n} = \mu, \text{ and}$$

$$\text{Variance, } \sigma^2_{\bar{X}} = \frac{\sigma^2 + \sigma^2 + \sigma^2 + \dots + \sigma^2}{n^2} = \frac{n.\sigma^2}{n^2} = \frac{\sigma^2}{n}$$

Thus, when we sample from a population of unknown distribution, whether finite or infinite, the sampling distribution of $\bar{X}$ will still be approximately normal with mean μ, and variance provided the sample size is large (i.e greater than or equal to 30, i.e. n ≥ 30).

5.3.1 Central Limit Theorem (CLT)

Regardless of the population distribution, if the sample size, n, is large enough (Winkler and Hays, 1975), the normal distribution is a good approximation to the sampling distribution of the mean.

THEOREM 5.3.1:

If random samples of n observations are drawn from a population with finite mean, μ, and standard deviation, σ, then, when n is large, the sample mean, $\bar{X}$, will be approximately normally distributed with mean equal to $\mu_{\bar{X}}$ and standard deviation, $\frac{\sigma}{\sqrt{n}}$. The approximation will become more and more accurate as n becomes large (Mendenhall, 1975).

The Central Limit Theorem (CLT) could be restated to apply to:

the sum of the sample measurements $(X_1 + X_2 + X_3 + \ldots + X_n)$, which would also tend to possess a normal distribution, in repeated sampling, with mean equal to μ and standard deviation, $\frac{\sigma}{\sqrt{n}}$, as n becomes large.

The CLT states that under rather general conditions, sums and means of samples of random measurements drawn from a population tend to possess approximately, a bell-shaped distribution, in repeated sampling.

Let us illustrate these with some examples:

5.3.2 Examples of Sampling with Replacement

Example 5.3.1: Suppose we are given a discrete uniform population $f(x) \approx \frac{1}{4}$, x = 0, 1, 2, 3, find the probability that a random sample of size 36, selected with replacement, will yield a sample mean, $\bar{X}$, greater than 1.4, but less than 1.8, if the mean is measured to the nearest tenth.

Solution: Calculate the mean and variance of the uniform distribution as:

$$\text{Mean} \quad \mu = \frac{0+1+2+3}{4} = \frac{6}{4} = \frac{3}{2} = \underline{1.5}$$

$$\text{Variance,} \quad \sigma^2 = \Sigma(X - \mu)^2 = \Sigma(X_i - \mu)^2 P(x)$$

$$= \frac{\left(0-\frac{3}{2}\right)^2 + \left(1-\frac{3}{2}\right)^2 + \left(2-\frac{3}{2}\right)^2 + \left(3-\frac{3}{2}\right)^2}{4}$$

$$\frac{\frac{9}{4}+\frac{1}{4}+\frac{1}{4}+\frac{9}{4}}{4} = \left(\frac{20}{4}\right)\Big/4 = \frac{5}{4} = 1.25$$

The sampling distribution of $\bar{X}$ may be approximated by the normal distribution with mean, $\mu_{\bar{X}} = \frac{3}{4}$ and variance, $\sigma_{\bar{X}}^2 = \frac{\sigma^2}{n}$.

n = 36; $\bar{x}_1 = 1.4$; $\bar{x}_2 = 1.8$

Find P(1.4 < X < 1.8)

First calculate $\sigma_{\bar{X}}^2 = \frac{\sigma^2}{n} = \left(\frac{5}{4}\right)\Big/36 = \frac{5}{4} x \frac{1}{36} = \frac{5}{144}$

$$\sigma_{\bar{X}} = \sqrt{\frac{5}{144}} = \frac{\sqrt{5}}{12} = \frac{2.236}{12} = \underline{0.1863}$$

Since X is discrete, we must apply the correction for continuity as:

$\bar{X}_1 = 1.4 + 0.05 = 1.45$

$\bar{X}_2 = 1.8 - 0.05 = 1.75$ (because $\bar{X}$ is less than 1.8)

$\therefore$ P(1.4 < X < 1.8) = P(1.45 < X < 1.75) is given by the shaded portion of the diagram:

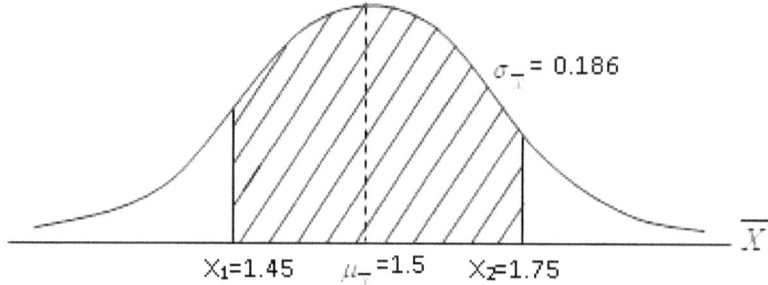

Figure: 5.3.1: P(1.4<X<1.8)

$$Z_1 = \frac{\bar{X}_1 - \mu}{\sigma_{\bar{X}}} = \frac{1.45 - 1.5}{0.186} = \frac{-0.05}{0.186} = -0.2688$$

$$Z_2 = \frac{\bar{X}_2 - \mu}{\sigma_{\bar{X}}} = \frac{1.75 - 1.5}{0.186} = \frac{0.25}{0.186} = 1.3441$$

P(1.4 < X < 1.8) ≈ P(-0.269 < Z < 1.344)

= P(Z < 1.344) – P(Z < -0.269)

= 0.9105 – 0.3932

= 0.5173

Example 5.3.1.2:

The average life-span of electrical bulbs is 800 hours with a standard deviation of 40 hours. What is the probability that a random sample of 16 bulbs will have an average life-span of less than 775 hours?

Solution: n=16; $\mu_{\bar{X}}$ = 800

$\sigma = 40 \therefore$

$$\sigma_{\bar{X}} = \frac{\sigma}{\sqrt{n}} = \frac{40}{\sqrt{16}} = \frac{40}{4} = 10$$

Sample size = 16 bulbs

$$\bar{X} P(\bar{X} < 775) = P\left(\frac{\bar{X} - \mu_{\bar{X}}}{\sigma_{\bar{X}}}\right) < \frac{775 - 800}{10}) = P\left(Z < \frac{-25}{10}\right) = P(Z < -2.5) = 0.006.$$

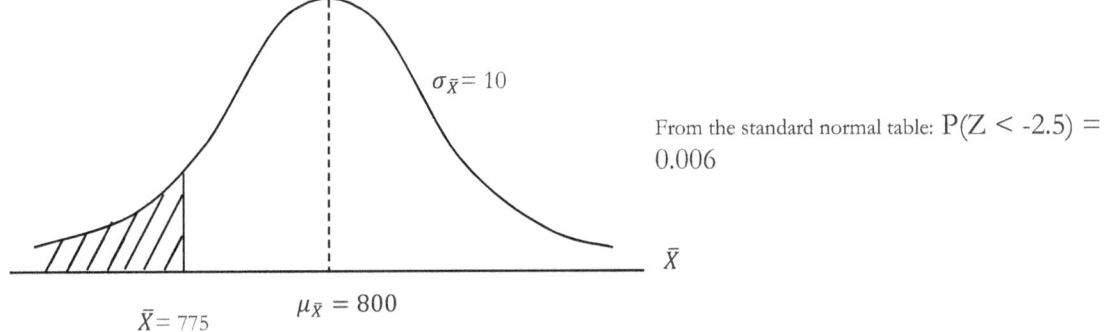

From the standard normal table: $P(Z < -2.5) = 0.006$

Figure 5.3.2: Probability that $\bar{X}$ is less than 775 hours

THEOREM 5.3.2:

If all possible random samples of size n are drawn, with replacement, from a finite population of size, N, with mean, μ, and standard deviation, σ, then, the sampling distribution of the mean, $\bar{X}$, will be approximately normally distributed with mean, $\mu_{\bar{X}} = \mu$ and standard deviation, $\sigma = \frac{\sigma}{\sqrt{n}}$. Hence: $z = \frac{\bar{X} - \mu}{\sigma/\sqrt{n}}$ is a value of a standard normal variable Z (Walpole, 1974).

5.3.3 Examples of Sampling without Replacement

Suppose the data in Table 5.3.2 are obtained from sampling a small finite sample of n = 2 from a population of size N = 4 without replacement, estimate the mean and the variance of $\bar{X}$.

Table 5.3.2: Sampling Distribution of $\bar{X}$ Without Replacement

$\bar{X}$	f	$f(\bar{X})$	
0.5	2	$^1/_6$	N = 4
1.0	2	$^1/_6$	n = 2
1.5	4	$^1/_3$	
2.0	2	$^1/_6$	
2.5	2	$^1/_6$	

Solution:

Use the sample mean to estimate the population mean: $\mu_{\bar{X}} = \mu$

$$\mu_{\bar{X}} = \sum \bar{X} f(\bar{X}) = (0.5)(1/6) + (1.0)\left(^1/_6\right) + (1.5)(1/3) + (2.0)\left(^1/_6\right) + (2.5)\left(^1/_6\right)$$

$$= \frac{0.5}{6} + \frac{1.0}{6} + \frac{1.5}{3} + \frac{2.0}{6} + \frac{2.5}{6}$$

$$= \frac{0.5 + 1.0 + 3.0 + 2.0 + 2.5}{6}$$

$$= \frac{9.0}{6} = \frac{3}{2} = \mu = 15$$

76

$$\text{Variance} = \sigma_{\overline{X}}^2 = \sum(\overline{X}_i - \mu)^2 \cdot f(\overline{X})$$

$$= \left(0.5 - \tfrac{3}{2}\right)^2 \left(\tfrac{1}{6}\right) + \left(1 - \tfrac{3}{2}\right)^2 \left(\tfrac{1}{6}\right) + \left(1.5 - \tfrac{3}{2}\right)^2 \left(\tfrac{1}{3}\right) + \left(2 - \tfrac{3}{2}\right)^2 \left(\tfrac{1}{6}\right) + \left(2.5 - \tfrac{3}{2}\right)^2 \left(\tfrac{1}{6}\right)$$

$$= (-1)^2 \left(\tfrac{1}{6}\right) + \left(\tfrac{1}{2}\right)^2 \left(\tfrac{1}{6}\right) + (0)^2 \left(\tfrac{1}{3}\right) + \left(\tfrac{1}{2}\right)^2 \left(\tfrac{1}{6}\right) + (1)^2 \left(\tfrac{1}{6}\right)$$

$$= \tfrac{1}{6} + \tfrac{1}{24} + 0 + \tfrac{1}{24} + \tfrac{1}{6}$$

$$= \frac{4 + 1 + 0 + 1 + 4}{24} = \frac{10}{24} = \frac{5}{12}$$

$$= \frac{5}{12} = \left(\tfrac{1}{6}\right)/n \left(\frac{N - n}{N - 1}\right)$$

$$= \left(\tfrac{5}{4}\right)/2 \left(\frac{4 - 2}{4 - 1}\right) = \left(\tfrac{5}{4}\right)/2 \left(\tfrac{2}{3}\right) = \frac{5}{4} x \frac{2}{1} x \frac{2}{3} = \frac{5}{3}$$

THEOREM 5.3.3:

If all possible random samples of size n are drawn, without replacement, from a finite population of size N with mean, μ, and standard deviation, σ, then the sampling distribution of the sample mean, $\overline{X}$, will be approximately normally distributed with a mean and standard deviation given by:

$$\mu_{\overline{X}} = \mu$$

$$\sigma_{\overline{X}} = \frac{\sigma}{\sqrt{n}} \sqrt{\frac{N-n}{N-1}}.$$ where, $\sqrt{\frac{N-n}{N-1}} = $ Finite population correction factor which approaches 1, if N is large, or infinite.

5.3.3 The Sampling Distribution of the Differences between Means

Walpole and Myers (1972) and Walpole (1974) have described the distribution of the differences between two sets of independent sample means, $\overline{x}_1 - \overline{x}_2$, as the sampling distribution of the statistic, $\overline{X}_1 - \overline{X}_2$. These authors also pointed out that, if both n1 and n2 are greater than, or equal to 30, the normal approximation for the distribution of $\overline{X}_1 - \overline{X}_2$ is very good - - as stated in theorem 5.3.3.

THEOREM 5.3.4:

If independent samples of size n1 and n2 are drawn from two large or infinite populations, discrete or continuous, with means $\mu_1 - \mu_2$, and variances $\sigma_1 - \sigma_2$, respectively, then the sampling distribution of the differences of means, $\overline{X}_1 - \overline{X}_2$, is approximately normally distributed with mean and standard deviation given by:

$$\mu_{\overline{X}_1 - \overline{X}_2} = \mu_1 - \mu_2$$

$$\sigma_{\overline{X}_1 - \overline{X}_2}^2 = \frac{\sigma_1^2}{n_1} + \frac{\sigma_2^2}{n_2}$$

$$\sigma_{\overline{X}_1-\overline{X}_2} = \sqrt{\frac{\sigma_1^2}{n_1} + \frac{\sigma_2^2}{n_2}}$$

Hence, $Z = \dfrac{(\overline{X}_1-\overline{X}_2)-(\mu_1-\mu_2)}{\sqrt{(\sigma_1^2/n_1+\sigma_2^2/n_2)}}$, is a value of standard normal variable Z.

Example 5.3.3:

Suppose we are given the following data:

$\mu_1 = 6.5$	$\mu_2 = 6.0$
$\sigma_1 = 0.9$	$\sigma_2 = 0.8$
$n_1 = 36$	$n_1 = 49$

What is the probability that the difference between the two means will be greater than 1.0?

Solution:

We apply the Theorem 5.3.3 to solve the problem:

$$\mu_{\overline{X}_1-\overline{X}_2} = \mu_1 - \mu_2 = 6.5 - 6.0 = 0.5$$

$$\sigma_{\overline{X}_1-\overline{X}_2} = \sqrt{\frac{\sigma_1^2}{n_1} + \frac{\sigma_2^2}{n_2}} = \sqrt{\frac{0.81}{36} + \frac{0.64}{49}}$$

$$= \sqrt{0.0225 + 0.013061224} = \sqrt{0.035561224} = 0.188576839 = 0.1886$$

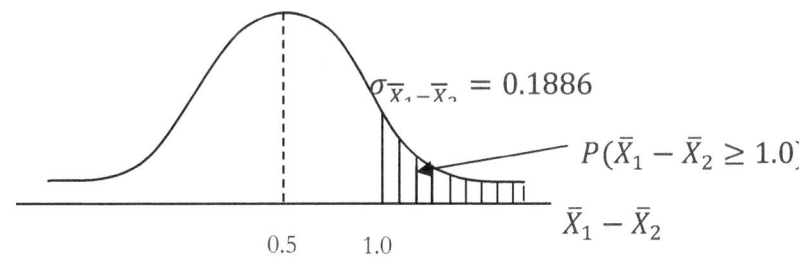

Figure 5.3.3: $P(\overline{X}_1 - \overline{X}_2 \geq 1.0)$

$P(\overline{X}_1 - \overline{X}_2 \geq 1.0) = 1 - P(\overline{X}_1 - \overline{X}_2 < 1.0)$

$$Z = \frac{(\overline{X}_1 - \overline{X}_2) - (\mu_1 - \mu_2)}{\sqrt{(\sigma_1^2/n_1 + \sigma_2^2/n_2)}} = \frac{1.0 - 0.5}{0.1886} = \frac{0.5}{0.1886} = 2.6511$$

$P(Z \geq 2.6511) = 1 - P(Z < 2.6511)$

$= 1 - 0.9960$

$= 0.004$

THEOREM 5.3.3.1:

If the random variables X and Y are independent and normally distributed with means μx and μy and variances σ_x^2 and σ_y^2, respectively, then the distribution of the difference, X – Y, is normally distributed with mean, μx-y = μx – μy and variance, $\sigma_{x-y}^2 = \sigma_x^2 + \sigma_y^2$.

5.3.4 Student's t-Distribution

For samples of size n ≥ 30, S2, the sample variance, is a good and unbiased estimate of σ2, the population variance. The deviations $(\overline{X} - \mu)$ of sample means from the parametric mean of a normal distribution are themselves normally distributed. If these deviations are divided by the parametric standard deviation, $(\overline{X} - \mu)/\sigma_{\overline{X}}$, they are still normally distributed, with μ=0, σ=1, i.e., they form a standard normal distribution.

If on the other hand, we calculate the variance, S_i^2 of each sample and the deviation for each mean $\overline{X}_i$ as $(\overline{X} - \mu)/S_{\overline{X}}$, where $S_{\overline{X}}$ is the estimate of the standard error of the mean of the ith sample, we find the distribution of the deviations to be wider and flatter than the normal distribution.

The new distribution $\left((\overline{X} - \mu)/S_{\overline{X}}\right)$ ranges wider than the corresponding normal distribution because the denominator is the sample standard error, rather than the parametric standard error, $\left(\frac{\overline{X}-\mu}{\sigma_{\overline{X}}}\right)$, and will sometimes be smaller, and at other times, greater than expected. This increased variation will be reflected in the greater variance of the ratio $((\overline{X} - \mu)/S_{\overline{X}})$. The expected distribution of this ratio is called the **"t-distribution"**, also known as **"STUDENT'S," distribution after W.S. Gossett, who first described it in 1908, publishing under the pseudonym "Student".**

The t-distribution is a function with a complicated mathematical formula that will not be presented here.

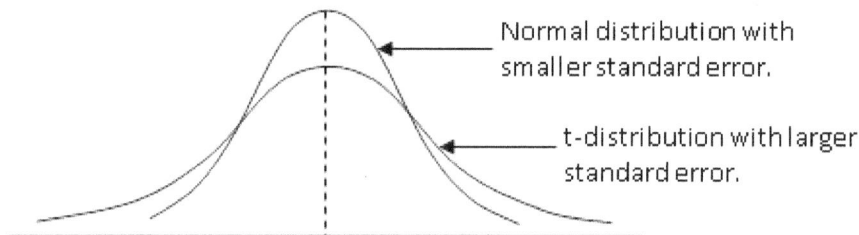

Normal distribution with smaller standard error.

t-distribution with larger standard error.

Figure. 5: The Normal and t-distributions contrasted

The t-distribution shares with the normal distribution the properties of being:

- symmetric (bell-shaped), extending from negative to positive infinity.
- But differs from the normal distribution because it assumes different shapes, depending on the number of degrees of freedom (v = n – 1), or the sample size.
- as the number of degrees of freedom increases, student's t-distribution approaches the shape of the standard normal distribution (μ = 0; σ = 1) ever more closely.
- t-distribution of degrees of freedom equal to 30, is essentially indistinguishable from a normal distribution.
- as the degree of freedom approaches infinity (∞), the t-distribution is the same as the normal distribution.
- like the normal distribution, there is also a Table for cumulative t-distribution for various degrees of freedom (v = n – 1).

THEOREM 5.4.1:

If $\overline{X}$ and s^2 are the mean and variance, respectively, of a random sample of size n, taken from a normal population having the mean, μ, and unknown variance, σ^2, then:

$$t = \frac{\bar{X} - \mu}{S/\sqrt{n}}$$

is a value of a random variable, T, having the t-distribution with

v = n – 1 degrees of freedom.

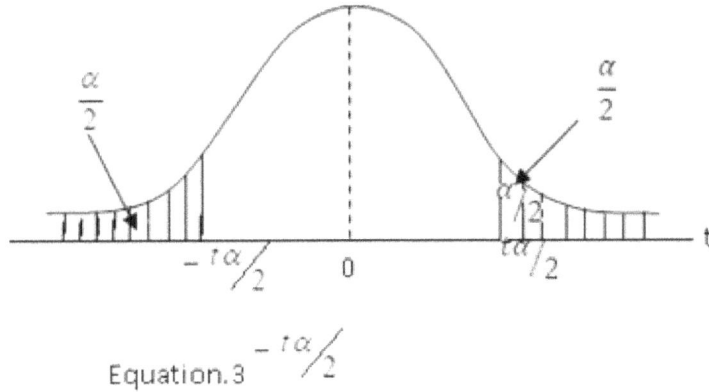

Equation.3

Figure. 5.4.2: Symmetric Property of the t-distribution

5.5 CHI-SQUARED DISTRIBUTION

According to Ross (2005), "if Z_1, Z_2, …. Z_n are independent standard normal random variables, then the random variable, $\sum_{i=1}^{n} Z_i^2$, is said to be a chi-squared random variable with n degrees of freedom. Note that for a standard normal random variable Z:

Var (Z) = 1

$$= E(Z^2) - (E[Z]^2).$$

$$= E(Z^2) \text{ since } E[Z] = 0.$$

Hence, $E[Z]^2 = 1$.

Therefore, $E[\sum_{i=1}^{n} Z_i^2] = \sum_{i=1}^{n} E[Z_i^2] = n$

Thus, the expected value of a chi-squared random variable is equal to the number of degrees of freedom.

If a random sample of size n (X_1, X_2, X_3, …, X_n) is drawn from a normal population with mean, μ, and variance, σ^2, and the sample variance, S^2, is computed as:

$S^2 = \frac{\sum_{i=1}^{n}(X_i - \bar{X})^2}{n-1}$, we obtain the value of the statistic, S^2.

Consider the distribution of the random variable, $\frac{(n-1)S^2}{\sigma^2}$,

$\frac{(n-1)S^2}{\sigma^2}$, has a chi-squared distribution with (n – 1) degrees of freedom.

Note that the standardized variables, (X_i–μ)/σ, i = 1, 2, …, n, where μ is the population mean, are independent standard normals, it follows that the sum of their squares:

$\frac{\sum_{i=1}^{n}(X_i - \mu)^2}{\sigma^2}$, has a chi-squared distribution with n degrees of freedom. If we substitute the sample mean, $\bar{X}$,

80

for the population mean, μ, the new quantity:

$\frac{\sum_{i=1}^{n}(X_i-\overline{X})^2}{\sigma^2}$, will remain a chi-squared random variable with n – 1 degrees of freedom because the population mean, μ, is replaced with its estimator (the sample mean, $\overline{X}$).

Basically, a chi-squared is the ratio of a sample variance multiplied by its degrees of freedom to the population variance:

$$x^2 = \frac{(n-1)S^2}{\sigma^2}$$

THEOREM 5.5:

If S2 is the variance of a random sample of size n taken from a normal population having the variance, $\sigma 2$, then, the random variable:

$$x^2 = \frac{(n-1)S^2}{\sigma^2},$$ has a chi-squared distribution with v = n – 1 degrees of freedom.

The values of the random variable x^2 are calculated from each sample (Walpole and Myers, 1972) by the formula:

$$x^2 = \frac{(n-1)S^2}{\sigma^2}.$$

The probability that a random sample produces a x^2 value greater than some specified value is equal to the area under the curve to the right of this value.

Thus, x^2 is a ratio of a sample variance (S^2) multiplied by its degree of freedom (n – 1) over the population variance (σ^2): $x^2 = \frac{(n-1)S^2}{\sigma^2}$, where S^2 = sample variance with

(n – 1) degrees of freedom; and

σ^2 = population variance.

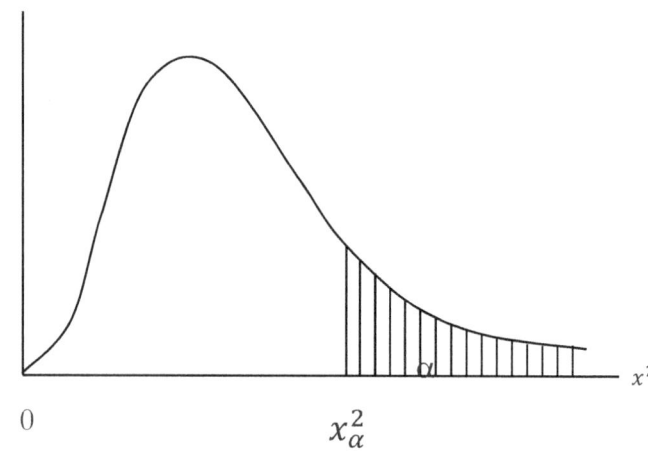

Figure. 5.5.1: Tabulated values of the chi-squared distribution

81

The values of x_α^2 for various values of α and $v = n-1$ are presented in Statistical Tables for the x^2 distribution. The areas, α, are the column headings, the degrees of freedom, $v = n-1$, are given in the left column, the table entries are the x^2 values.

Exactly 95% of a chi-squared distribution with $n-1$ degrees of freedom lies between $x_{0.975}^2$ and $x_{0.025}^2$. A x^2 value falling to the right of $x_{0.025}^2$ is not likely to occur, unless our assumed value of σ^2 is too small. Likewise, a x^2 value falling to the left of $x_{0.975}^2$ is unlikely, unless our assumed value of σ^2 is too large.

As in the case of t-distribution, there is not merely one x^2-distrubition, but there is one distribution for each number of degrees of freedom. Therefore, x^2-distribution is a function of v, the number of degrees of freedom.

The chi-squared distribution is a probability density function where values range from zero to positive infinity. Chi-squared values cannot be negative. As the degrees of freedom $(n-1)$ increases, the x^2-distribution approaches a normal distribution.

The chi-squared distribution finds its major applications in:

- Testing for the goodness-of-fit - - where observed values agree with the theoretical distribution:

$$x^2 = \Sigma \frac{(o_i - e_i)^2}{e_i}, \text{ where: } o_i = \text{observed values; } e_i = \text{expected values.}$$

- Testing for degree of independence;
- Testing for the homogeneity of variances.

Example 5.5.1:

A manufacturer of car batteries guarantees that his batteries will last, on the average, 3 years, with a standard deviation of one year. If five of these batteries last 1.9, 2.4, 3.0, 3.5 and 4.2 years, is the manufacturer still convinced that his batteries have a standard deviation of one year?

Solution:

$$\text{Estimate } S^2 = \frac{\sum_{i=1}^{n} X_i^2 - \frac{(\Sigma X_i)^2}{n}}{n-1} = \frac{48.26 - 45.0}{4} = \frac{3.26}{4} = 0.815$$

$$x^2 = \frac{(n-1)S^2}{\sigma^2} = \frac{(4)(0.815)}{1} = 3.26$$

Thus, from x2-Table (Table A.6 – Walpole (1974)), $x_{(4)}^2 = 0.484$.

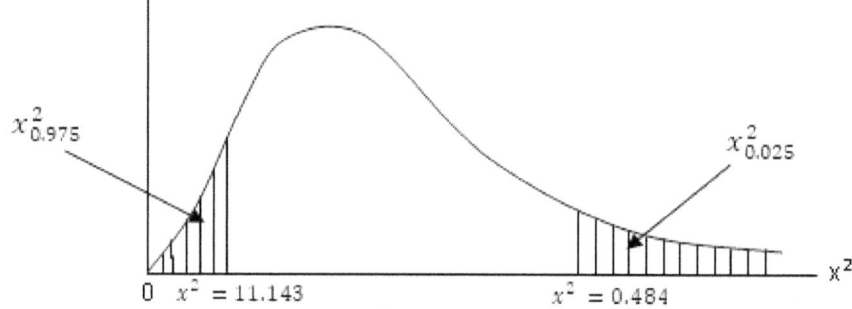

Figure 5.5.2: 95% Confidence Interval for a Chi-squared.

95% of X^2 value to the left or to the right is given as:

$$X^2_{0.975} > \sigma^2 > X^2_{0.025} = \frac{(n-1)S^2}{X^2_{\frac{\alpha}{2}}} < \sigma^2 < \frac{(n-1)S^2}{X^2_{1-\frac{\alpha}{2}}}$$

$$= 0.484 < \sigma^2 < 11.143$$

Since 95% of the estimated $x^2_{(4)}$ contains 1.0 standard deviation, the manufacturer has no reason to doubt his standard deviation of one year.

5.6 F-DISTRIBUTION

The F-distribution is one of the most important distributions in applied statistics (Walpole, 1974). Theoretically, we might define the F-distribution as the ratio of two independent chi-squared distributions, each divided by their degrees of freedom. Hence, if f is a value of the random variable, F, we have:

$$f = \frac{X_1^2/V_1}{X_2^2/V_2} = \frac{S_1^2/\sigma_1 2}{S_2^2/\sigma_2^2} = \frac{\sigma_2^2 S_1^2}{\sigma_1^2 S_2^2}$$

where X^2 is a value of a chi-squared distribution with $v = n - 1$ degrees of freedom.

To obtain an f-value, first select a random sample of size n_1 from a normal population having a variance σ_1^2 and compute S_1^2/σ_1^2. Then, select an independent sample of size n_2 from a second normal population with variance σ_2^2 and compute S_2^2/σ_2^2. The ratio of the two quantities s_1^2/σ_1^2 and S_2^2/σ_2^2 produces an f-value.

The distribution of all possible f-values, where s_1^2/σ_1^2 is the numerator and S_2^2/σ_2^2 is the denominator is called the F-distribution with $v_1(n_1-1)$ and $v_2(n_2-1)$ degrees of freedom.

Thus, the curve of the F-distribution depends, not only on the two parameters v_1 and v_2, but also on the order in which we state them i.e. $(v_1, v_2) \neq (v_2, v_1)$.

Like the normal, x^2 and t-distribution, F-distribution also has tabulated values of f_α only for $\alpha = 0.05$ and $\alpha = 0.01$ for various combinations of the degrees of freedom, v_1 and v_2.

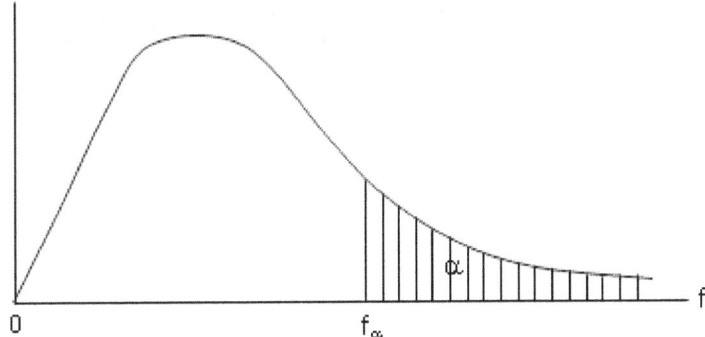

Figure. 5.6.1: Tabulated values of the F-distribution

Once the values of v_1 and v_2 are given, the f-values are easily identifiable from the Table A.7 (Walpole, 1974). Hence, the f-value with 6 and 10 degrees of freedom, leaving an area of 0.05 to the right, is $f_{0.05, (6,10)}$ = 3.22. The Table can be used to find values of $f_{0.95}$ and $f_{0.99}$.

Writing $f_{\alpha(V_1, V_2)}$ for f_α with V_1 and V_2 degrees of freedom, then:

$$f_{1-\alpha}(v_1, v_2) = \frac{1}{f_{\alpha(V_2, V_1)}}$$

$$f_{0.95}(6,10) = \frac{1}{f_{0.05}(10,6)} = \frac{1}{4.06} = 0.246$$

The F-distribution is primarily applied in analysis of variance (ANOVA) to test simultaneously the equality of several means.

Walpole and Myers (1972) stated the F-distribution in a theorem presented below:

THEOREM 5.6:

If S_1^2 and S_2^2 are the variances of independent random samples of sizes n_1 and n_2 taken from normal populations with variances σ_1^2 and σ_2^2, respectively, then:

$$F = \frac{S_1^2/\sigma_1^2}{S_2^2/\sigma_2^2} = \frac{\sigma_2^2 S_1^2}{\sigma_1^2 S_2^2}$$

has an F-distribution with $v_1 = n_1 - 1$ and $v_2 = n_2 - 1$ degrees of freedom.

6 ESTIMATION THEORY

6.1 INTRODUCTION

The theory of Statistical Inference (Walpole, 1974; Ross, 2005), which has become known as the **DECISION THEORY**, may be defined as those methods by which one makes inferences, or generalizations, about a population based on the information obtained from samples selected from the population. Here, we shall consider inferences about unknown population parameters such as the mean, proportion, difference between means and their standard deviation; by computing statistics from random samples and applying the theory of sampling distributions.

The study of statistics is the study of decision making in the presence of uncertainty. The decisions made are almost always based on probability. Thus, a solid grounding in the basics of probability is essential to understanding the statistical procedures. Decision theory may be divided into two major areas:

1. Estimation.
2. Tests of Hypotheses.

Ross (2005) defined an **ESTIMATOR** as **"a statistic whose value depends on the particular sample drawn"**, while the value of the estimator is called the **ESTIMATE** which is used to predict the value of a population parameter.

An estimate of a population parameter may be given as a **POINT ESTIMATE**, or as an **INTERVAL ESTIMATE**. The parameter may be a Mean, a Variance, a Proportion, a Total, or any other population characteristic (Walpole, 1974).

The drawback to point estimation is that it yields a single value for the parameter. How good is the value generated? What is the assurance that it is even close to the actual value of the parameter being estimated?

To get an idea of not only the value of the parameter, μ, being estimated, but also the accuracy of the estimate, researchers turn to the method of **INTERVAL ESTIMATION**, **OR CONFIDENCE INTERVALS**.

6.2 A POINT ESTIMATOR OF A POPULATION MEAN

According to Walpole and Myers (1972) and Ross (2005), a point estimator of the unknown population mean,μ, is given by the statistic, $\overline{X}$, whose sampling distribution is centered at μ and in most applications, the variance is smaller than that of any other estimator.

An estimator whose value is equal to the parameter is said to be **an UNBIASED estimator of that parameter**. Thus, the sample mean,$\overline{X}$, is an unbiased estimator of μ, because:

$$E(\overline{X}) = \mu.$$

While its standard deviation is:

$$\sigma_{\overline{X}} = \frac{\sigma}{\sqrt{n}}$$

which is called the STANDARD ERROR of the mean $(\overline{X})$.

Example 6.2.1:

Suppose samples of four readings are given as: 3.6, 3.9, 3.4 and 3.5. Estimate the mean and the standard error of the mean.

Sample Mean =

$$\frac{\sum X_i}{n} = \bar{X} = 3.6 + 3.9 + 3.4 + \frac{3.5}{4} = \frac{14.4}{4} = 3.6$$

$$S_X = \sigma_X = \sqrt{\frac{\sum X_i^2 - \frac{(\sum X_i)^2}{n}}{n-1}} = \sigma_X = \left(51.98 - \frac{207.36}{4}\right)/3 = \sqrt{\frac{51.98 - 51.84}{3}} = \frac{0.14}{3}$$

$$\sigma_{\bar{X}} = \frac{\sigma}{\sqrt{n}} = \frac{0.2160}{2} = 0.11$$

Therefore, the true population mean, μ, will not differ from 3.6 by more than 0.11.

6.3 A POINT ESTIMATOR OF A POPULATION PROPORTION

A point estimator of the population proportion, p, in a binomial experiment, is given by the statistic

$$\hat{P} = \frac{X}{n}$$.

The sample proportion, $\hat{P} = \frac{X}{n}$, will be used as the point estimate for the parameter, p. If n is sufficiently large, the distribution of $\hat{P}$ is approximately normally distributed with mean:

$$\mu_p = E(\hat{P}) = E\left(\frac{X}{n}\right) = \frac{np}{n} = p$$

and variance:

Variance $(p) = Var\left(\frac{X}{n}\right) = \frac{1}{n^2} Var(X)$

Variance $(p) = npq$

$$\sigma_{\hat{p}}^2 = \sigma_{\bar{X}/n}^2 = \frac{\sigma_{\bar{X}}^2}{n^2} = \frac{npq}{n^2} = \frac{pq}{n}$$

Standard Error of the Proportion = $\sigma_{\hat{p}} = \sqrt{\frac{\hat{p}\hat{q}}{n}}$

OR E(X) = np: Variance $S_{\bar{X}}^2 = np(1-p) = npq$

$$S_X = \sqrt{np(1-p)} = \sqrt{npq}$$

Since $\bar{X}$, the proportion of the sample that has the characteristic, is equal to $\frac{X}{n}$, we see that:

$$E(\bar{X}) = \frac{E(X)}{n} = \frac{np}{n} = p \text{ , and}$$

$$S_{\bar{X}} = \frac{S_X}{n} = \sqrt{\frac{p(1-p)}{n}} = \sqrt{\frac{pq}{n}}$$

Thus, for large n, the Central Limit Theorem states that:

$$Z = \frac{\frac{X}{n} - p}{\sqrt{\frac{p(1-p)}{n}}} = \frac{X - np}{\sqrt{np(1-p)}} = \frac{X - \mu}{\sqrt{npq}}$$

$$Z = \frac{\hat{p} - p}{\sqrt{\frac{pq}{n}}} = \frac{X - np}{\sqrt{npq}}$$

Therefore, $P(-Z_{\frac{\alpha}{2}} \leq Z \leq Z_{\frac{\alpha}{2}}) = 1 - \alpha$

Substitute for Z:

$$P(-Z_{\frac{\alpha}{2}} \leq \frac{\hat{p} - p}{\sqrt{\frac{Pq}{n}}} \leq Z_{\frac{\alpha}{2}}) = 1 - \alpha$$

Multiplying each term by $\sqrt{\frac{Pq}{n}}$ then subtracting $\hat{p}$ and multiplying by -1, we obtain:

$$P\left(\hat{p} - Z_{\frac{\alpha}{2}}\sqrt{\frac{pq}{n}} \leq p \leq \hat{p} + Z_{\frac{\alpha}{2}}\sqrt{\frac{pq}{n}}\right) = 1 - \alpha$$

By substituting the point estimate $\hat{p} = \frac{X}{n}$ for p:

$$P\left(\hat{p} - Z_{\frac{\alpha}{2}}\sqrt{\frac{\hat{p}\hat{q}}{n}} \leq p \leq \hat{p} + Z_{\frac{\alpha}{2}}\sqrt{\frac{\hat{p}\hat{q}}{n}}\right) \approx 1 - \alpha$$

Therefore, $(1-\alpha)$ 100% confidence interval for p is given by:

$$\hat{p} - Z_{\frac{\alpha}{2}}\sqrt{\frac{\hat{p}\hat{q}}{n}} \leq p \leq \hat{p} + Z_{\frac{\alpha}{2}}\sqrt{\frac{\hat{p}\hat{q}}{n}}$$

Example 6.3.1:

The Faculty of Law, **UNIBEN,** has introduced students' dress code. To determine students' reaction to the dress code, the school authority randomly selected a sample of 50 students and questioned them to see those in favor of the dress code. If 20 students favored the dress code;

What is the proportion of all students in favor of the dress code?

What is the standard error of this proportion?

Solution:

P = $\underline{\text{No. of students in favor of the dress code}}$ = $\underline{20}$ = 0.40

No. of students in the sample 50

Standard Error of the estimate = $\sigma_p = \sqrt{\frac{pq}{n}} = \sqrt{\frac{(0.4)(0.6)}{50}}$

$$= \sqrt{\frac{0.24}{50}} = \sqrt{0.0048} = 0.069282 \approx \underline{0.0693}$$

6.4 ESTIMATING A POPULATION VARIANCE

In Section 5.0, we stated that the characteristics of populations – mean,μ, and variance, σ^2 - - are usually unknown and we must therefore use sample data to estimate them. An unbiased point estimate of the unknown population variance, σ^2, is provided by the sample variance, S^2. Hence the statistic, S^2, is called an estimator of σ^2 defined by:

$$S^2 = \frac{\sum_{i=1}^{n}(X_i - \overline{X})^2}{n-1},$$

if the population mean, μ, is unknown. But, if the population mean, μ, is known, the appropriate estimator of σ^2 would be (Ross, 2005):

$$\sigma^2 = \frac{\sum_{i=1}^{n}(X_i - \mu)^2}{n-1},$$

since the population variance is the average of the expected squared difference between an observation and the population mean,μ as:

$$\sigma^2 = E[(X_i - \mu)^2]$$

Example 6.4.1:

A Sociologist took a survey to learn the number of hours that an average government worker works per week. The data he obtained were: 48, 65, 72, 37, 22, 19, 55, 28, 49, 60. Estimate the population mean and its standard deviation.

Solution:

The population mean, μ, is estimated by:

$$\overline{X} = \frac{\sum_{i=1}^{n} X_i}{n} = \frac{48+65+...+49+60}{10} = \frac{455}{10} = 45.5$$

Population variance, σ^2, is estimated by:

$$\sigma^2 = S^2 = \frac{\sum_{i=1}^{n} X_i^2 - \frac{(\Sigma X_i)^2}{n}}{n-1} = \left(23{,}737 - \frac{207{,}025}{10}\right)\Big/9$$

$$\frac{23{,}737 - 20{,}702.5}{9} = \frac{3{,}034.5}{9} = 337.166667$$

$$S = \sigma = \sqrt{337.166667} = 18.36209865 = \underline{18.3621}$$

6.5 ESTIMATING THE DIFFERENCE BETWEEN TWO MEANS $\bar{X}_1 - \bar{X}_2$: A CASE WHERE σ_1^2 AND σ_2^2 ARE KNOWN

In Section 5.3.3, we discussed the sampling distribution of the statistic, $\bar{X}_1 - \bar{X}_2$, the difference between two sample means. If we have two independent normal populations with means, μ_1 and μ_2, and variances, σ_1^2 andσ_2^2, respectively (Walpole and Myers (1972) and Walpole (1974)), a point estimator of the difference between, $\mu_1 - \mu_2$, is given by the statistic, $\bar{X}_1 - \bar{X}_2$.

To obtain a point estimate of μ_1 and μ_2:

- select two independent random samples of sizes n_1 and n_2;
- compute the two sample means, $\bar{X}_1$ and $\bar{X}_2$;
- compute the difference between the two sample means as: $\bar{X}_1 - \bar{X}_2$.
- provided that the two populations are normal;
- or, provided that the sample sizes are greater than, or equal to 30 (n_1 and $n_2 \geq 30$) – i.e. large samples.

Then: $\bar{X}_1 - \bar{X}_2$ is approximately normally distributed with mean, $\mu_{\bar{X}_1 - \bar{X}_2}$ and standard deviation,

$$\sigma_{\bar{X}_1 - \bar{X}_2} = \sqrt{\frac{\sigma_1^2}{n_1} + \frac{\sigma_1^2}{n_2}}$$

Hence:

$$Z = \frac{(\bar{X}_1 - \bar{X}_2) - (\mu_1 - \mu_2)}{\sqrt{\left(\frac{\sigma_1^2}{n_1} + \frac{\sigma_2^2}{n_2}\right)}} \text{ is a standard normal variable.}$$

Example 6.5.1:

A post-graduate Biostatistics class comprising 50 women and 75 men was given a test in which the women made an average grade of 76% with a standard deviation of 6, while the men made an average grade of 82%, with a standard deviation of 8. Test the hypothesis that there is no difference between the average grades of the men and women, versus the alternative hypothesis that there is a difference between the two grades at $\alpha = 0.05$.

Solution:

$$H_0: \mu_1 = \mu_2 \text{ or } \mu_1 - \mu_2 = 0$$

$$H_1: \mu_1 \neq \mu_2 \text{ or } \mu_1 - \mu_2 \neq 0$$

$$\alpha = 0.05$$

Critical level: $Z_{0.025}$ i.e. all z-scores that are less than $-Z_{0.025}$ and greater than $+ Z_{0.025}$.

$Z_{0.025} = \pm 1.96$ (From the normal Table A.4).

Compute:, $Z = \dfrac{(\bar{X}_1 - \bar{X}_2) - (\mu_1 - \mu_2)}{\sqrt{\frac{\sigma}{n_1} + \frac{\sigma_2^2}{n_2}}}$

$$= \frac{(82 - 76) - (0)}{\sqrt{\frac{36}{50} + \frac{64}{75}}} = \frac{6}{0.84853 + 0.92376} = \frac{6}{1.772288}$$

$$= \frac{6}{1.772288} = \underline{3.38545}$$

Conclusion: Since z-calculated = 3.385 is greater than the tabled value = 1.960, reject H_0 and conclude that $\mu_1 \neq \mu_2$ – the men had a significantly better grade than the women.

6.6 ESTIMATING THE DIFFERENCE BETWEEN TWO MEANS $(\bar{X}_1 - \bar{X}_2)$: A CASE WHERE σ_1^2 AND σ_2^2 ARE UNKNOWN BUT EQUAL

If the two independent populations are approximately normally distributed where σ_1^2 and σ_2^2 are unknown but assumed equal ($\sigma_1^2 = \sigma_2^2 = \sigma^2$), and the sample sizes, n_1 and n_2, are small, (n_1, $n_2 < 30$), we must turn to the t-distribution to test the difference between the two sample means.

Since the common unknown population variance (σ^2) must be estimated from the two small samples (n_1, $n_2 < 30$) to obtain a point estimate for a standard normal variable of the form:

$$Z = \frac{(\bar{X}_1 - \bar{X}_2) - (\mu_1 - \mu_2)}{\sqrt{\sigma^2\left[\left(\frac{1}{n_1}\right) + \left(\frac{1}{n_2}\right)\right]}},$$

where σ^2 is estimated by "pooling" or combining the two sample variances for the point estimate as

$$\sigma^2 = S_p^2:$$

$$S_p^2 = \frac{(n_1 - 1)S_1^2 + (n_2 - 1)S_2^2}{n_1 + n_2 - 2}$$

Substitute S_p^2 for σ^2 and obtain:

$$T = \frac{(\bar{X}_1 - \bar{X}_2) - (\mu_1 - \mu_2)}{\sqrt{S_p^2\left(\frac{1}{n_1}\right) + \left(\frac{1}{n_1}\right)}},$$

which has a t-distribution with $n_1 + n_2 - 2$ degrees of freedom for checking the value of T from a t-distribution Table (A.6: Walpole, 1974).

Example 6.6.1:

Suppose 10 boys and 12 girls were tested in an Ecology class and their average grades were 81 and 85, respectively, for the boys and the girls, with a standard deviation of 5 and 4. Test the hypothesis at 95% significance level that there is no difference between the two means versus the alternative that the difference is significant.

Solution:

(i) H_0: $\mu_1 = \mu_2$ or $\mu_1 - \mu_2 = 0$

H_1: $\mu_1 \neq \mu_2$ or $\mu_1 - \mu_2 \neq 0$

$\alpha = 0.05$

(ii)   **Critical level:** $t_{(n_1+n_2-2, 0.025)}$ i.e. $t_{10+12-2, 0.025} t_{(20, 0.025)} = \pm 2.086$ from the t-Table.

(iii) **Decision Rule:** Reject H_0, if t-calculated is greater than or equal to t-tabulated.

(iv) **Compute S_p^2**, the point estimate for the common unknown variance:

$$S_p^2 = \frac{(n_1 - 1)S_1^2 + (n_2 - 1)S_2^2}{n_1 + n_2 - 2} = \frac{(n_1 - 1)S_1^2 + (n_2 - 1)S_2^2}{n_1 + n_2 - 2}$$

$$= \frac{(10 - 1)(5^2) + (12 -)(4^2)}{10 + 12 - 2} = \frac{(9)(25) + (11)(16)}{20}$$

$$= \frac{225 + 176}{20} = \frac{401}{20} = 20.05$$

$$S_p = \sqrt{20.05} = 4.477723$$

(v) **Compute** $T = \dfrac{(\overline{X}_1 - \overline{X}_2) - (\mu_1 - \mu_2)}{s\sqrt{\left(\frac{1}{n_1} + \frac{1}{n_2}\right)}} = \dfrac{(85 - 81) - (0)}{4.4777\sqrt{\frac{1}{10} + \frac{1}{12}}}$

Substitute

$$S_p = \frac{4}{4.4777\sqrt{(0.1 + 0.083)}} = \frac{4}{4.4777\sqrt{(0.1833)}} = \frac{4}{(4.4777)(0.428135492)}$$

$$= \frac{4}{1.917062295} = 2.0865$$

(vi) **Conclusion:** Since the calculated t = 2.0865 is equal to or less than $t_{(20, 0.025)} = 2.086$, we conclude that there is no difference between the two means. Therefore, accept H_0 and conclude that $\mu_1 = \mu_2$. From the Decision Rule above, we could also reject H_0, since t-calculated is equal to t-tabulated.

6.7 ESTIMATING THE DIFFERENCE BETWEEN TWO MEANS$(\overline{X}_1 - \overline{X}_2)$: A CASE WHERE σ_1^2 AND σ_2^2 ARE UNKNOWN AND UNEQUAL

The differences between two means$(\overline{X}_1 - \overline{X}_2)$, from two independent small samples, where the two variances are unknown and therefore not equal, we still obtain good results when the populations are normal, provided that n1 = n2. But for small samples where n1 and n2 are of unknown and of unequal population variances, and it is impossible to select equal sample sizes, the statistic most often used (Walpole and Myers, 1972) to test differences between two such means, has an approximate t-distribution:

$$T' = \frac{(\overline{X}_1 - \overline{X}_2) - (\mu_1 - \mu_2)}{\sqrt{\left(\frac{S_1^2}{n_1} + \frac{S_2^2}{n_2}\right)}}$$

with approximate number of degrees of freedom (v) estimated from the formula:

$$V = \frac{\left(S_1^2/n_1 + S_2^2/n_2\right)^2}{\left[\left(S_1^2/n_1\right)^2/(n_1 - 1)\right] + \left[\left(S_2^2/n_2\right)^2/(n_2 - 1)\right]},$$

which is rarely an integer, or a whole number, which must be approximated to the nearest whole number.

Example 6.7.1:

The 15 years' records of rainfall in Nifor (Edo State of Nigeria) have an average of 1.94 inches with a standard deviation of 0.45 inches. In Udo area of Edo State, the average rainfall was 1.04 inches with a standard deviation of 0.26 inches, in the last 10 years. Test the hypothesis at $\alpha = 0.05$, that there is no difference in the average rainfalls in these two locations, assuming that the observations came from normal populations with different variances.

Solution:

$$\overline{X}_1 = 1.94 \ \overline{X}_2 = 1.04$$

$$S_1 = 0.45 \ S_2 = 0.26$$

$$n_1 = 15 \ n_2 = 10$$

H0: μ1 = μ2

H1: μ1 ≠ μ2

α = 0.05

i. First estimate the degrees of freedom (v) for the approximate t-test:

$$V = \frac{(S_1^2/n_1 + S_2^2/n_2)^2}{[(S_1^2/n_1)^2/(n_1 - 1)] + [(S_2^2/n_2)^2/(n_2 - 1)]}$$

Substitute the data provided into this formula:

$$= \frac{(0.2025/15 + 0.0676/10)^2}{[(0.2025/15)^2/14] + [(0.0676/10)^2/9]}$$

$$= \frac{(0.0135 + 0.00676)^2}{[(0.0135)^2/14] + [(0.00676)^2/9]}$$

$$= \frac{(0.02026)^2}{[0.00018225/14] + [0.0000456976/9]} = \frac{0.0004104676}{0.00001301785714 + 0.00000507751111}$$

$$= \frac{0.0004104676}{0.00001809536825} = 22.68357263 = 22.7 \approx \underline{23}$$

Therefore, t(23,0.025) = 2.069.

Compute $T' = \dfrac{(\overline{X}_1 - \overline{X}_2) - (\mu_1 - \mu_2)}{\sqrt{\left(\dfrac{S_1^2}{n_1}\right) + \left(\dfrac{S_2^2}{n_2}\right)}}$

Substitute the given data: $T' = \dfrac{(1.94 - 1.04) - (0)}{\sqrt{0.2025/15 + 0.0676/10}}$

$$= \frac{0.90}{\sqrt{0.0135 + 0.00676}} = \frac{0.90}{\sqrt{0.02026}}$$

$$= \frac{0.90}{0.142337626} = 6.32299 = 6.323$$

Conclusion: Since T' = 6.323 >> $t_{0.025,23}$ = 2.069, we reject H0 and conclude that the difference between the average rainfalls of Nifor and Udo is substantially significant.

Note: In testing the differences between two means, the difference between the two population means ($\mu_1 - \mu_2$) is set equal to zero. But if the data for the two population means, μ_1 and μ_2, are available, we must substitute them in the formulas for estimating the statistics of the difference.

6.8 ESTIMATING THE DIFFERENCE BETWEEN TWO PROPORTIONS

Independent samples of size n_1 and n_2 are selected at random from two binomial populations with means, n_1p_1 and n_2p_2, and variances, $n_1p_1q_1$ and $n_2p_2q_2$, respectively. Suppose the proportion of successes in each sample are estimated by $\hat{p}_1$ and $\hat{p}_2$, a point estimator of the difference between the two proportions, $p_1 - p_2$, is given by the statistic, $\hat{p}_1 - \hat{p}_2$.

If n_1 and n_2 are sufficiently large, $\hat{p}_1$ and $\hat{p}_2$ are each approximately normally distributed with mean p_1 and p_2 and variance $\frac{p_1q_1}{n_1}$ and $\frac{p_2q_2}{n_2}$ respectively. As we stated in Section 5.3.3, the difference between two independent means from normally distributed populations will also be normally distributed. Hence, selecting independent samples from two populations, the variables $\hat{p}_1$ and $\hat{p}_2$ will be independent and by the reproductive property of the normal distribution, $\hat{p}_1 - \hat{p}_2$ is approximately normally distributed with mean:

$$\mu_{\hat{p}_1 - \hat{p}_2} = \hat{p}_1 - \hat{p}_2 \text{ and}$$

Variance: $$\sigma^2_{\hat{p}_1 - \hat{p}_2} = \frac{\hat{p}_1\hat{q}_1}{n_1} + \frac{\hat{p}_2\hat{q}_2}{n_2}$$

Therefore, $$Z = \frac{(\hat{p}_1 - \hat{p}_2) - (p_1 - p_2)}{\sqrt{\left(\frac{\hat{p}_1\hat{q}_1}{n_1}\right) + \left(\frac{\hat{p}_2\hat{q}_2}{n_2}\right)}}$$ is a standard normal variable.

Example 6.8.1:

The Forest Nursery of UNIBEN has 400 seedlings of ***Irvingia gabonensis*** and 100 seedlings of ***Garcinia cola***. A sample of 20 *Irvingia gabonensis* and another sample of 30 *Garcinia cola* were independently selected and the two proportions were estimated. Is there any difference in the two proportions at α = 0.05 level of significance?

Solution:

n_1 = 400: X_1 = 20

n_2 = 100: X_2 = 30

α = 0.05. Critical region $Z_{0.025}$ = ± 1.96

H0: p_1 = p_2 or $p_1 - p_2$ = 0

H1: $p_1 \neq p_2$ or $p_1 - p_2 \neq 0$

$$\hat{p}_1 = \frac{X_1}{n1} = \frac{20}{400} = 0.05$$

$$\hat{p}_2 = \frac{X_2}{n_2} = \frac{30}{100} = 0.3$$

Decision Rule: Reject H$_0$, if z-calculated is greater than, or equal to z-tabulated.

$$\mu_{\hat{p}_1-\hat{p}_2} = p_1 - p_2 = 0.05 - 0.3 = -0.25$$

$$\sigma^2_{\hat{p}_1-\hat{p}_2} = \frac{\hat{p}_1\hat{q}_1}{n_1} + \frac{\hat{p}_2\hat{q}_2}{n_2} = \frac{(0.05)(0.95)}{400} + \frac{(0.3)(0.7)}{100}$$

$$= \frac{0.0475}{400} + 1.\frac{0.21}{100} = 0.00011875 + 0.0021 = 0.00221875$$

$$\therefore |Z| = \frac{(\hat{p}_1-\hat{p}_2)-(p_1-p_2)}{\sqrt{\left(\frac{\hat{p}_1-\hat{q}_1}{n_1}\right)+\left(\frac{\hat{p}_2-\hat{q}_2}{n_2}\right)}} = \frac{-0.25}{\sqrt{0.00221875}} = \frac{-0.25}{0.047103609} = -5.307449$$

Conclusion: Since $Z_{0.025}$ = -1.96 >> than z-cal = -5.31, we reject H$_0$ and conclude that the difference between the two proportions is highly significant.

6.9 ESTIMATING THE DIFFERENCE IN PAIRED OBSERVATIONS

To estimate the difference of two means when the samples are not independent and the variances of the two populations are not necessarily equal, like the observations in the two samples made on the same individual, we use the difference of paired observations (Walpole and Myers, 1972).

The observations in the two samples made on the same individual are related and therefore form a pair. The difference between a pair of observations is designated, di. Such differences are the values of a random sample D_1, D_2, D_3, ..., D_n, from a population which we assume to be normal with mean, μD, and unknown variance, σ^2_D. We estimate σ^2_D by S^2_d, the variance of the differences constituting the sample.

Therefore, S^2_d is a value of the statistic S^2_d which fluctuates from sample to sample. The point estimator of $\mu 1 - \mu 2 = \mu D$ is given by $\overline{D}$ and is tested by:

$$\text{Mean, } \mu_D, \text{ estimated by: } \bar{d} = \frac{\sum_{i=1}^{n} d_i}{n}$$

$$\text{Variance, } \sigma^2_D \text{ estimated by: } S^2_d = \frac{\sum_{i=1}^{n} d_i^2 - \frac{(\Sigma d_i)^2}{n}}{n-1}$$

$$S^2_{\bar{d}} = \frac{S^2_d}{n} \text{ i.e. } S_{\bar{d}} = \sqrt{\frac{S^2_d}{n}} = \frac{S_d}{\sqrt{n}}$$

$$\text{Test statistic} = T = \frac{\overline{D}-\mu_D}{\frac{S_d}{\sqrt{n}}} \text{ estimated by: } t = \frac{\bar{d}-\mu_D}{\frac{S_d}{\sqrt{n}}}$$

Example 6.9.1:

The diastolic blood pressure of 10 300-level Agric students was taken before and after a prescribed exercise. Is there any difference in the diastolic blood pressure before and after the exercise in the data below?

Table 6.9: Diastolic BP of 10 300-level Agric Students Before and After Exercise

	Before	After	d_i (Before − After)	d_i^2
1.	76	81	-5	25
2.	60	52	8	64
3.	85	87	-2	4
4.	58	70	-12	144
5.	91	86	5	25
6.	75	77	-2	4
7.	82	90	-8	64
8.	64	63	1	1
9.	79	85	-6	36
10.	88	83	5	25

$$\Sigma d_i^2 = 392$$

Solution:

H0: $\mu_D = 0$

H1: $\mu_D \neq 0$

$\alpha = 0.05$

Critical region at $\alpha = 0.05 = t_{0.025,9} = \pm 2.262$, since the sample size is only 10.

df = 9.

Calculate Mean difference = $\bar{d} = \dfrac{\Sigma_{i=1}^{n} d_i}{n}$

$$= \frac{-16}{10} = -1.6$$

Calculate Variance = $S_d^2 = \dfrac{\Sigma_{i=1}^{n} d_i^2 - \dfrac{(\Sigma d_i)^2}{n}}{n-1} = \dfrac{392 - 25.6}{9}$

$$= \frac{366.4}{9} = 40.711111$$

$S_d = \sqrt{40.711111} = 6.380525927$

$S_{\bar{d}} = \dfrac{S_d}{\sqrt{n}} = \dfrac{6.380525927}{\sqrt{10}} = \dfrac{6.380525927}{3.16227766} = 2.01769946$

$t* = \dfrac{\bar{d} - \mu_D}{S_{\bar{d}}} = \dfrac{-1.6}{2.01769946} = -0.79298232$

Conclusion: Since $t^* = -0.793 \gg t_{0.025,9} = -2.262$, we reject H_0 and conclude there is a significant difference in diastolic BP before and after exercise.

7 DETERMINING THE NECESSARY SAMPLE SIZE

The sampling procedure, or experimental design, according to Mendenhall (1975), affects the quantity of information per measurement. The sampling procedure, along with the sample size, n, controls the total amount of relevant information in a sample. One of the most frequently asked questions is: How many measurements should be included in the sample?

In estimation, we would like to know how accurate the experimenter wishes his estimate to be by specifying a bound on the error of estimation and an associated confidence level, $(1-\alpha)$ 100%.

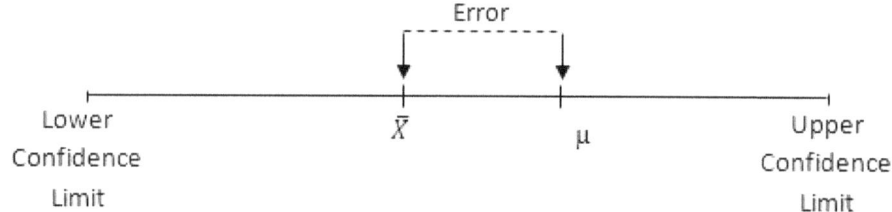

Figure 7.0: Error in Estimating μ by $\bar{X}$

$$\bar{X} - Z_{\frac{\alpha}{2}} \frac{\sigma}{\sqrt{n}} \qquad\qquad \bar{X} + Z_{\frac{\alpha}{2}} \frac{\sigma}{\sqrt{n}}$$

Therefore,

Walpole and Myers (1972) and Walpole (1974) asserted that the $(1-\alpha)$ 100% confidence interval provides an estimate of the accuracy of our point estimate. If μ is the center value of the interval, then $\bar{X}$ estimates μ without error. This is seldom the case. The size of the error is the absolute difference between μ and $\bar{X}$, and we can be $(1-\alpha)$ 100% confident that this difference will be less than $Z_{\frac{\alpha}{2}} . \frac{\sigma}{\sqrt{n}}$.

7.1 DETERMINING SAMPLE SIZE WHEN σ IS KNOWN, OR ESTIMATED FROM n ≥ 30

Walpole (1974) stated that: "If $\bar{X}$ is used as an estimate of μ, we can be $(1-\alpha)$100% confident that the error will be less than $Z_{\frac{\alpha}{2}} . \frac{\sigma}{\sqrt{n}}$ i.e., $e = z_{\frac{\alpha}{2}} . \frac{\sigma}{\sqrt{n}}$.

Walpole (1974) further stated that: "If $\bar{X}$ is used as an estimate of μ, we can be $(1-\alpha)$100%

confident that the error (e) will be less than a specified amount, e, when the sample size is:

$$n = \left(\frac{Z_{\alpha/2}\sigma}{e}\right)^2$$

These formulas (formulae) assume that, either the population variance, σ_1^2 is known, or a preliminary estimate, S_1^2 is made from a sample size, n ≥ 30.

Example 7.1.1:

The mean and standard deviation for the quality point averages of a random sample of 36 college students are calculated to be 2.6 and 0.3, respectively. How large a sample is required, if we want to be 95% confident that our estimate of μ is off by less than 0.05?

Solution:

Sample standard deviation, S = 0.3 obtained from the preliminary sample of size 36 will be used for

σ; e = 0.05; CI = 95%.

$$\therefore n = \left(\frac{Z_{\alpha/2}\sigma}{e}\right)^2$$

$Z0.025 = 1.96$

$e = 0.05$

$\sigma = S = 0.3$

$$n = \left(\frac{(1.96)(0.3)}{0.05}\right)^2 = \left(\frac{0.588}{0.05}\right)^2 = (11.76)^2 = 138.2976$$

$n = 138.3 \approx 139$

If there is no information on σ, or s is available, Mendenhall (1974) suggested the use of the Range (R) in which the measurements fall because the range is approximately equal to 4σ, i.e., Range = 4σ and hence

$$\frac{R}{4} = \sigma.$$

Thus, $n = \frac{\sigma^2}{4}$.

7.2 DETERMINING SAMPLE SIZE WHEN σ IS UNKNOWN AND A PRELIMINARY SAMPLE OF n < 30

From our discussion of t-distribution in Section 5.4, we saw that the t-distribution approaches the normal distribution, as the sample size increases. Thus, each sample of size, n, would generate a t-distribution which has its area tabulated in the T-table. All we need to know in estimating the sample size is the t-values corresponding to the degree of confidence (0.95, 0.98, 0.99) of $t = \frac{\bar{X}-\mu}{\frac{S}{\sqrt{n}}}$ from a preliminary sample of size n.

For sample sizes less than 30, the maximum error, e, can be calculated by: $|t|\frac{S}{\sqrt{n}}$, where $|t|$ is the absolute value of t and hence the sample size, n, can be estimated as:

$$n = \left(\frac{t_{\alpha/2}S}{e}\right)^2 = \frac{\left(t_{\alpha/2}\right)^2 S^2}{e^2}$$

Table 7.2: Values of $t_{n-1,\alpha}$

n−1	0.10	0.05	α 0.025	0.01	0.005
6	1.440	1.943	2.447	3.193	3.707
7	1.415	1.895	2.365	2.998	3.499
8	1.397	1.860	2.306	2.896	3.355
9	1.383	1.838	2.262	2.821	3.250
10	1.372	1.812	2.228	2.764	3.169

7.3 DETERMINING THE SAMPLE SIZE FOR ESTIMATING A PROPORTION

If p is the center value of a $(1-\alpha)$ 100% confidence interval, then $\hat{p}$ estimates p without error. Most of the time, $\hat{p}$ will not be exactly equal to p and the point estimate is in error (Walpole and Myers, 1972). The size of this error will be the difference between p and $\hat{p}$, and we can be $(1-\alpha)$ 100% confident that this difference will be less than $Z_{\alpha/2}\sqrt{\dfrac{\hat{p}\hat{q}}{n}}$.

Thus,

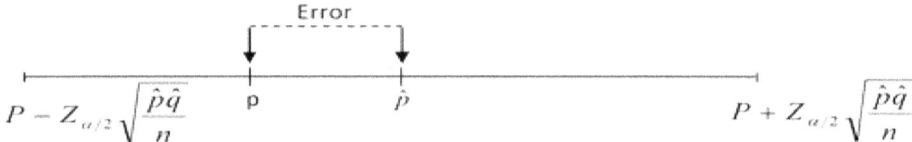

Figure. 7.3: Error in Estimating p by $\widehat{p}$

These authors observed that: **"If $\widehat{p}$ is used as an estimate of p, we can be $(1-\alpha)$ 100% confident that the error will be less than $Z_{\alpha/2}\sqrt{\dfrac{\hat{p}\hat{q}}{n}}$," i.e.,**

$$e = Z_{\alpha/2}\sqrt{\frac{\hat{p}\hat{q}}{n}}$$

Furthermore, **"if $\widehat{p}$ is used as an estimate of p, we can be $(1-\alpha)$ 100% confident that the error will be less than a specified amount, e, when the sample size is:**

$$n = \frac{Z^2_{\alpha/2}\hat{p}\hat{q}}{e^2}$$

A crude estimate of p could be used for $\hat{p}$ to estimate n; or a preliminary sample of size n ≥ 30 could be taken to estimate p.

An upper bound for n can be established for any degree of confidence by noting that $\hat{p}\hat{q} = \hat{p}(1-\hat{p})$, which must be at most equal to $\dfrac{1}{4}$, since $\hat{p}$ must lie between 0 and 1; when $\hat{p} = 1/2$, $\hat{p} = \dfrac{1}{2}$, $\hat{p}\hat{q} =$

$\left(\frac{1}{2}\right)\left(\frac{1}{2}\right) = \left(\frac{1}{4}\right)$. Hence, Walpole (1974) stated that: **"If $\hat{p}$ is used as an estimate of p, we can be, at least, $(1-\alpha)$ 100% confident that the error will be less than a specified amount, e, when the sample size is:**

$$n = \frac{Z_{\alpha/2}^2}{4e^2}$$

Example 7.3.1:

In a random sample of 500 fresh students of UNIBEN, 160 have cheated in exams. How large a sample is required, if we want to be 95% confident that our estimate of p is off by less than (1) 0.02 and (2) at least 95% confident?

Solution:

1. Let us treat the 500 fresh students as a preliminary sample. Proportion estimate,

$\hat{p} = 160/500 = 0.32$.

e = 0.02

$\alpha = 0.05$

$Z_{\alpha/2} = Z0.025 = 1.960$

$$n = \frac{Z_{\alpha/2}^2 \hat{p}\hat{q}}{e^2} = \frac{(1.96)^2(0.32)(0.68)}{(0.02)^2}$$

$$= \frac{(3.8416)(0.2176)}{(0.0004)} = \frac{0.83593216}{0.0004} = 2089.8304 \approx \underline{2{,}090}$$

We can be at least 95% confident that our sample proportion will not differ from the true proportion by more than 0.02 if we select a sample size:

$$n = \frac{Z_{\alpha/2}^2}{4e^2} = \frac{(1.96)^2}{4(0.02)^2} = \frac{3.8416}{4(0.0004)}$$

$$= \frac{3.8416}{0.0016} = \underline{2401}$$

Example 7.3.2:

If an experimenter wishes to estimate the proportion, p, that a person will react in favor of a proposal, how many persons must be included in the experiment? Assume that he will be satisfied, if the error of estimation is less than 0.04, with a probability equal to 0.90; assume also that p lies somewhere around 0.6.

Solution:

Since $(1 - \alpha) = 0.9$, $\alpha = 0.1$ and $\frac{a}{2} = 0.05$

$Z_{\alpha/2} = Z_{0.05} = 1.645$

e = 0.04

$$\hat{p} = 0.6$$

$$n = \frac{Z_{\alpha/2}^2 \hat{p}\hat{q}}{e^2} = \frac{(1.645)^2(0.6)(0.4)}{(0.04)^2}$$

$$= \frac{(2.706025)(0.24)}{0.0016} = \frac{0.649446}{0.0016}$$

$$= 405.90375 \approx \underline{406}$$

Example 7.3.3:

A sample of a certain hybrid strain of corn will be examined to see what proportion exhibits a certain genetic characteristic. What sample size should be examined in order to justify generalizing the result to the entire population, accurate to within 0.01, with a probability of 0.95? Assume (a) it is reasonable to use a figure of 0.20 as an estimate for p and (b) nothing is known about p.

Solution:

a) e = 0.01. For 95% confidence, $\alpha = 0.05$

$$Z_{\alpha/2} = Z_{0.025} = 1.96$$

Assuming p = 0.2

$$n = \frac{Z_{\alpha/2}^2 \hat{p}\hat{q}}{e^2} = \frac{(1.96)^2(0.2)(0.8)}{(0.01)^2} = \frac{(3.8416)(0.16)}{0.0001}$$

$$\frac{0.614656}{0.0001} = 6,146.56 \approx 6,147$$

b) Assuming nothing is known about p, we must use the maximum value of p = 0.5.

Hence:

$$n = \frac{Z_{\alpha/2}^2 \hat{p}\hat{q}}{e^2} = \frac{(1.96)^2(0.5)(0.5)}{(0.01)^2} = \frac{(3.8416)(0.25)}{0.0001}$$

$$= \frac{.9604}{0.0001} = \underline{9,604}$$

7.4 INTERVAL ESTIMATION: CONFIDENCE INTERVALS

Ross (2005) defined an **"INTERVAL ESTIMATOR"** or **CONFIDENCE INTERVAL** of a **population parameter as an interval that is predicted to contain the parameter, while the CONFIDENCE we ascribe to the interval is the probability that it will contain the parameter.**

To determine an interval estimator of a population parameter, we use the sampling distribution of the point estimator of that parameter. It is possible that, $\overline{X}$ the point estimator of the population parameter, μ, may lie either above or below the population mean. But we would not expect it to deviate by more than approximately $2\sigma\overline{x}$ (i.e., $\frac{2\sigma}{\sqrt{n}}$) from μ (Mendenhall, 1975).

In Section 7.1, we saw that the error, when $\overline{X}$ is used to estimate the population mean, μ, is: $e = \frac{Z_{\alpha/2}\sigma}{\sqrt{n}}$.

Hence, if we choose $\left(\overline{X} - Z_{\alpha/2}\sigma/\sqrt{n}\right)$ as the lower point of the interval, called the **LOWER CONFIDENCE LIMIT**, and $\left(\overline{X} + Z_{\alpha/2}\sigma/\sqrt{n}\right)$ as the upper point, or **UPPER CONFIDENCE LIMIT**, the interval most probably, will enclose the true population mean. This interval, computed from the selected random sample, is called a **95% CONFIDENCE INTERVAL**. In other words, we are 95% confident that our computed interval does, in fact, contain the population parameter, μ.

The interval computed from a particular sample is then called $(1-\alpha)$ 100% **CONFIDENCE INTERVAL**. The fraction, $(1-\alpha)$ is called the **CONFIDENCE COEFFICIENT** and the end points are called **CONFIDENCE LIMITS, or FIDUCIAL LIMITS**.

The longer the confidence interval, the more confident we can be that the given interval contains the unknown parameter. Ideally, we prefer a short interval with a high degree of confidence.

Thus,

$$P\left(|\overline{X} - \mu| \leq Z_{\alpha/2}\frac{\sigma}{\sqrt{n}}\right) = 1 - \alpha$$

i.e.

$$P\left(\overline{X} - Z_{\alpha/2}\frac{\sigma}{\sqrt{n}} \leq \mu \leq \overline{X} + Z_{\alpha/2}\frac{\sigma}{\sqrt{n}}\right) = 0.95$$

$$P\left(\overline{X} - 1.96\frac{\sigma}{\sqrt{n}} \leq \mu \leq \overline{X} + 1.96\frac{\sigma}{\sqrt{n}}\right) = 0.95$$

That is, with 95% probability, the interval $\overline{X} \pm 1.96\sigma/\sqrt{n}$, will contain the population mean, μ.

It can be seen that the length of the interval will also depend on the confidence level (confidence coefficient), $(1-\alpha)$, as seen in Table 7.4.

Table 7.4: Confidence Limits for μ

Confidence Coefficient $(1-\alpha)$	Corresponding Value of α	$Z_{\frac{\alpha}{2}}$	Confidence Interval
0.90	0.10	$Z_{0.05} = 1.645$	$\overline{X} \pm 1.645\sigma/\sqrt{n}$
0.95	0.05	$Z_{0.025} = 1.960$	$\overline{X} \pm 1.960\sigma/\sqrt{n}$
0.99	0.01	$Z_{0.005} = 2.576$	$\overline{X} \pm 2.576\sigma/\sqrt{n}$

7.4.1A $(1-\alpha)$ 100% Confidence Interval for μ, when σ is Known

When the population is normal with a known variance, σ_1^2 or the sample size, n, is greater than, or equal to 30 ($n \geq 30$), and $\overline{X}$ is the mean of the sample, $Z_{\alpha/2}$ is the value of a standard normal, then $(1-\alpha)$ 100% C.I. is given by:

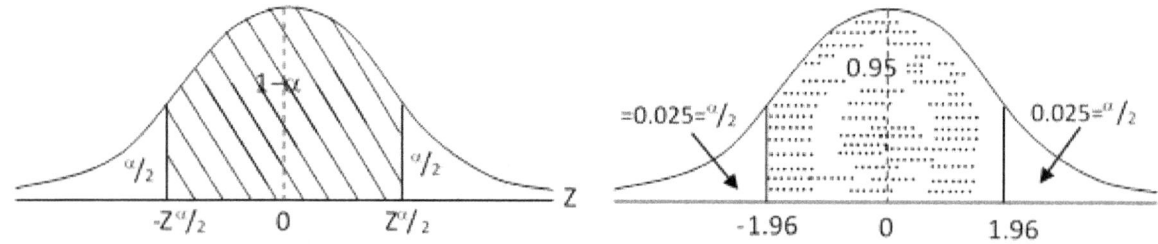

Figure. 7.4.1: P($-Z_{\frac{\alpha}{2}} \leq Z \leq Z_{\frac{\alpha}{2}}$) = 1–$\alpha$ = P(-1.96 $\leq$ Z $\leq$ 1.96) =0.95

Example 7.4.1:

Suppose the sample mean, $\bar{X}$, based on n=50, is found to be 871, with a standard deviation of 21. Find a 90% confidence interval for the population mean, μ.

Solution:

$\bar{X} = 871$

$S = 21$

$\alpha = 0.10$: $^{\alpha}/_{2} = 0.10/2 = 0.05$

$Z_{0.05} = 1.645$

$n = 50$

Since the sample size is n = 50, we can use s to estimate σ as the population variance:

$\therefore$ (1–α) 100% CI = $\bar{X} \pm Z_{\frac{\alpha}{2}} \frac{\sigma}{\sqrt{n}}$

$= 871 \pm (1.645) \frac{21}{\sqrt{50}} = 871 \pm \frac{34.545}{7.071067812}$

$= 871 \pm 4.8854$

$= 866.1146 \leq \mu \leq 875.8854$

7.4.2 A (1–α) 100% CONFIDENCE INTERVAL FOR μ WHEN σ IS UNKNOWN AND n < 30

When the sample size is small, or less than 30, but the population is approximately normal with unknown variance, σ^2, we use the sample data - - $\bar{X}$ and S - - and base our confidence interval on the statistic T_{n-1} given by:

$$T_{n-1} = \frac{\bar{X} - \mu}{S/\sqrt{n}}$$

where: S = Sample standard deviation.

T_{n-1} = is a random variable, having n-1 degrees of freedom.

Due to the symmetric distribution of the t-random variable,

P ($T_n > t_{n,\alpha}$) = α, which has n degrees of freedom.

103

$P(T_n < t_{n,\alpha}) = 1-\alpha.$

Hence, $P(|T_n| \leq t_{n,\frac{\alpha}{2}}) = P(-t_{n,\frac{\alpha}{2}} \leq T_n \leq t_{n,\frac{\alpha}{2}}) = 1-\alpha$

But the quantity, $\frac{\bar{X}-\mu}{S/\sqrt{n}}$ has a t-distribution with n-1 degree of freedom because, upon estimating the sample mean, $\bar{X}$, we lose one degree of freedom. The quantity $t_{n,\alpha}$ is analogous to the quantity Z_α of the standard normal distribution.

In the same manner, when σ is unknown, we can show that:

$P(\bar{X} - t_{n-1,\frac{\alpha}{2}} \frac{S}{\sqrt{n}} \leq \mu \leq \bar{X} + t_{n-1,\frac{\alpha}{2}} \frac{S}{\sqrt{n}}) = 1-\alpha$, is a $(1-\alpha)$ 100% confidence interval – i.e.,

$\bar{X} \pm t_{n-1,\frac{\alpha}{2}} \frac{S}{\sqrt{n}}$ is $(1-\alpha)$ 100% confidence interval.

Example 7.4.2:

To investigate the amount (ppm) of a toxic chemical, PCB, content in the breast milk of cows, an animal nutritionist took a sample of 20 cows in a farm and obtained the following data:

16, 0, 0, 2, 3, 6, 8, 2, 5, 0, 12, 10, 5, 7, 2, 3, 8, 17, 9, 1.

Obtain:

(i)   95% confidence interval.
(ii)   99% confidence interval of the average PCB content in the milk of cows.

Solution:

Calculate the sample mean, $\bar{X}$ and S^2, the sample variance, as:

$$\bar{X} = \frac{\Sigma X_i}{n} = \frac{116}{20} = 5.8$$

$$\alpha = 0.05; \alpha = 0.01; \frac{\alpha}{2} = \frac{0.05}{2} = 0.025; \frac{0.01}{2} = 0.005$$

$$S^2 = \frac{\Sigma X_i^2 - \frac{(\Sigma X_i)^2}{n}}{n-1} = [1,164 - \frac{(116)^2}{20}]/19 = \frac{1164-672.8}{19}$$

$$= \frac{491.2}{19} = 25.85263158$$

$$S = \sqrt{25.85263158} = 5.084548316$$

$$t_{19,0.025} = 2.093.$$

$$t_{19,0.005} = 2.861.$$

(i)   100 $(1-\alpha)$ CI at $\alpha = 0.05$: $\bar{X} \pm (2.093)\frac{(5.084548316)}{\sqrt{20}} = (2.093)\frac{(5.084548316)}{\sqrt{4.472135955}}$

$$= 5.0845 \pm (2.093)(1.136939567).$$

$$= 5.0845 \pm 2.3796 = 2.7049 \le \mu \le 7.4641.$$

(ii) $100\,(1-\alpha)$ CI at $\alpha = 0.01$: $\bar{X} \pm -t_{19,0.005}\frac{s}{\sqrt{n}} = 5.0845 \pm (2.861)(1.136939567)$

$$= 5.0845 \pm 3.25278.$$

$$= 1.83172 \le \mu \le 8.33728.$$

7.4.3 Confidence Interval for the Difference between Two Means, when σ_1^2 and σ_2^2 are Known

In Section 6.5, we discussed the point estimator for the difference between two independent normal populations with means, μ_1 and μ_2, and known variances, σ_1^2 and σ_2^2 The conditions for the point estimator hold true for the confidence interval for the difference between two sample means, $\bar{X}_1 - \bar{X}_2$ and it is estimated by:

$$(\bar{X}_1 - \bar{X}_2) \pm Z\alpha_{/2}\sqrt{\frac{\sigma_1^2}{n_1} + \frac{\sigma_2^2}{n_2}}$$

since its sampling distribution is normal with mean $\mu_{(\bar{X}_1 - \bar{X}_2)}$

and variance, $\sigma_{(\bar{X}_1 - \bar{X}_2)}^2 = \frac{\sigma_1^2}{n_1} + \frac{\sigma_2^2}{n_2}$

and standard deviation, $\sqrt{\frac{\sigma_1^2}{n_1} + \frac{\sigma_2^2}{n_2}}$

A $(1-\alpha)100\%$ CI for $\mu_1 - \mu_2$ is:

$$(\bar{X}_1 - \bar{X}_2) - Z\alpha_{/2}\sqrt{\frac{\sigma_1^2}{n_1} \pm \frac{\sigma_2^2}{n_2}} \le \mu_1 - \mu_2 \le (\bar{X}_1 - \bar{X}_2) + Z\alpha_{/2}\sqrt{\frac{\sigma_1^2}{n_1} \pm \frac{\sigma_2^2}{n_2}}$$

Example 7.4.3:

Using the data in Example 6.5.1, find 96% confidence interval for the difference, $\mu_1 - \mu_2$, where μ_1, is the mean for men and μ_2 is the mean for women.

Solution:

$Z_{\alpha/2} = Z_{0.04/2} = Z_{0.02}$. From standard normal Table (A.4 – Walpole, 1974), we see that we cannot obtain

exactly $Z_{0.02}$. Hence, we must combine the values of columns 0.05 and 0.06 as $(0.05+0.06)/2 = 0.055$.

Thus, $Z_{0.02} = 2.055$ or the values $\frac{(0.0202 + 0.0197)}{2} = \frac{0.399}{2} = 0.01995 \approx 0.02$

$\bar{X}_1 - \bar{X}_2 = 82 - 76 = 6$

Since $n_1 = 75$; $n_2 = 50$ are both large, we can substitute $S_1 = 8$, $S_2 = 6$ for σ_1 and σ_2, respectively.

$100\,(1-\alpha)$ CI for $\mu_1 - \mu_2$ is:

$$(\bar{X}_1 - \bar{X}_2) \pm Z_{\frac{\alpha}{2}} \sqrt{\frac{S_1^2}{n_1} + \frac{S_2^2}{n_2}} = 6 \pm 2.055 \sqrt{\frac{64}{75} + \frac{36}{50}} = 6 \pm 2.58 = 3.42 \le \mu_1 - \mu_2 \le 8.58$$

7.4.4 Confidence Interval for μ1–μ2, when $\sigma_1^2 = \sigma_2^2$, but Unknown

The (1–α) 100% confidence interval for $\mu_1 - \mu_2$, for small samples of n_1 and $n_2 < 30$, but from normal populations, whose variances, σ_1^2 and σ_2^2, are equal but unknown, is given by:

$$(\bar{X}_1 - \bar{X}_2) - t_{\frac{\alpha}{2}} S_p \sqrt{\frac{1}{n_1} + \frac{1}{n_2}} \le \mu_1 - \mu_2 \le (\bar{X}_1 - \bar{X}_2) + t_{\frac{\alpha}{2}} S_p \sqrt{\frac{1}{n_1} + \frac{1}{n_2}}$$

where:

Pooled variance $= S_p^2 = \dfrac{(n_1-1)S_1^2 + (n_2-1)S_2^2}{n_1+n_2-2}$

If the population variances are considerably different, we will still obtain good results when the populations are normal, provided $n_1 = n_2$ (Walpole, 1974).

7.4.5 CONFIDENCE INTERVAL FOR μ₁ – μ₂, WHEN $\sigma_1^2 \ne \sigma_2^2$ AND ARE UNKNOWN

The (1–α) 100% confidence interval for μ₁ – μ₂, for small samples (n_1, $n_2 < 30$) from normal populations whose variances are unequal and unknown is given by:

$$(\bar{X}_1 - \bar{X}_2) - t_{\frac{\alpha}{2}} \sqrt{\frac{S_1^2}{n_1} + \frac{S_2^2}{n_2}} \le \mu_1 - \mu_2 \le (\bar{X}_1 - \bar{X}_2) + t_{\frac{\alpha}{2}} \sqrt{\frac{S_1^2}{n_1} + \frac{S_2^2}{n_2}}$$

The $t_{\alpha/2}$, in this case, has approximate number of degrees of freedom that must be estimated from the formula:

$$V = \frac{(S_1^2/n_1 + S_2^2/n_2)^2}{[(S_1^2/n_1)^2/(n_1-1)] + [(S_2^2/n_2)^2/(n_2-1)]},$$ which may or may not be an integer but must be

approximated to the nearest integer.

Example 7.4.5:

Use the data in Example 6.7.1 and find a 95% confidence interval for the difference between the true average rainfalls in these two locations, assuming the observations came from normal populations with different variances:

$$\bar{X}_1 = 1.94: \bar{X}_2 = 1.04$$

$$S_1 = 0.45: S_2 = 0.26$$

$$n_1 = 15: n_2 = 10$$

Solution:

First estimate the degrees of freedom for the $t\alpha/2$ test as:

$$V = \frac{(S_1^2/n_1 + S_2^2/n_2)^2}{[(S_1^2/n_1)^2/(n_1-1)] + [(S_2^2/n_2)^2/(n_2-1)]} = \frac{(0.2025/15 + 0.0676/10)^2}{[(0.2025/15)^2/14] + [(0.0676/10)^2/9]}$$

$$= \frac{(0.0135 + 0.00676)^2}{(0.00001301785714 + 0.000005077511111)} = \frac{0.0004104676}{0.00001809536825}$$

$$= 22.68357263 \approx 23$$

$$\therefore\ t_{23,\,0.025} = 2.069$$

$$\bar{X}_1 - \bar{X}_2 = 1.94 - 1.04 = 0.90$$

$(1-\alpha)$ 100% CI

$$= (\bar{X}_1 - \bar{X}_2) - t_{\alpha/2}(23)\sqrt{\frac{S_1^2}{n_1} + \frac{S_2^2}{n_2}} \leq \mu_1 - \mu_2 \leq (\bar{X}_1 - \bar{X}_2) + t_{\alpha/2}(23)\sqrt{\frac{S_1^2}{n_1} + \frac{S_2^2}{n_2}}$$

$$= 0.90 \pm 2.069\sqrt{\frac{0.2025}{15} + \frac{0.0676}{10}} = 0.90 \pm (2.069)(0.14233762)$$

$$= 0.90 \pm 0.2944965$$

$$= 0.61 \leq \mu_1 - \mu_2 \leq 1.19$$

7.4.6 CONFIDENCE INTERVAL FOR PROPORTION, P, WHEN n ≥ 30

For large samples, when n is greater than or equal to 30, the binomial parameter, p, is approximately normal. Therefore, the $(1-\alpha)$ 100% confidence interval for p is given by:

$$\hat{p} \pm Z_{\frac{\alpha}{2}}\sqrt{\frac{\hat{p}\hat{q}}{n}}$$

That is:

$$\hat{p} - Z_{\frac{\alpha}{2}}\sqrt{\frac{\hat{p}\hat{q}}{n}} \leq p \leq \hat{p} + Z_{\frac{\alpha}{2}}\sqrt{\frac{\hat{p}\hat{q}}{n}}$$

Example 7.4.6:

A random sample of 100 undergraduates shows that 82 were non-smokers. Find the 99% confidence interval for **p**, the proportion of all the students who are non-smokers.

Solution:

$$\alpha = 0.01: \ {}^{\alpha}/_2 = 0.005: \ Z_{0.005} = 2.576: \ \hat{p} = 0.82$$

$$(1-\alpha) \ 100\% \ \text{CI for p:} \ \hat{p} \pm Z_{0.005}\sqrt{\frac{(0.82)(0.18)}{100}}$$

$$= 0.82 \pm (2.576)\sqrt{\frac{0.1476}{100}} = 0.82 \pm (2.576)\sqrt{0.00001476}$$

$$= 0.82 \pm (2.576)(0.038418745)$$

$$= 0.82 \pm 0.098966687$$

$$= 0.721033313 \leq p \leq 0.91897$$

$$= 0.721 \leq p \leq 0.919$$

7.4.7 CONFIDENCE INTERVAL FOR P, WHEN σ IS UNKNOWN AND n < 30

If the sample size, n, is small and the standard deviation is unknown, the $(1-\alpha)$ 100% confidence interval is given by:

$$\hat{p} \pm t_{n-1,\frac{\alpha}{2}}\sqrt{\frac{\hat{p}\hat{q}}{n}}$$

$(1-\alpha)$ 100% CI for p is: $\hat{p} - t_{n-1,\frac{\alpha}{2}}\sqrt{\frac{\hat{p}\hat{q}}{n}} \leq p \leq \hat{p} + t_{n-1,\frac{\alpha}{2}}\sqrt{\frac{\hat{p}\hat{q}}{n}}$

7.4.8 CONFIDENCE INTERVAL FOR THE DIFFERENCE BETWEEN TWO PROPORTIONS, $P_1 - P_2$

The $(1-\alpha)$ 100% confidence interval for the difference between two proportions will be considered in view of the two preceding sections:

(i)   When the sample sizes, n_1 and n_2, are greater than 30;
(ii)   When the sample sizes, n_1 and n_2 are less than 30.

(i) When the sample sizes are greater than 30, we use the standard normal distribution as:

$(1-\alpha)$ 100% CI:

$$(\hat{p}_1 - \hat{p}_2) - Z_{\alpha/2}\sqrt{\frac{\hat{p}_1\hat{q}_1}{n_1} + \frac{\hat{p}_2\hat{q}_2}{n_2}} \leq p_1 - p_2 \leq (\hat{p}_1 - \hat{p}_2) + Z_{\alpha/2}\sqrt{\frac{\hat{p}_1\hat{q}_1}{n_1} + \frac{\hat{p}_2\hat{q}_2}{n_2}}$$

When the sample sizes are small, or less than 30, we use the t-distribution as:

(1–α) 100% CI:

$$(\hat{p}_1 - \hat{p}_2) - t_{n-1,\frac{\alpha}{2}}\sqrt{\frac{\hat{p}_1\hat{q}_1}{n_1} + \frac{\hat{p}_2\hat{q}_2}{n_2}} \le p_1 - p_2 \le (\hat{p}_1 - \hat{p}_2) + t_{n-1,\frac{\alpha}{2}}\sqrt{\frac{\hat{p}_1\hat{q}_1}{n_1} + \frac{\hat{p}_2\hat{q}_2}{n_2}}$$

$$(\hat{p}_1 - \hat{p}_2) - t_{n-1,\frac{\alpha}{2}}\sqrt{\frac{\hat{p}_1\hat{q}_1}{n_1} + \frac{\hat{p}_2\hat{q}_2}{n_2}} \le p_1 - p_2 \le (\hat{p}_1 - \hat{p}_2) + t_{n-1,\frac{\alpha}{2}}\sqrt{\frac{\hat{p}_1\hat{q}_1}{n_1} + \frac{\hat{p}_2\hat{q}_2}{n_2}}$$

8 TESTING STATISTICAL HYPOTHESES

The testing of statistical hypotheses is perhaps the most important decision theory (Walpole, 1974). In a hypothesis testing problem, there is a pre-conceived theory, or an assumption, concerning the population characteristic under study. It is often stated in terms of a population parameter.

The reasoning employed in testing a hypothesis bears a striking resemblance to the procedure used in a law court trial where the court assumes that the accused is innocent until proved guilty. Thus, the prosecution presents all the available evidence in an attempt to contradict the "not guilty" hypothesis and so obtain a conviction.

Similarly, to test a statistical hypothesis, we must decide whether that hypothesis appears to be **"consistent"** with the data of the sample or **"not consistent"**. This implies that there are two theories, or hypotheses, involved in any statistical study: (a) the hypothesis being proposed by the experimenter is called the **NULL HYPOTHESIS** and is symbolized by H_0, which may be of the form: $H_0: \mu = \mu_0$ i.e., the population parameter is equal to a specified value, and (b) the negation of this hypothesis, called the **ALTERNATIVE HYPOTHESIS,** designated by H_1 can take one of these three forms:

(i) $H_1: \mu < \mu_0$ i.e., μ is less than the specified value – a one-tailed test.
(ii) $H_1: \mu > \mu_0$ i.e., μ is greater than the specified value – a one-tailed test.
(iii) $H_1: \mu \neq \mu_0$ i.e., μ is not equal to the specified value – a two-tailed test.

The alternative hypothesis is what the experimenter is interested in confirming, in order to establish the decision of **"not consistent" or "guilty"**.

Therefore, the experimenter, or statistician, must test the hypothesis in an objective manner to determine whether the sample data are "consistent" with the Null hypothesis. If the sample data are consistent with the null hypothesis, we accept the null hypothesis. Otherwise, we reject the null hypothesis and accept the alternative hypothesis.

The decision of whether to reject or accept the null hypothesis is based on the value of a **"Decision Maker"** called the **TEST STATISTIC**. The value of the test statistic is divided into two groups called:

1. the **REJECTION REGION, OR CRITICAL REGION** – which contains the set of values of the test statistic that will lead to the rejection of the null hypothesis;
2. the **ACCEPTANCE REGION** – which contains the set of values of the test statistic that will lead to the acceptance of the null hypothesis.

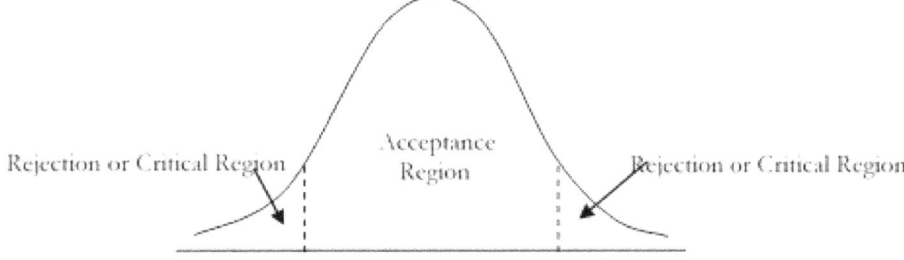

Figure. 8.1: Acceptance and Rejection Regions of a Two-Tailed Test

8.1 THE TWO TYPES OF POSSIBLE DECISION ERRORS

From the foregoing, we notice that, after the calculation of the test statistic, the experimenter is still faced with two possible types of errors in his decision:

a) He may erroneously reject the null hypothesis (H_0), when it is true, and falsely support the alternative hypothesis (H_1), when it is wrong. This type of error is called a **TYPE I ERROR**, denoted by the symbol α.

Definition (Mendenhall, 1975):

"Rejecting the null hypothesis when it is true is called a TYPE I ERROR for a statistical test. The probability of making a Type I Error is denoted by the symbol α".

Mathematically stated: **P (Reject H_0, when H_0 is true) = α = Type I**

i.e.: **P (Reject H_0/H_0 is true) = α.**

The value, α, is called **the LEVEL OF SIGNIFICANCE of the test and is usually under the control of the experimenter, and it is set in advance, with commonly chosen values of α = 0.05 and 0.01.** The classical procedure for testing a null hypothesis is to fix a small significance level, α, in advance, and then require that the probability of rejecting H_0, when H_0 is true, is less than or equal to α, i.e.,

P (Reject H_0/H_0 is true) $\leq \alpha$ = P (Type I Error) = Level of Significance = α.

This is the **Error that I call the "Pilate Error" because Pilate found no fault in Jesus Christ and declared: "for I find no fault in this man"**; and yet Pilate delivered Jesus to be crucified; thus declaring the innocent Jesus Christ guilty and subsequently sentenced him to death **(Matthew 27: 23-26; Mark 15:12-15; Luke 23: 4; 13-24; John 19: 4-6).**

Level of Significance (α):

A level of significance (α) of 1% signifies that when the population mean (μ) is correct as specified, the sample mean will fall in the critical areas (regions) of rejection only 1% of the time. Referring to Figure 8.1, 0.005 or 0.5% of the sample means will fall in each area of rejection and 99% of the means will fall in the region of acceptance.

If the sample size is large, statistical significance may occur, even for a difference which is of marginal practical importance.

b) He may not reject the null hypothesis (H_0) when it is false. This type of error is called a **TYPE II ERROR for the statistical test and its probability is denoted by the symbol, β.**

DEFINITION (Mendenhall, 1975):

"Accepting the null hypothesis (H_0), when it is false, is called a TYPE II ERROR for a statistical test. The probability of making a Type II error, when some specific alternative is true, is denoted by the symbol, β".

i.e. **P (Accept H_0, when H_0 is false) = Type II Error = β**

or **P (Accept H_0/H_0 is false) = β**

P (Type II Error) = β

For a fixed sample size, n, α and β are inversely related – as one increases, the other decreases. Increasing the sample size provides more information upon which to base the decision and hence reduces both α and

β.

Under an experimental situation, the probabilities of Type I and Type II errors for a test measure the risk of making an incorrect decision; **the more serious the consequences of making an error, the smaller the level of significance which is used.**

It is desirable to have both types of error small. However, the two types of error are not independent – the smaller numerically the Type I Error, then the larger numerically the potential Type II Error.

While Type I Error i.e. the level of significance (α), is directly under the control of the experimenter, the Type II Error (β) is a little harder to handle than α because for a particular test, **β depends on which particular alternative is true. That is, β can be found only if a particular value of the alternative is specified.** The Types I and II Errors and possible decisions are presented in Table 8.1:

Table 8.1: The Null Hypothesis and the Errors Committed

Decision	Null Hypothesis	
	H_0 True	H_1 True
Reject H_0	Type I Error Probability = α	Correct Decision (No Error) = $1 - \beta$ = Power
Accept H_0 (Fail to Reject H_0)	Correct Decision No (Error) = $1 - \alpha$	Type II Error Probability = β

Note: The probability of committing a Type I Error is the probability that the test statistic will fall into the critical (or rejection) region by chance, even though the null hypothesis is true.

8.2 PROCEDURE FOR TESTING HYPOTHESES

The procedure for testing a statistical hypothesis includes the following steps:

1. State the null hypothesis and its assumptions i.e. H_0: $\mu = \mu_0$.
2. State the alternative hypothesis: H_1: $\mu < \mu_0$; $\mu > \mu_0$; $\mu \neq \mu_0$.
3. State the level of significance chosen i.e. $\alpha = 0.05$ or 0.01.
4. Select the appropriate test statistic and establish the critical region.
5. Compute the value of the statistic from a random sample of size n.
6. State the conclusion: Reject H_0, if the statistic has a value in the critical region; otherwise, accept H_0.

8.3 STATING THE CONCLUSION

Generally, we use a small level of significance (α) and thus provide protection against rejecting a true hypothesis. If the value of the test statistic observed is in the critical region, we reject the null hypothesis (H_0), or say that **"the result is significant at the specified level"** (Dixon and Massey, 1969). Other levels of significance (α) may not lead to the rejection of H_0.

If the value of the test statistic is observed in the **"acceptance region"**, we say **"the result is not significant"**, or **"there is no sufficient evidence to reject H_0, the null hypothesis"**, or **"we fail to reject H_0"**, rather than saying "we accept H_0" (Mendenhall, 1975: Ross, 2005). Dixon and Massey (1969) suggested that we should avoid a wording that implies that the hypothesis has been proved to be true.

8.4 "P-VALUE"

Many professional journals and articles report a **"p-value"**, or a "probability level" which usually indicates (Dixon and Massey, 1969) the smallest value of α that would lead to a rejection of the null hypothesis. Ross (2005) has defined **the p-value "as the smallest significance level at which the data would lead to a rejection of the null hypothesis"**. It gives the probability that data as unsupportive of H_0 as those observed, will occur when H_0 is true.

To test the null hypothesis, H_0: $\mu \geq \mu_0$ versus H_1: $\mu < \mu_0$ at $\alpha = 0.05$, the test statistic is: $\dfrac{\sqrt{n}(\overline{X} - \mu_0)}{\sigma}$ and we reject

H_0, if $\dfrac{\sqrt{n}(\overline{X} - \mu_0)}{\sigma} < Z\alpha$ Therefore,

$$P\text{-value} = P\left(\frac{\sqrt{n}(\overline{X} - \mu_0)}{\sigma}\right) \leq P(Z \leq Z\alpha) - \text{for a one-tailed test; or}$$

$$P\text{-value} = P\left(\frac{\sqrt{n}(\overline{X} - \mu_0)}{\sigma}\right) \geq P(Z \geq Z_{\alpha/2}) - \text{for a two-tailed test.}$$

Example 8.4.1:

A firm wanted to test the hypothesis that the mean of their product was at least 1.5mm with a standard deviation of 0.7mm, versus the alternative that it is less. They took a sample of 20 items and found the average to be 1.42mm. (i) What conclusion could be drawn from the data? (ii) What is the p-value?

Solution:

H_0: $\mu \geq 1.5$mm

H_1: $\mu < 1.5$mm

$\sigma = 0.7$mm

$\alpha = 0.05$: $Z_{0.05} = 1.645$

$\overline{X} = 1.42$mm

$$Z = \frac{\sqrt{n}(\overline{X} - \mu_0)}{\sigma} = \frac{\sqrt{20}(1.42 - 1.5)}{0.7} = \frac{4.472135955(-0.08)}{0.7}$$

$$= \frac{-0.357770876}{0.7}$$

$$= -0.511101252$$

$P(Z \leq -0.5111) = 0.3050 - \text{From Table A.4 (Walpole, 1974).}$

$P\text{-value} = P(Z \leq -0.5111) = 0.3050.$

Conclusion: Do not reject H_0 because the mean, μ, is really less than 1.5, and the p-value = 0.3050 which is larger than $\alpha = 0.05$ and therefore the evidence is not strong enough to prove the claim.

8.5 POWER OF THE TEST FOR $H_0: \mu = \mu_0$

An important concept in connection with hypothesis testing is **the power of a test** which is $1-\beta$, the complement of β, which is the probability of rejecting the null hypothesis when, in fact, it is false (Sokal and Rohlf, 1969) and the alternative hypothesis is correct. **Power = $1-\beta$**.

For any given test, we would like to have the quantity, $1-\beta$, as large as possible and β as small as possible. Values of β can only be calculated, if we know the actual (or true) population parameter. **Therefore, we have to calculate β or $1-\beta$ for a continuum of alternative values**. When $1 - \beta$ is plotted against the possible alternative values of μ, the graph is called **a POWER CURVE** (Sokal and Rohlf, 1969), or **OPERATING CHARACTERISTIC CURVE (O–C)** (Dixon and Massey, 1969; Byrkit, 1975; Mendenhall, 1975) for the test under consideration.

The power of a test $(1 - \beta)$ is influenced by:

1. As the sample size is increased, α and β decrease.
2. As α increases, β decreases, and as α decreases, β increases.
3. The value of β is large, if the true value of μ, is close to μ_0, and small if the true value is very different from μ_0.

To plot the power $(1-\beta)$ curve against the alternatives, μ, we use the standard error of $\overline{X} = \sigma / \sqrt{n}$, the sample size n, α and $Z_{\alpha/2}$, as well as the confidence intervals on μ_0 under different sample sizes:

Example 8.5.1:

$H_0: \mu = 67; H_0: \mu = 68$

$H_1: \mu \neq 67$

$\alpha = 0.05$

$\sigma = 3.0$

n = 25 and n = 100

Solution:

Compute CI for the various alternatives of μ, z_1, z_2, β and $1-\beta$

$$\text{CI:} \overline{X} \pm Z_{\alpha/2} \frac{\sigma}{\sqrt{n}}; Z = \frac{\overline{X} - \mu}{\sigma}; z_1 - z_2; n = 25; n = 100$$

For n = 25: LCL =

UCL67 + 1.76 = 68.18.

$\begin{pmatrix} \text{where} & : & \text{LCL} & = & \text{Lower} & \text{Confidence} & \text{Limit} \\ & & \text{UCL} & = & \text{Upper} & \text{Confidence} & \text{Limit} \end{pmatrix}$

For n = 100:

$$LCL = \overline{X} \pm Z_{\alpha/2} \frac{\sigma}{\sqrt{n}} = 68 - 1.96\frac{3}{10} = 0.588 = 67.41$$

$$UCL = 68 + 0.588 = 68.59$$

Using these lower and upper confidence intervals when n = 25 and n = 100, we compile this table:

$$H_0: \mu = 67 \qquad H_0: \mu = 68$$

$$n = 25 \qquad n = 100$$

μ	z_1	z_2	P(z_1)-P(z_2) β	Power (1–β)	z_1	z_2	P(z_1) -P(z_2) β	1–β
68.5	-4.47	-0.53	0.29	0.71	-6.97	-3.04	0.00	1.00
68	-3.63	0.30	0.62	0.38	-5.30	-1.37	0.09	0.91
67.5	-2.80	1.13	0.87	0.13	-3.68	0.30	0.62	0.38
67	-1.96	1.96	0.95	0.05	-1.96	1.96	0.95	0.05
66.5	-1.13	2.80	0.87	0.13	-0.30	3.63	0.62	0.38
66	-0.30	3.63	0.62	0.38	1.37	5.30	0.09	0.91
65.5	0.53	4.47	0.29	0.71	3.04	6.97	0.00	1.00

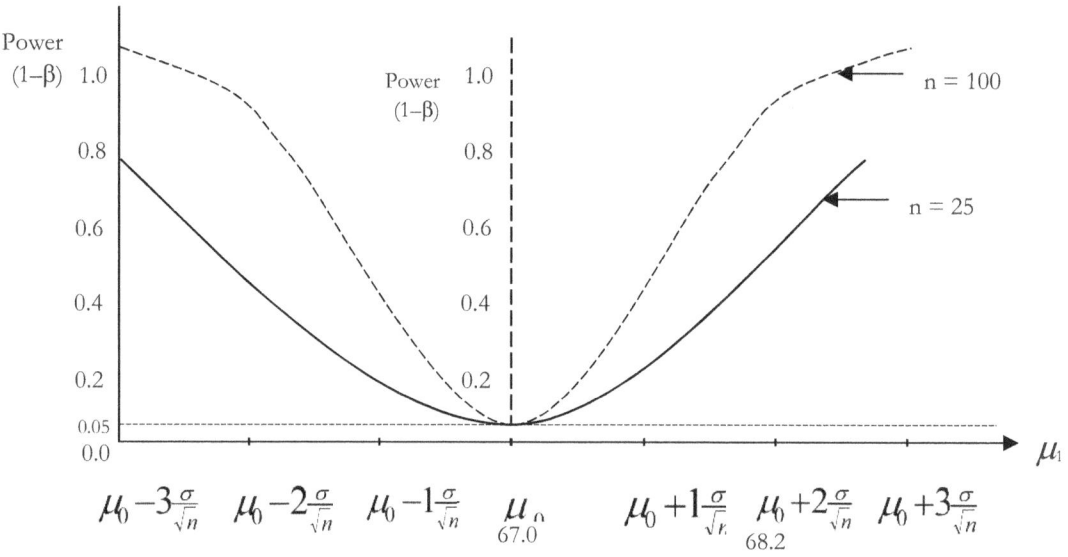

Figure 8.5.1 Power of a Two-Tailed Single Sample Test of $H_0: \mu = 67$ and $H_0: \mu = 68$

Once the power curve is drawn, it can be used to read the value of 1–β for a given alternative, μ; and an alternative μ can be read for a given 1–β.

8.6 ONE-SIDED (ONE-TAILED) TESTS OF HYPOTHESES

One-sided or one-tailed tests of hypotheses are generally of two types and can only be detected from the Alternative hypothesis (H_1). The location of the one-side of interest i.e., the critical region, is determined from the H_1 stated. Once H_1 is stated, at a given significance level, α, the location of the critical region is determined. **These one-tailed tests are called DIRECTIONAL REJECTION REGION.**

115

For example, if we want to test whether a new herbicide is better than those already in the market, we will state the one-tailed hypothesis as:

(a) H_0: $\mu \leq \mu_0$ i.e.the new herbicide (μ) is less than or equal to the ones (μ_0) already

in market.

H_1: $\mu > \mu_0$ i.e. the new herbicide (μ) is better than the ones (μ_0) already in market.

$\alpha = 0.05$

Select the Test statistic and establish the critical region:

$Z_\alpha = Z_{0.05} = 1.65$.

Compute the statistic as: Z^* or $Z_{cal} = \dfrac{\sqrt{n}(\bar{X} - \mu_0)}{\sigma}$

Pictorially, the critical region is shown in Figure 8.6.1:

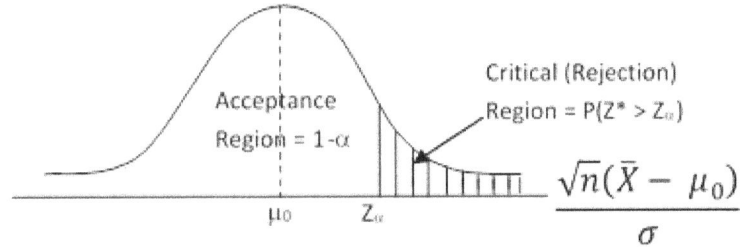

Figure 8.6.1: One-tailed (Right-hand) Test of H_1: $\mu > \mu_0$

In this example, we are only interested in z-values much greater than $Z_\alpha = 1.65$. Thus, we reject H_0, if the calculated test statistic, Z^*, in this case, is greater than $Z_\alpha = Z_{0.05} = 1.65$ - - that is, if the sample mean $\bar{X}$ is greater than μ_0.

Alternatively, we could test the hypothesis as:

(b) H_0: $\mu \geq \mu_0$ i.e. the new herbicide is as good as, or better, than those already in the market.

H_1: $\mu < \mu_0$ i.e.the new herbicide is not as good as those already in market.

$\alpha = 0.05$.

Select the Test statistic and establish the critical region:

$-Z_\alpha = -Z_{0.05} = -1.65$

Decision Rule: Reject H_0, if $Z^* \geq Z_{0.05} = -1.65$

Calculate Z^* as: $-Z_{cal} = \dfrac{-\sqrt{n}(\bar{X} - \mu_0)}{\sigma}$ pictorially represented as shown in Figure 8.5.2.

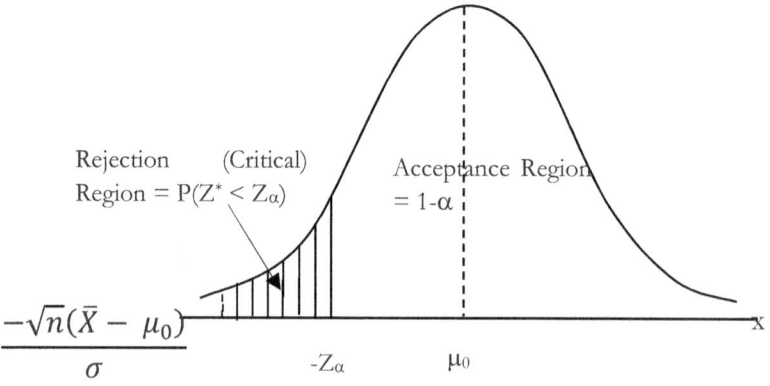

$$\frac{-\sqrt{n}(\bar{X} - \mu_0)}{\sigma}$$

Figure 8.6.2: One-tailed (Left-hand) Test of H₁: $\mu < \mu_0$

In this example, we are interested in z-values (or t-values, or χ^2, or F- values) which are much smaller than $-Z_\alpha = -Z_{0.05} = -1.65$ - - that is, if the sample mean, $\bar{X}$, is smaller than the hypothesized value, μ_0.

Thus, directional hypotheses are implied when the basic question involves terms like **"more than", "better than", "increased", "less than", or "declined"**. Therefore, directional tests arise from already available knowledge, or information, about the current situation in the population of interest and the value of μ_0 under H₀.

From such available knowledge on the population of interest, or from past experience, it is known that:

i. the population is normally distributed.
ii. the sample size, n, is large, or greater than 30 ($n \geq 30$).
iii. the population variance, σ^2, is known.
iv. the level of significance, α, is small i.e. $\alpha = 0.05$ or 0.01.
v. the statistical tests, z or t, can be applied due to the Central Limit Theorem.

But when no information is available on the direction of the difference to expect, if H₀ is false, the basic question is: "Is there any difference?" or "Is there any change?" Such questions (Whinkler and Hays (1975)) take us to test techniques for non-directional hypothesis testing - - a two-tailed test.

8.7 TWO-SIDED (TWO-TAILED) TESTS OF HYPOTHESES

When there is no information on the true nature of the population of interest, the rejection region should be arranged so that H₀ will be rejected when there are extreme departures from expectation - - whether too low or too high, occur. This calls for a rejection in both tails, or regions and it is called a **TWO-TAILED, or TWO-SIDED TEST**. The hypothesis is stated as:

H₀: $\mu = \mu_0$ i.e. the population mean, μ, is equal to a specified value, μ_0.

H₁: $\mu \neq \mu_0$ i.e. the population mean, μ, is not equal to a specified value, μ_0.

$\alpha = 0.05$ or 0.01

Select the test statistic and establish the critical regions as:

$$\pm Z_{\alpha/2} = Z_{0.025} = \pm 1.960$$

Compute the statistic $Z_{cal} = \dfrac{\sqrt{n}(\bar{X} - \mu_0)}{\sigma}$ and compare it to $Z_{0.025} = \pm 1.96$

Decision Rule: Reject H_0 if $|Z_{cal}| \geq Z_{0.025} = \pm 1.960$ i.e. if the absolute value of the computed statistic is greater than or equal to the tabled value.

Pictorially, a two-tailed test is represented in Figure 8.7.1:

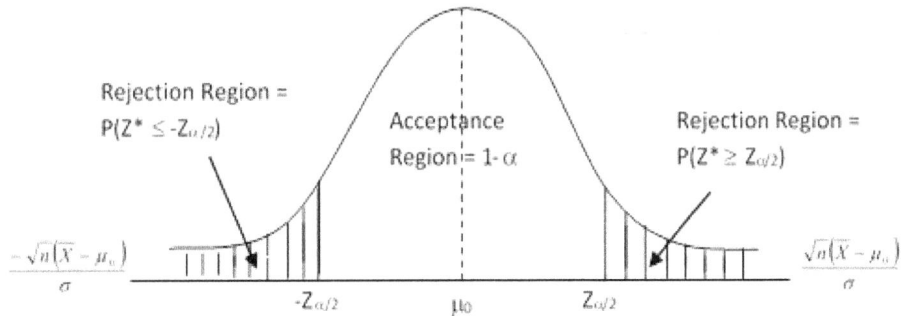

Figure 8.7.1: A Two-tailed Rejection Region (H_1: $\mu \neq \mu_0$)

Testing a two-tailed hypothesis is equivalent to setting a $(1-\alpha)$ 100% confidence interval on μ discussed in Section 7.4.1 and 7.4.2: $(1-\alpha)$ 100% CI:

$$\bar{X} - Z_{\frac{\alpha}{2}} \frac{\sigma}{\sqrt{n}} \leq \mu \leq \bar{X} + Z_{\frac{\alpha}{2}} \frac{\sigma}{\sqrt{n}}$$ and accepting H_0, if μ lies in the internal and rejecting H_0, if μ lies outside the internal.

Therefore, a two-tailed test of hypothesis may be achieved by testing the hypothesis, as above, or setting confidence intervals on μ. In both cases, we use the same statistic:

$$Z* = \frac{\sqrt{n}(\bar{X} - \mu_0)}{\sigma}$$, or other test statistics like t, χ^2 or F.

8.8 THE POWER OF A STATISTICAL TEST

If H_0: $\mu = \mu_0$, P(Reject $H_0/\mu = \mu_1$) = P(Reject H_0/H_0 is true) = α. But P(Reject H_0/H_1 is true) = 1– P(Reject H_0/H_1 is true) = 1–β = Power. For any value of μ_1, then, the probability, P(Reject $H_0/\mu = \mu_1$) can be computed, and this probability is called the power of the test of H_0 against the alternative H_1 (Winkler and Hays, 1975).

The power of a test of H_0 is the ability of a decision rule to detect, from the evidence, that the true situation differs from a hypothetical one. Winkler and Hays (1975) emphasized that a high-powered test of H_0 insures us of detecting when H_0 is false. The larger the departure of H_0 from the true situation, H_1, the more powerful is the test of H_0 -- all things being equal.

Large values of $\bar{X}$ favor H_1, while small values favor H_0 and the rejection region is of the form: $\bar{X} \geq C$. For $\alpha = 0.05$, the critical value (C) is: $\mu_0 + 1.65 \dfrac{\sigma}{\sqrt{n}}$ and the power of the test can be computed for any value of μ_1 as:

118

$$\mu_1 = \mu_0 + \frac{\sigma}{\sqrt{n}}$$

The power is then: $P(\text{Reject } H_0 / \mu_1 = \mu_0 + \frac{\sigma}{\sqrt{n}}) =$

$$P\left(\bar{X} \geq \mu_0 + 1.65\frac{\sigma}{\sqrt{n}} \mid \mu_1 = \mu_0 + \frac{\sigma}{\sqrt{n}}\right) =$$

$$= P\left(\frac{Z \geq \left(\mu_0 + 1.65\frac{\sigma}{\sqrt{n}}\right) - \left(\mu_0 + \frac{\sigma}{\sqrt{n}}\right)}{\sigma/\sqrt{n}}\right)$$

$$= P\left(Z \geq \mu_0 + 1.65\frac{\sigma}{\sqrt{n}} - \mu_0 + \frac{\sigma}{\sqrt{n}}\right) = P(Z \geq 0.65) = \beta$$

$$= 1 - P(Z < 0.65) = 1 - \beta = Power$$

$$= 1 - 0.7422 = 0.2578$$

This is illustrated in Figure 8.8.1, where the shaded area is β, while the unshaded area of the curve is $1-\beta$, the power.

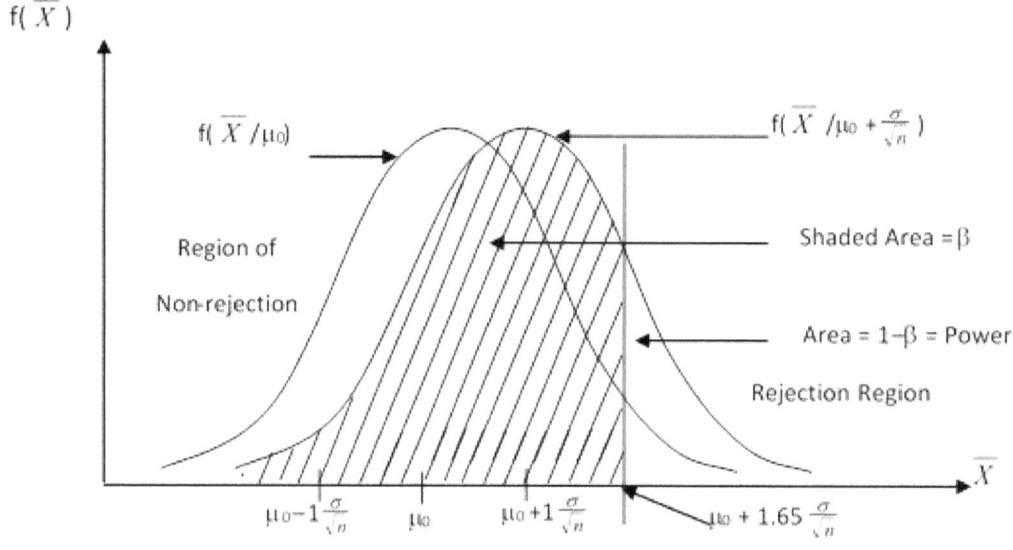

Figure 8.8.1: Power as Area under a Sampling Distribution

Of course, a different alternative hypothesis will produce a different power. If $\alpha=0.05$, $z=1.65$.

If $\mu_1 = \mu_0 + 3\frac{\sigma}{\sqrt{n}}$, then: $P(\text{Reject } H_0 / \mu_1 = \mu_0 + 3\frac{\sigma}{\sqrt{n}}) = \beta$

Hence, $P(\text{Reject } H_0 / \mu_1 = \mu_0 + 3\frac{\sigma}{\sqrt{n}}) = P(\bar{X} \geq \mu_0 + 1.65\frac{\sigma}{\sqrt{n}} \mid \mu_1 = \mu_0 + 3\frac{\sigma}{\sqrt{n}})$

$$= P(Z \geq (\mu_0 + 1.65\frac{\sigma}{\sqrt{n}}) - (\mu_0 + 3\frac{\sigma}{\sqrt{n}}))$$

$$= P(Z \geq \mu_0 + 1.65\frac{\sigma}{\sqrt{n}} - \mu_0 - 3\frac{\sigma}{\sqrt{n}})$$

$$= P(Z \geq 1.65\frac{\sigma}{\sqrt{n}} - 3\frac{\sigma}{\sqrt{n}}) = P(Z \geq -1.35) = \underline{0.911492} = \beta.$$

$$= 1 - P(Z < 1.35) = \text{Power} \text{ - - From Standard Normal Table A.4.}$$

$$= 1 - 0.911492 = 0.0885$$

$$= 1 - \beta = 0.0885$$

Therefore, we see that, if the true value, μ_1, is sufficiently greater than μ is actually $\mu_0 + 3\frac{\sigma}{\sqrt{n}}$, the test has a high probability.

The relation of power to the true value of μ_1 is called the POWER FUNCTIONS, OR, POWER

CURVES - - where the horizontal axis gives the possible values of μ_1 in terms of μ_0 and $\frac{\sigma}{\sqrt{n}}$, and the

vertical axis indicates the power of that alternative, rising for increasing values of μ_1 and approaches

1.00 for very large values – see Figure 8.8.2.

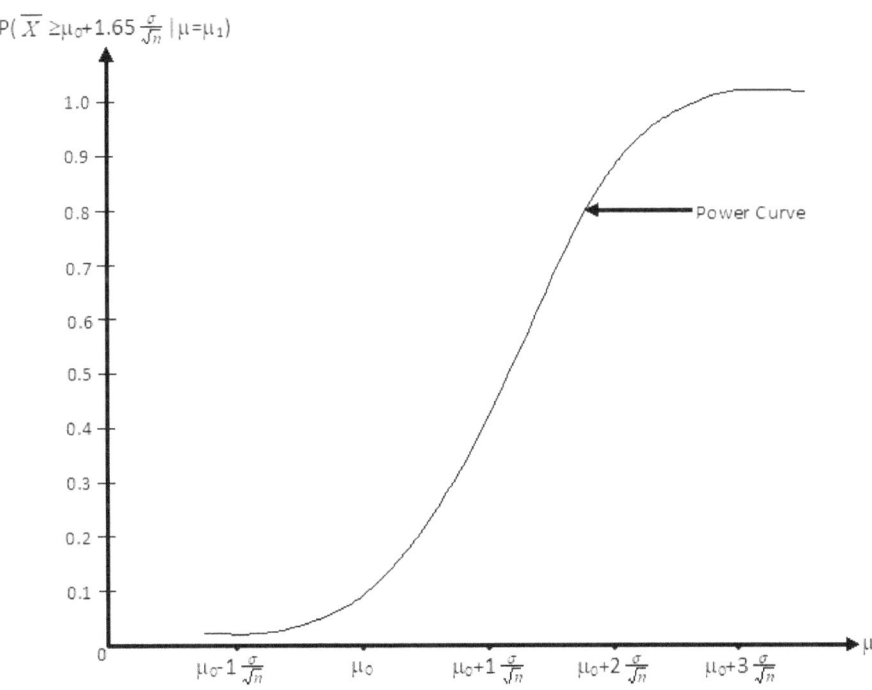

Figure 8.8.2: Power Curve

Winkler and Hays (1975) observed that the greater the discrepancy between the tested hypothesis and the true situation, the greater the power, and a larger α (0.1, 0.2, 0.3) makes for a relatively more powerful test of H_0.

Example 8.8.1:

Suppose that for a normally distributed population, $\mu_0 = 138$, $\sigma = 2$ and a sample of size, n = 100, and $\mu_1 = 142$ i.e.,

120

$\mu_0 = 138$

$\mu_1 = 142$

$\alpha = 0.05$

$\alpha = 2.0$

Thus, under H_0, $\overline{X}$ is normal with a value of 138, and standard deviation of 2; under H_1, $\overline{X}$ is normal with a value of 142 and standard deviation of 2. Determine the critical value (C) of $\overline{X}$ and the decision rule of the form: P(Reject H_0 if $\overline{X} \geq$ C) and hence we need to determine C.

Solution:

$\alpha = $ P(Reject H_0/H_0 is true) = P($\overline{X} \geq$ C /$\mu = 138$) = 0.05

But $\alpha = 0.05$

But $P(X \geq C \mid \mu = 138) = P(Z \geq \dfrac{c-138}{2})$

Hence we want: $P(Z \geq \dfrac{C-138}{2}) = 0.05$

From the standard normal Table, the probability of 0.05 has a Z-value = 1.65.

$\therefore$ P(Z $\geq$ 1.65) = 0.05

Hence: $\dfrac{C-138}{2} = 1.65$

$\therefore$ C – 138 = 2(1.65) = C = 138 + 2(1.65) = 141.30

$\overline{X}$ has a critical value = 141.30

β can be calculated as: β = P(Accept H_0 H_1 is true)

= P($\overline{X} <$ 141.30$\mu = 142$).

$P(Z < \dfrac{141.3-142}{2}) = P(Z < \dfrac{-0.7}{2}) = P(Z < -0.35).$

From the Z-Table, Z = -0.35 is 0.3632.

$\therefore$ $\beta = 0.3632$.Power = 1 - β = 0.6368.

Pictorially, this can be represented as:

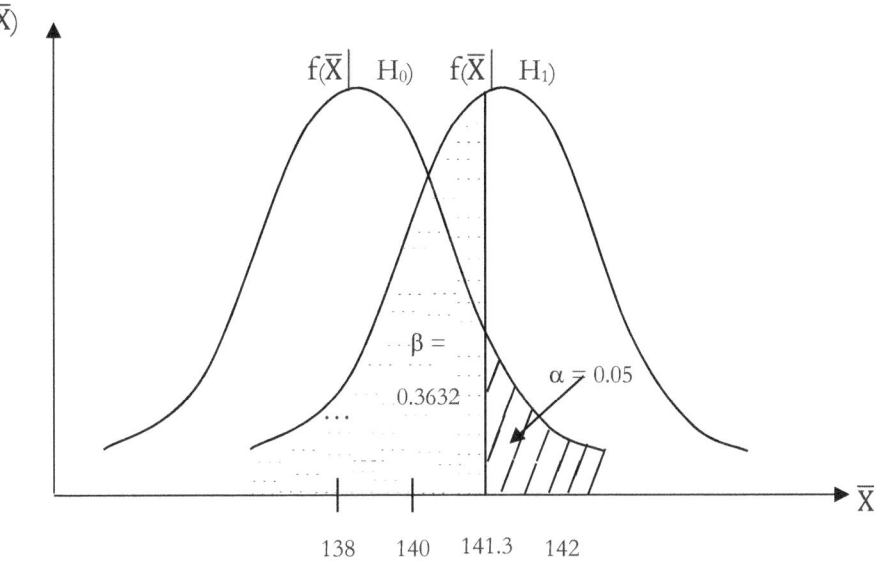

Figure 8.8.3: Probabilities of Error as Areas Under Sampling Distributions for C = 141.3; n = 100

8.9 POWER AND OPERATING CHARACTERISTIC CURVES FOR ONE-TAILED AND TWO-TAILED TESTS

According to Winkler and Hays (1975), the powers of one and two-tailed tests of the same hypothesis will be different, given the same α-level and the same true alternative. The one-tailed test is more powerful than a two-tailed test, if the true alternative is in the direction of the rejection region, over all such possible values of μ.

We noted in the previous section that the power function is the probability of rejecting H_0 as a function of the parameter of interest. The power curve is:

$$P(\text{Reject } H_0 \mu = \mu_0 + \frac{\sigma}{\sqrt{n}}) \text{ - - as shown in figure 8.9.1.}$$

The Operating Characteristic Curve is:

$$P(\text{Accept } H_0 = 1 - P(\text{Reject } H_0\ \mu) = \text{Power Curve – as shown in figure 8.9.1.}$$

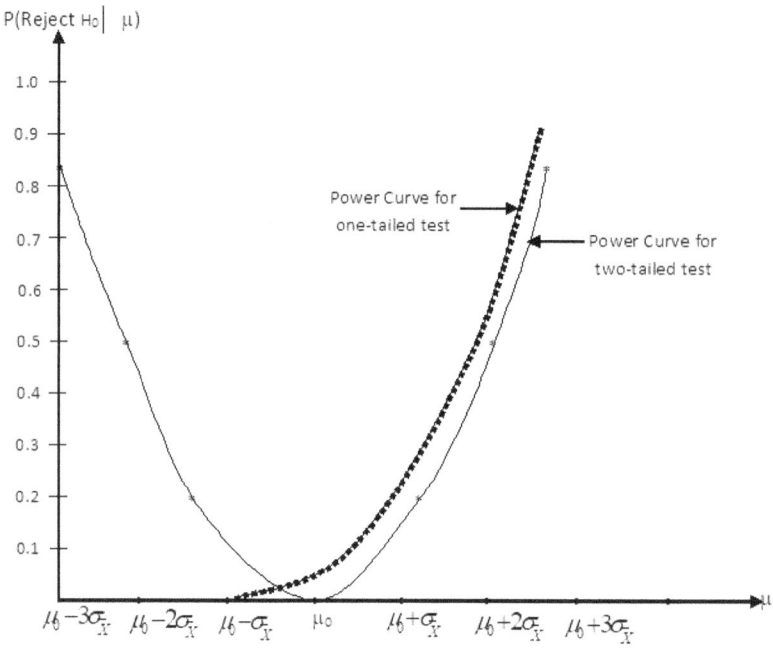

Figure 8.9.1: Power Curves for One-tailed and Two-tailed Tests

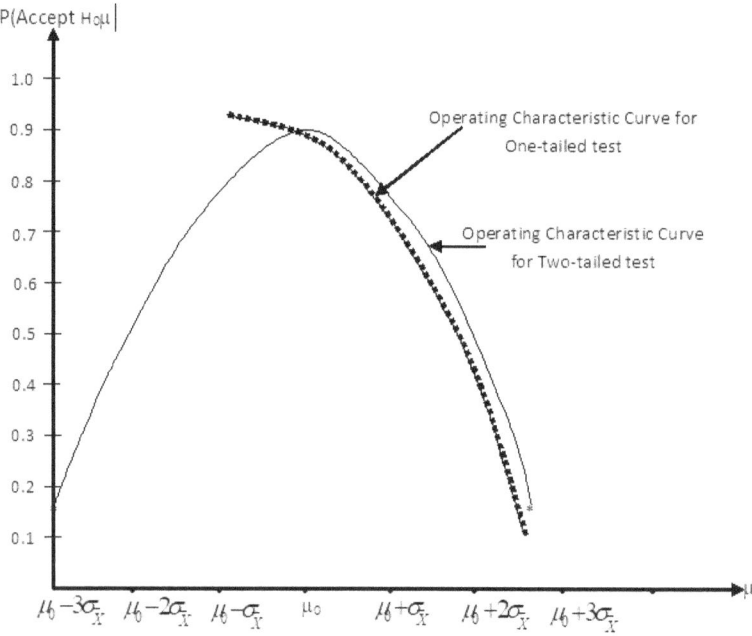

Figure 8.9.2: Operating Characteristic Curves for One-tailed and Two-tailed Tests

8.10.1 One-Sided Tests of a Single Large Population Mean and Proportion

In this section, we will consider one-sided tests of hypotheses that concern the means and proportions of a single population value. The conditions for applying these tests include:

1.   The population is normally distributed, or

123

2. The sample size is large i.e. n ≥ 30

3. The population standard deviation, σ, is known.

A. If these conditions hold, then we must use Z as our test statistic as:

i. $Z = \dfrac{(\bar{X}-\mu)}{\frac{\sigma}{\sqrt{n}}}$ - - for testing a single mean against a specified value.

ii. $Z = \dfrac{(\hat{P}-P_0)}{\sqrt{\frac{\hat{p}\hat{q}}{n}}}$ - - for testing a single proportion against a specified value.

The following examples illustrate the two types of one-sided tests:

Example 8.10.1:

A Crop Scientist raised cucumber fruits in UNIBEN farm and weighed each fruit which had a mean weight of 12.5 grams with a standard deviation of 1.5grams. (a) Cucumbers that weighed less than this mean are not marketed; (b) Cucumbers are only marketed, if they weigh more than the average. A sample of 25 fruits yielded a mean of 13.0 grams and α=0.05.

Solution

(a) H_0: μ ≤ 12.5

H_1: μ > 12.5

(implies one-tailed on the right side as shown in diagram)

σ = 1.5

$\bar{X}$ = 13.0

α = 0.05

$Z_{0.05}$ = 1.650 = Critical region - - from Z-Table A.4

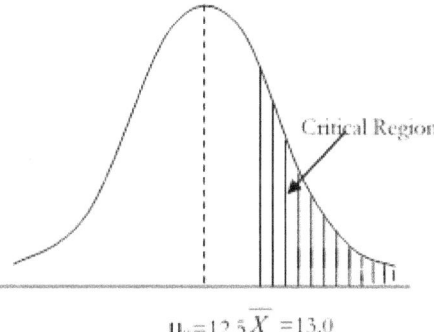

Decision Rule: Reject H_0, if Z_{cal} (Z*) is greater than 1.650.

Figure 8.10.1: P(X>13.0).

Compute the test statistic $Z^* = \dfrac{\bar{X}-\mu}{\frac{\sigma}{\sqrt{n}}}$

$Z* = \dfrac{13.5 - 12.5}{1.5/5} = \dfrac{1.0}{0.3} = 3.3333$

Note: Z* or Z_{cal}, or t* or t_{cal} means the calculated, or estimated Z-value, or t-value.

Conclusion: Since Z_{cal} = 3.333 >> $Z_{0.05}$ = 1.650, reject H_0 and conclude that the cucumbers have means that are significantly larger than 12.5 grams.

(b) H_0: $\mu \geq 12.5$

H_1: $\mu < 12.5$

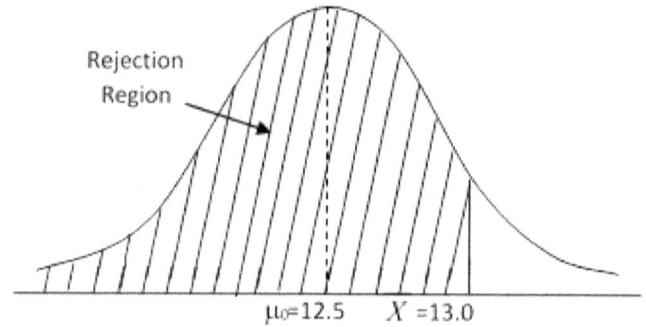

Figure 8.10.2: P(X<13).

(implies one-tail on the left side as shown in the diagram)

$\sigma = 1.5$

$\overline{X} = 13.0$

$\alpha = 0.01$

Critical Region = $Z_{0.01}$ = -2.33 - - from Z-Table.

Decision Rule: Reject H_0, if Z_{cal} (Z^*) < -2.33

Compute test statistic as Z_{cal} (Z^*) = $\dfrac{\overline{X} - \mu}{\dfrac{\sigma}{\sqrt{n}}}$

$$Z* = \frac{13.0 - 12.5}{1.5/5} = \frac{0.5}{0.3} = 1.667$$

Conclusion: Since $Z^* = 1.667 \gg -Z_{0.01} = -2.33$, do not reject H_0. We do not have sufficient information to reject H_0. We need additional research on cucumbers.

Example 8.10.2:

A Wildlife Expert in UNIBEN raises African Giant snails for eight weeks and claims that the average weight of these snails is 21 grams. A post-graduate student took a sample of 100 of the snails and weighed each one and obtained an average of 20.5 grams with a standard deviation of 2.0 grams. Does the PG student's data support the Wildlife Expert's claim?

Solution:

$H_0: \mu = 21$

$H_1: \mu < 21$

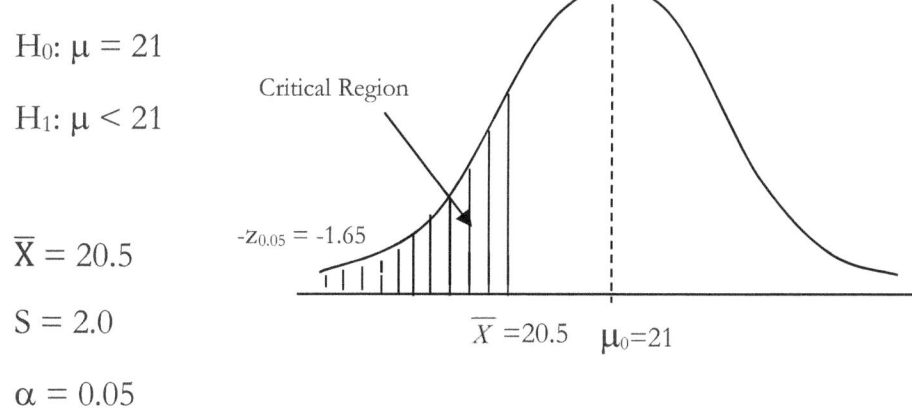

$\overline{X} = 20.5$

$S = 2.0$

$\alpha = 0.05$

Figure 8.10.3: P(X<21).

It is always advisable to draw or sketch the diagram of the critical region of interest.

$Z_{0.05} = 1.65$ - - From the standard normal Table A.4 (Walpole, 1974).

Compute Test Statistic as:

$$z = \frac{\overline{X} - \mu_0}{\frac{\sigma}{\sqrt{n}}}$$

Note: We use z-test statistic here because the sample size, n = 100, is large and so we can use the standard normal distribution because of the Central Limit Theorem. We could still use the formula:

$z = \frac{\overline{X} - \mu_0}{\frac{\sigma}{\sqrt{n}}}$ and obtain the same result. But if the sample size, n, is less than 30, we would have used

t-test.

$$t = \frac{\overline{X} - \mu_0}{\frac{S}{\sqrt{n}}}.$$

$$z* = \frac{20.5 - 21}{\frac{2}{10}} = \frac{-0.5}{0.2} = -2.5$$

Conclusion: Since $Z* = -2.5 < Z_{0.05} = -1.65$, reject H_0 and conclude that the data do not support the Wildlife Expert's claim.

Example 8.10:3:

Suppose in Example 8.6.2, the Wildlife Expert's claims that at least 70% of his eight weeks old snails attain a body weight of 21 grams. A sample of 100 snails shows that only 60% of them attain a body weight of 21 grams. Does this result support the Expert's claim at $\alpha = 0.05$?

Solution:

$H_0: P \geq 0.7$

$H_1: P < 0.7$

$\alpha = 0.05$

$\hat{p} = 0.6$

$n = 100$

Critical Region $= Z_{0.05} = 1.65$

Decision Rule: Reject H_0, if $Z^* < Z_{0.05}$

Compute the test statistic as:

$$Z* = \frac{\hat{P} - P_0}{\sqrt{\frac{\hat{p}\hat{q}}{n}}} = \frac{0.6 - 0.7}{\sqrt{\frac{(0.6)(0.4)}{100}}} = \frac{-0.1}{\sqrt{\frac{0.24}{100}}} = \frac{-0.1}{\sqrt{\frac{0.24}{10}}}$$

$$= \frac{-0.1}{\sqrt{0.024}} = -\frac{0.1}{0.15492} = -0.6455$$

Conclusion: Since $Z^* = -0.646 << 1.65$ (or since $Z^* = -0.646 > Z_{0.05} = -1.65$), reject H_0. The result does not support the expert's claim.

8.10.4 One-Tailed Test of Means and Proportions of One Small Sample

B If the requirements of:

1. Sample size is small i.e. $n < 30$,
2. The population standard deviation, σ, is unknown,
3. The population is assumed normal,

hold, we can no longer use Z as the test statistic.

We must now use the t-test statistic as:

(i) $\quad t* = \dfrac{\bar{X} - \mu_0}{\frac{S}{\sqrt{n}}}$ -- for testing a single Mean against a specified alternative.

(ii) $\quad t* = \dfrac{\hat{P} - P_0}{\sqrt{\frac{\hat{p}\hat{q}}{n}}}$ -- for testing a single proportion against a specified alternative.

Here, we must now use the T-Table (Table A.5 – Walpole 1974).

Example 8.10.4:

Suppose we have the following data:

n = 25

$\bar{X}$ = 26

s = 6

α = 0.05

μ = 30

Test the hypothesis that the mean, μ is greater than or equal to 30, versus the alternative that μ is less than 30.

Solution:

H_0: $\mu \geq 30$

H_1: $\mu < 30$

n = 25

$\bar{X}$ = 26

s = 6

α = 0.05

Critical Region = $t_{0.05, n-1}$ = $t_{0.05, 24}$ = 1.711

$$\frac{S}{\sqrt{n}} = \frac{6}{5} = 1.2$$

Decision Rule: Reject H_0 if $|t| < t_{0.05} = 1.71$

Compute test statistic as:

$$t* = \frac{\bar{X} - \mu_0}{\frac{S}{\sqrt{n}}} = \frac{26.30}{\frac{6}{\sqrt{25}}} = \frac{-4.}{1.2} = -3.333$$

Conclusion: Since (i) $t*$ = -3.33 << $t_{0.05}$ = 1.711, reject H_0.

Absolute value of $t*> t_{0.05} = 1.71$.

Example 8.10.5:

The Silviculturist in UNIBEN claims that her forest nursery contains about 75% seedlings of Irvingia spp. A forest farmer randomly selected 15 seedlings from the forest nursery and found that 9 of the seedlings were Irvingia spp. Does this support the Silviculturist's claim at 5% level of significance, if the seedlings are assumed to be normally distributed?

Solution:

$H_0: p \geq 0.75$

$H_1: p < 0.75$

$\alpha = 0.05$

$n = 15$

$X = 9: = \dfrac{x}{n} = \dfrac{9}{15} = 0.6: \hat{q} = 1 - 0.6 = 0.4$

Critical Region $= t_{0.05},\ n\text{-}1 = t_{0.05},\ 14 = 1.761$.

Decision Rule: Reject H_0, if $|t_{cal}|$ is less than 1.761.

Compute the test statistic as: $t^* = \dfrac{\hat{P} - P_0}{\sqrt{\dfrac{\hat{p}\hat{q}}{n}}}$

Substitute the data above: $t^* = \dfrac{0.6 - 0.75}{\sqrt{\dfrac{(0.6)(0.4)}{15}}} = \dfrac{-0.15}{\sqrt{\dfrac{0.24}{15}}} = \dfrac{-0.15}{\sqrt{0.016}}$

$= \dfrac{-0.15}{0.12649} = -1.18585 = -1.186$

Conclusion: Since $|t^*| = 1.186 < t_{0.05} = 1.761$, reject H_0 and conclude that the observed data do not support the Silviculturist's claim.

8.10.5 Tests of Hypotheses Concerning Variance (σ^2) or Standard Deviation (σ)

Frequently, it is important to study the variability of some measurements. To test the hypothesis that the variance of a normal population, σ^2, has some specified value, σ_0^2, we must adopt the following procedure:

(i) $H_0: \sigma^2 \geq \sigma_0^2$

$H_1: \sigma^2 < \sigma_0^2$

(ii) Choose the level of significance, $\alpha = 0.05$ or 0.01.
(iii) Select a random sample of size, n.

Compute the point estimate, S^2, of the population variance, σ^2.

(iv) Compute the statistic, S^2, as the estimator of σ^2.
(v) Using the statistic, S^2, compute another statistic, χ^2, as:

$\chi^2 = \dfrac{(n-1)S^2}{\sigma^2}$ - - which has a χ^2 distribution with $(n - 1)$ degrees of freedom.

(vi) Decision Rule: Reject H_0, if $\dfrac{(n-1)S^2}{\sigma^2} < \chi^2_{0.05},\ (n\text{-}1)$.

Example 8.10.5.1:

Suppose that a sample of size, $n = 10$, yields a standard deviation of 11.2 and it is known that $\sigma^2 \leq 100$. Can we claim that the sample variance is significantly greater than 100?

Solution:

H_0: $\sigma^2 \leq \sigma_0^2$ i.e. $\sigma^2 \leq 100.0$

H_1: $\sigma^2 > \sigma_0^2$ i.e. $\sigma^2 > 100.0$

$\alpha = 0.0516$

$S^2 = (11.2)^2 = 125.44$

$$\chi^2 = \frac{(n-1)S^2}{\sigma^2} = \frac{(9)(125.44)}{100.0} = \frac{1,128.96}{100} = 11.2896 \approx 11.290$$

$\chi^2_{0.05, 9} = 16.919$ - - From χ^2-Table (Walpole (1974) – Table A.6)

Conclusion: Since $\chi^2_{cal} = 11.290 < X^2_{0.05, 9} = 16.92$, do not reject H_0. There is no sufficient information to reject H_0, and so, we conclude that the sample variance is actually less than, or equal to 100.

8.11.0 A Two-Tailed Test on a Single Population Mean, Proportion and Variance, Using the Normal, T and Chi-Squared Distributions

In section 8.6, we stated that **"when we have no information about the true nature of the population of interest, the rejection region should be so arranged that H_0 will be rejected when there are extreme departures from expectation - - whether too low or too high"**. This calls for a rejection in both tails (or regions) and it is called **a TWO-TAILED OR TWO-SIDED TEST**.

The hypothesis is stated as:

H_0: $\mu = \mu_0$ i.e. the population mean,μ, is equal to a specific value, μ_0.

H_1: $\mu \neq \mu_0$ i.e. the population mean, μ, is not equal to a specific value, μ_0.

$\alpha = 0.05$ or 0.01: Under a two-tailed test, α is divided into two as: $\alpha = \dfrac{\alpha}{2} = \dfrac{0.05}{2} = 0.025$ or $\dfrac{0.01}{2} = 0.005$.

The key for determining whether a hypothesis is one-tailed or two-tailed is in the alternative hypothesis, H_1: $\mu \neq \mu_0$, i.e. once we see the statement: "not equal to" ($\mu \neq \mu_0$), it immediately tells us that we have a two-tailed hypothesis.

Choose $\alpha = 0.05$, or 0.01 and divide it by 2: $\dfrac{0.05}{2} = 0.025$ or

$$\dfrac{0.01}{2} = 0.005$$

Establish critical regions as: $Z_{0.025} = \pm 1.96$; $Z_{0.005} = \pm 2.576$

Pictorially, a two-tailed test is represented as:

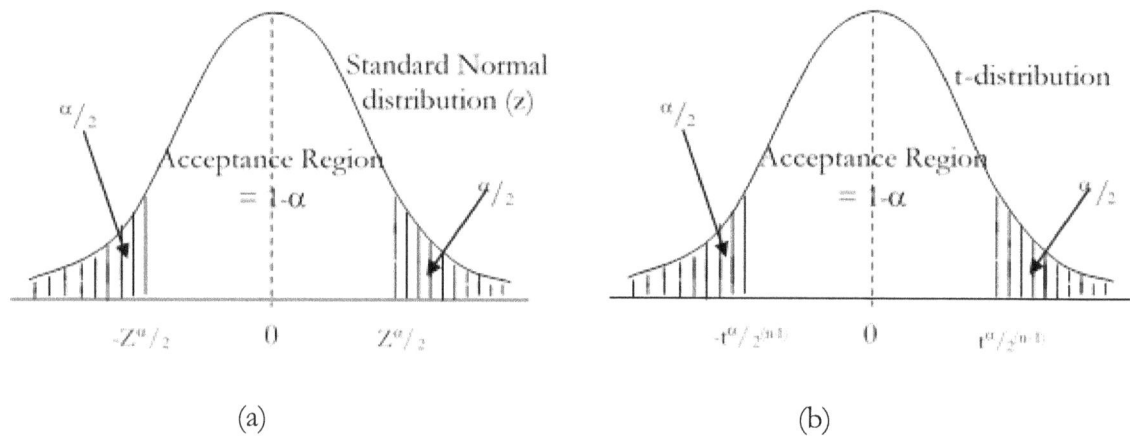

(a)   (b)

Figure 8.11.1:Critical Region for the Alternative Hypothesis: $\mu \neq \mu_0$: Under (a) Standard Normal and (b) and t-distribution

A. We apply Figure 8.11.1a if:

   i. σ^2 is known, or
   ii.   Population is normal ($X_i \sim N (\mu, \sigma^2)$)
   iii.   Sample size, n, is large i.e. $n \geq 30$

   iv.   Use $Z_{\frac{\alpha}{2}}$ $Z_{\frac{\alpha}{2}}$: $Z_{0.025} = \pm 1.96$ as the critical value, when $\alpha = 0.05$.

          $Z_{0.005} = \pm 2.576$ as the critical value, when $\alpha = 0.01$.

   v.   To test means and proportions:

The Test statistic for testing Means is: $Z * = \dfrac{\overline{X} - \mu_0}{\frac{\sigma}{\sqrt{n}}} = \dfrac{\sqrt{n}}{\sigma}\left|\overline{X} - \mu_0\right|$ - - meaning we use the absolute

value of the difference between the estimated mean,$\overline{X}$, and the hypothesized value, μ_0 as: $\left|\overline{X} - \mu_0\right|$.

The same test statistic: $\dfrac{\hat{p} - p_0}{\sqrt{\frac{\hat{p}\hat{q}}{n}}}$ is used to test a proportion.

The calculated $Z*$ is then compared with $Z_{\frac{\alpha}{2}} = Z_{0.025} = \pm 1.96$ or $Z_{\frac{\alpha}{2}} = Z_{0.005} = \pm 2.58$.

Decision Rule is: Reject H_0 if $Z*$ is less than $-Z_{0.025} = -1.96$ or greater than $+ 1.96$ at $\alpha = 0.05$.

B. We apply Figure 8.11.1b if:

   i. σ^2 is unknown and is estimated as S^2.
   ii.   Population is normal ($X_i \sim N (\mu, \sigma^2)$).
   iii.   Sample size, n, is small i.e. $n < 30$.

iv.   Use $t_{\alpha/2,\ (n-1)}$ – is normal, but depends on the number of degrees of freedom (n-1). Under the same $\alpha = 0.05$ or 0.01, different degrees of freedom in the t-Table will give different values. See Table 7.2.

v.   To test Means and Proportions:

The Test Statistic for testing a Mean against a specified value is: $t^* = \dfrac{\overline{X} - \mu_0}{\frac{S}{\sqrt{n}}}$ - - compared with $t_{(n-1),\alpha/2}$

from Table A.5 (Walpole, 1974).

The Test Statistic for testing a proportion against a specified value is:

$t^* = \dfrac{\hat{p} - p_0}{\sqrt{\frac{\hat{p}\hat{q}}{n}}}$ - - compared with $t_{(n-1),\alpha/2}$ from Table A.5 (Walpole, 1974).

C.   For a two-tailed test of a Sample Variance against a specified value, the hypothesis is stated as:

H_0: $\sigma^2 = \sigma^2$

H_1: $\sigma^2 \neq \sigma_0^2$

$\alpha = 0.05$ or 0.01

The Test Statistic is χ^2 whose critical region is $\chi^2_{\frac{\alpha}{2}}\left(\chi^2_{0.025,n-1}\right)$ as shown below:

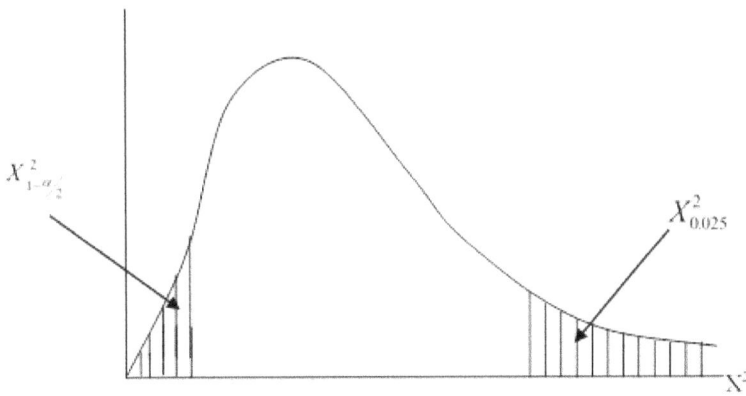

Figure 8.11.2: Testing H_1: $\sigma^2 \neq \sigma_0^2$ Using χ^2 distribution.

The Test Statistic is $\chi^2 = \dfrac{(n-1)S^2}{\sigma^2}$ which is compared with:

$$\chi^2_{0.025,(n-1)} \text{ and } \chi^2_{0.975}$$

Decision Rule: Reject H_0, if $\chi^2_{cal} > \chi^2_{0.025,(n-1)}$ or $\chi^2_{cal} < \chi^2_{0.925,(n-1)}$.

Thus, testing a sample statistic, $\overline{X}, \hat{p}$, or S^2 against a parametric value μ_0, P_0, or σ_0^2, in a two-tailed hypothesis, is equivalent to setting confidence limits on the parameter of interest: under the same conditions as in sections 8.7 A, B and C: $(1-\alpha)$ 100% CI is:

i. $\overline{X} \pm Z_{\alpha/2} \dfrac{\sigma}{\sqrt{n}}$

ii. $\hat{P} \pm Z_{\alpha/2} \sqrt{\dfrac{\hat{p}\hat{q}}{n}}$

iii. $\overline{X} \pm t_{(n-1),\alpha/2} \dfrac{S}{\sqrt{n}}$

iv. $\hat{P} \pm t_{(n-1),\alpha/2} \sqrt{\dfrac{\hat{p}\hat{q}}{n}}$

v. $\dfrac{(n-1)S^2}{\chi_{\alpha/2}^2} < \sigma^2 < \dfrac{(n-1)S^2}{\chi_{1-\alpha/2}^2}$

9 TESTS OF HYPOTHESES CONCERNING THE MEANS OF TWO POPULATIONS

Frequently, we are interested in testing the hypothesis that two independent population means are equal, when a random sample from each population is available. Due to the various physical conditions of such populations, it is not possible to use the same test statistic.

Therefore, to apply any of the test statistics, we need to examine the prevailing conditions in the two populations that will determine the sampling distributions and hence the test statistics. Some of such prevailing conditions are whether or not:

- the population distributions are normal.
- the population variances are known, or assumed known.
- the population variances are not known but assumed equal.
- the population variances are unknown and unequal.
- the sample sizes, n_1, and n_2, are large, or n_1, $n_2 \geq 30$.
- the sample sizes, n_1 and n_2 are small, or n_1, $n_2 < 30$.
- the two samples are related (correlated) i.e. test on the same individuals before and after administration of a drug or physical exercise.

The two-tailed tests of hypotheses under the above conditions are similar to estimating differences between two means, proportions and variances, discussed in Section 6.5 to 6.9. The sampling distributions used for testing the differences of means, proportions and variances are exactly the same as those for testing two-tailed hypotheses.

Assuming that we randomly selected a sample of size n_1: $X_1, X_2, X_3, ..., X_{n_1}$, from a normal population with mean, μ_1 and variance σ_1^2, and also randomly selected another independent sample of size n_2: $X_1, X_2, X_3, ..., X_{n_1}$, from a normal population with mean, μ_2 and variance σ_2^2, and we want to test the null hypothesis that the two population means are equal against the alternative that they are not equal:

$$H_0: \mu_1 = \mu_2$$

$$H_1: \mu_1 \neq \mu_2$$

Critical Region = $Z_{0.025} = \pm 1.960$; $Z_{0.005} = \pm 2.576$ (Walpole, 1974: Table A.4).

(i)   Calculate the estimators of μ_1 and μ_2 as the sample means:

$$\bar{X}_1 = \frac{\Sigma X_{1i}}{n_1}$$

$$\bar{X}_2 = \frac{\Sigma X_{2i}}{n_2}$$

(ii)   Decide on $\alpha = 0.05$ or 0.01

(iii)   The difference between two normal random variables, $\bar{X}_1 - \bar{X}_2$, is also normally distributed with mean, 0:

$$E(\bar{X}_1 - \bar{X}_2) = E(\bar{X}_1) - E(\bar{X}_2) = \mu_1 - \mu_2 \text{ and}$$

$$\text{Variance: } Var(\bar{X}_1 - \bar{X}_2) = Var(\bar{X}_1) + Var(-\bar{X}_2)$$

$$= Var(\bar{X}_1) + (-1)^2 Var(\bar{X}_2)$$

$$= Var(\overline{X}_1) + Var(\overline{X}_2)$$

$$= \frac{\sigma_1^2}{n_1} + \frac{\sigma_2^2}{n_2}$$

(iv) Use the Test Statistic

$$z^* = \frac{\bar{X}_1 - \bar{X}_2}{\sqrt{\frac{\sigma_1^2}{n_1} + \frac{\sigma_2^2}{n_2}}} = \sim N(0,1)$$

(v) **The Critical Region** $|z^*| \geq Z_{\alpha/2}$

(vi) Compute the Test Statistic from the available data.

(vii) State the statistical conclusion.

9.1 TESTING THE EQUALITY OF MEANS OF TWO NORMAL POPULATIONS WHEN THE TWO VARIANCES, σ_1^2 AND σ_2^2 ARE KNOWN

It should be obvious that when we test for the equality of two means, proportions or variances, we are invariably testing the difference between the two means, proportions or variances. The conditions here are that:

- The two populations are normal.
- The two variances, σ_1^2 and σ_2^2, are known.

The hypothesis is:

H_0: $\mu_1 = \mu_2$: or H_0: $\mu_1 - \mu_2 = 0$ i.e. – the two means are equal; or there is no difference between the two means.

H_1: $\mu_1 \neq \mu_2$: H_1: $\mu_1 - \mu_2 \neq 0$ i.e. – the two means are not equal; or the difference between the two means is not equal to zero.

$$\alpha = 0.05 \text{ i.e. } {}^{\alpha}/_2 = 0.025$$

Critical Region: $Z_{\alpha/2} = Z_{0.025} = \pm 1.960$ – Obtained from the standard normal Table A.4 (Walpole, 1974).

Estimate $\bar{X}_1$ and X_2, and substitute them in the Test Statistic z^*.

$$\text{Test Statistic } z^* = \frac{\bar{X}_1 - \bar{X}_2 - (\mu_1 - \mu_2)}{\sqrt{\frac{\sigma_1^2}{n_1} + \frac{\sigma_2^2}{n_2}}} = \frac{\bar{X}_1 - \bar{X}_2}{\sqrt{\frac{\sigma_1^2}{n_1} + \frac{\sigma_2^2}{n_2}}}$$

z^* is then compared with Z-Table = ± 1.960, to know whether or not to reject H_0.

Example 9.1.1:

A sample of 50 employees from a Forest Nursery in FRIN, Benin-City, has a mean wage of 160 naira per week with a standard error of the mean of 1.44 naira. Another sample of 40 Forest Nursery employees in Sakponba has a mean weekly wage of 155 naira with a standard error of 1.50 naira. Test the difference between the mean wages at $\alpha = 0.05$.

Solution:

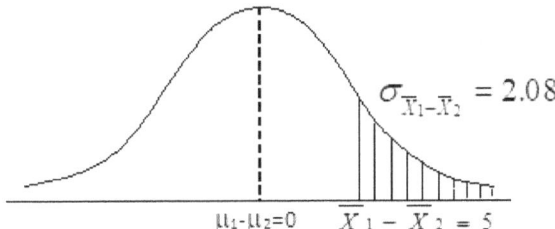

Figure 9.1: Testing the Difference between $\overline{X}_1 - \overline{X}_2$.

H_0: $\mu_1 - \mu_2 = 0$ i.e. The two wages are equal.

H_1: $\mu_1 - \mu_2 \neq 0$ i.e. The two wages are not equal..

$$\alpha = 0.05 \; i.e. \; \frac{0.05}{2} = 0.025$$

Critical Region: $Z_{0.025} = \pm 1.960$:

Decision Rule: Reject H_0, if $Z* > Z_{0.025} = 1.960$

$\overline{X}_1 = 160 : \frac{\sigma}{\sqrt{n_1}} = 1.44 : n_1 = 50 \therefore \frac{\sigma_1^2}{n_1} = \left(\frac{\sigma_1}{\sqrt{n_1}}\right)^2 = (1.44)^2 = 2.0736$

$\overline{X}_2 = 155 : \frac{\sigma}{\sqrt{n_2}} = 1.50 : n_2 = 40 : \frac{\sigma_2^2}{n_2} = \left(\frac{\sigma_2}{\sqrt{n_2}}\right)^2 = (1.50)^2 = 2.25.$

$\overline{X}_1 - \overline{X}_2 = 160 - 155 = 5.$

Substitute these data in:

$$Z* = \frac{\overline{X}_1 - \overline{X}_2 - (\mu_1 - \mu_2)}{\sqrt{\frac{\sigma_1^2}{n_1} + \frac{\sigma_2^2}{n_2}}} : \mu_1 - \mu_2 = 0, \text{ according to the null hypothesis}$$

$$= \frac{160 - 155 - 0}{\sqrt{(1.44)^2 + (1.50)^2}} = \frac{5}{\sqrt{(2.0736 + 2.25}}$$

$$= \frac{5}{\sqrt{4.3236}} = \frac{5}{2.079326814} = 2.404624403 \approx 2.405$$

136

Conclusion: Since $Z^* = 2.405 > Z_{0.025} = 1.960$, reject H_0 and conclude that there is a significant difference between the two wages at 5% level of significance.

Example 9.1.2:

A UNIBEN Sociologist wanted to test the difference between the average ages of 100-level students in UNIBEN and Ambrose Ali Universities. The Sociologist took a random sample of 120 new students from UNIBEN and had a mean age of 20.2 years, with a standard deviation of 1.2 years. A random sample of 100 new students from AAU had a mean age of 21 years with a standard deviation of 1.5 years. Can we conclude that the average age of new students from the two Universities are the same at $\alpha = 0.05$?

Solution:

H_0: $\mu_2 - \mu_1 = 0$ i.e. There is no difference between the two ages.

H_1: $\mu_2 - \mu_1 \neq 0$ i.e. There is a difference between the two ages.

$\alpha = 0.05$ i.e. $0.025 = \pm 1.960$ – From Table A.4 (Walpole, 1974).

Critical Region: $Z_{0.025} = \pm 1.960$.

Decision Rule: Reject H_0, if $|z^*| > Z_{0.025} = \pm 1.960$.

Data: $n_1 = 120$: $n_2 = 100$.

$$\overline{X}_1 = 20.2: \bar{X}_2 = 21.0: \bar{X}_2 - \bar{X}_1 = 0.8$$

$$S_1 = 1.2 \quad : S_2 = 1.5$$

The diagrammatic representation is:

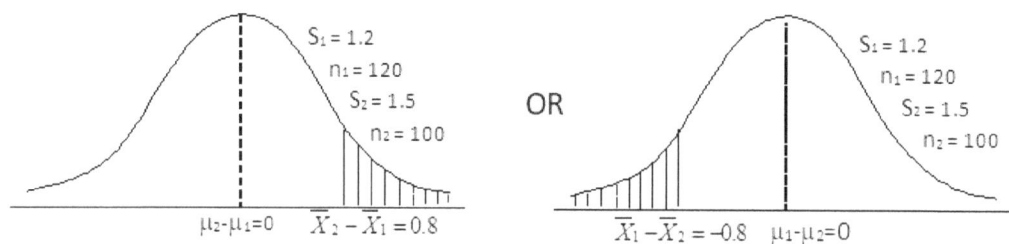

Figure 9.1.1: Testing the Equality of two Population Means $(\mu_1 = \mu_1)$

Since the population variances, σ_1^2 and σ_2^2 are not known but because the sample sizes; $n_1 = 120$; $n_2 = 100$ are large, we can use the sample variances in place of the unknown population variances as:

$$\text{Test Statistic } |z^*| = \frac{\overline{X}_2 - \overline{X}_1 - (\mu_2 - \mu_1)}{\sqrt{\frac{s_1^2}{n_1} + \frac{s_2^2}{n_2}}}$$

$$= \frac{S_1^2}{n_1} = \frac{(1.2)^2}{120} : \frac{S_2^2}{n_2} = \frac{(1.5)^2}{100}$$

137

Substitute these figures in the formula:

$$z^* = \frac{0.8 - 0}{\sqrt{\frac{(1.2)^2}{120} + \frac{(1.5)^2}{100}}} = \frac{0.8}{\sqrt{\frac{1.44}{120} + \frac{2.26}{100}}} = \frac{0.8}{\sqrt{(0.012 + 0.0225)}}$$

$$= \frac{0.8}{\sqrt{0.0345}} = \frac{0.8}{0.186} = 4.30$$

Conclusion: Since $Z^* = 4.30 >> Z_{0.025} = 1.960$, reject H_0 and conclude that the difference between the mean age of new students in UNIBEN and AAU is highly significant at $\alpha = 0.05$.

9.2 TESTING THE EQUALITY OF MEANS WHEN THE VARIANCES, σ_1^2 AND σ_2^2 ARE UNKNOWN, BUT THE SAMPLE SIZES ARE LARGE $(N_1, N_2 \geq 30)$

Unlike section 9.1, when the population parameters, μ_1 and μ_2 and σ_1^2 and σ_2^2, are unknown, but the sample sizes are large, the sample variances will be approximately equal to the population variances. Therefore, it is reasonable, due to the Central Limit Theorem, to substitute the sample variances, S_1^2 and S_2^2, for the population variances, σ_1^2 and σ_2^2, and make use of the same test statistic:

$z^* = \dfrac{\bar{X}_1 - \bar{X}_2 - (\mu_1 - \mu_2)}{\sqrt{\frac{S_1^2}{n_1} + \frac{S_2^2}{n_2}}}$ and substitute the values of S_1^2 and S_2^2:

Test Statistic: $z^* = \dfrac{\bar{X}_1 - \bar{X}_2 - (\mu_1 - \mu_2)}{\sqrt{\frac{S_1^2}{n_1} + \frac{S_2^2}{n_2}}}$.

Example 9.2.1:

A Toothpaste manufacturer, in Lagos, hypothesized that the addition of fluoride to the toothpaste **decreased** the number of teeth cavities in the users. To test his hypothesis, he produced two brands of toothpaste - - one with fluoride, the other without. He administered the two brands of toothpaste to 100 persons each for a specified period and then examined their teeth for cavities and obtained these data:

Fluoride Group	Regular Group
$n_1 = 100$	$n_2 = 100$
$\bar{X}_1 = 8$	$\bar{X}_2 = 9$
$S_1 = 3$	$S_2 = 4$

Can we conclude at 1% level of significance that the addition of fluoride improved the toothpaste's efficacy in reducing tooth decay?

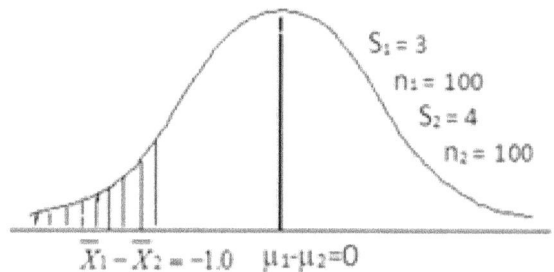

Figure 9.2.1: Testing a difference that is less than zero

Solution:

H_0: $\mu_1 - \mu_2 = 0$

H_1: $\mu_1 - \mu_2 \neq 0$

$\alpha = 0.01$

Critical Region: $Z_{0.005} = \pm 2.33$

Decision Rule: Reject H_0, if $|Z^*| > Z_{0.025} = 2.576$

Using the data above:

The Test Statistic $z^* = \dfrac{\overline{X}_1 - \overline{X}_2 - (\mu_1 - \mu_2)}{\sqrt{\dfrac{S_1^2}{n_1} + \dfrac{S_2^2}{n_2}}}$

Substituting the data, we have:

$$z^* = \frac{8 - 9 - 0}{\sqrt{\dfrac{9}{100} + \dfrac{16}{100}}} = \frac{-1.0}{\sqrt{\dfrac{25}{100}}} = \frac{-1.0}{\sqrt{0.25}} = \frac{-1.0}{0.5} = -2.0$$

$$= -2.0$$

Conclusion: Since $|Z^*| = 2.0 < Z_{0.005} = 2.33$, do not reject H_0 and conclude that there is no sufficient proof that the addition of fluoride decreases tooth decay.

9.3 TESTING THE EQUALITY OF MEANS WHEN SAMPLE SIZES ARE SMALL (N_1; $N_2 <$ 30) AND POPULATION VARIANCES, σ_1^2 AND σ_2^2, ARE UNKNOWN, BUT EQUAL ($\sigma_1^2 = \sigma_2^2 = \sigma^2$)

When testing the equality of means of small samples (n_1, $n_2 < 30$) whose population variances (σ_1^2, σ_2^2) are unknown, but approximately equal, or the small samples are obtained from the same population, we can assume that the unknown variances are equal to a common variance ($\sigma_1^2 = \sigma_2^2 = \sigma^2$). We obtain a standard normal variable in the form of:

139

$$Z = \frac{\bar{X}_1 - \bar{X}_2 - (\mu_1 - \mu_2)}{\sqrt{\frac{\sigma^2}{n_1} + \frac{\sigma^2}{n_2}}}$$

$$= \frac{\bar{X}_1 - \bar{X}_2 - (\mu_1 - \mu_2)}{\sqrt{\sigma^2 \left(\frac{1}{n_1} + \frac{1}{n_2}\right)}}$$

This common variance, σ^2, can be estimated by combining or "pooling" the sample variances. The pooled variance, denoted by S_p^2 is estimated as:

$$S_p^2 = \frac{(n_1 - 1)S_1^2 + (n_2 - 1)S_2^2}{n_1 + n_2 - 2}$$

Upon substituting S_p^2 for σ^2 in the above, the test statistic becomes:

$$T = \frac{\bar{X}_1 - \bar{X}_2 - (\mu_1 - \mu_2)}{\sqrt{S_p^2 \left(\frac{1}{n_1} + \frac{1}{n_2}\right)}}, \text{ with } (n_1 + n_2 - 2) \text{ degrees of freedom.}$$

This computed value (T) is then compared with a tabled value, $t_{0.025}$, with $(n_1 + n_2 - 2)$ degrees of freedom – in Table A.5 (Walpole, 1974).

Example 9.3.1:

To test the quality of Soybean-based feed and palm-kernel-meal feed, they were fed to two random samples of 12 pigs.

The weight-gains were as shown:

Soybean Feed	31	34	29	26	32	35	38	34	30	29	32	31
Palm-kernel Feed	26	24	28	29	30	29	32	26	31	29	32	28

Assuming that the populations are normal and their variances are equal: $\sigma_1^2 = \sigma_2^2 = \sigma^2$, are the weight-gains equal in the two groups, at $\alpha = 0.05$?

Solution:

H_0: $\mu_1 - \mu_2 = 0$ i.e. The weight gains under the two types of feeds are equal.

H_1: $\mu_1 \neq \mu_2 \neq 0$ i.e. The weight gains under the two types of feeds are not equal.

$\alpha = 0.05$

$\sigma_1^2 = \sigma_2^2 = \sigma^2$, but unknown and sample sizes $n_1 + n_2 < 30$

Critical Region: $t_{0.025, n_1 + n_2 - 2} = t_{0.025, 22} = \pm 2.074$

Decision Rule: Reject H_0, if $|t*| > 2.074$.

Computations:

$$\overline{X}_1 = \frac{\sum X_{1i}}{n_1}: \overline{X}_2 = \frac{\sum X_{2i}}{n_2}$$

$$\bar{X}_1 = \frac{381}{12} = 31.75: S_1^2 = \frac{112.25}{11} = 10.2045$$

$$\bar{X}_2 = \frac{344}{12} = 28.67: S_2^2 = \frac{66.64}{11} = 6.0606$$

$$S_p^2 = \frac{(n_1-1)S_1^2+(n_2-1)S_2^2}{n_1+n_2-2} = \frac{(11)(10.2045)+(11)(6.0606)}{12+12-2}$$

$$= \frac{112.2495 + 66.6666}{22} = \frac{178.9161}{22} = 8.13255$$

$$S_p^2 = 8.13255. \quad S_p = \sqrt{8.13255} = 2.85176.$$

$$t^* = \frac{\bar{x}_1 - \bar{x}_2}{S_p\sqrt{\frac{1}{12} + \frac{1}{12}}} = \frac{31.75 - 28.67}{2.85176\sqrt{0.166667}} = \frac{3.083}{(2.85176)(0.40824821)} = 2.648$$

Conclusion: Since $t^* > t_{0.025,22} = 2.074$, reject H_0 and conclude that the two types of feeds are significantly different in their effect on the weight-gains of pigs. In fact, the soybean feed is better than the palm kernel feed.

Example 9.3.2:

A private clinic in Benin-City studied two groups of babies that weighed approximately the same at birth and randomly assigned them to two methods of feeding for 6 weeks: (i) feeding on a formula alone and (ii) feeding on breast milk alone. The following data on the babies' weight-gains were obtained:

Formula Feeding: 5, 7, 8, 9, 6, 7, 10, 6.

Breast-Milk Feeding: 9, 10, 8, 6, 8, 7, 9.

Are the babies' weight-gains in the two methods of feeding the same, at $\alpha = 0.05$ level of significance?

Solution:

$H_0 = \mu_1 - \mu_2 = 0$.i.e. there is no difference in the babies' weight-gains under the two feeding methods.

$H_1: \mu_1 - \mu_2 \neq 0$.i.e.there is a difference in the babies' weight-gains under the two feeding methods.

$\alpha = 0.05: \frac{0.05}{2} = 0.025$.We use a t-distribution because $n_1 + n_2 < 30$.

Critical Region: $t_{0.025,n_1+n_2-2} = t_{0.025,9+7-2} = t_{0.025,14} = 2.145$.

Decision Rule: Reject H_0, if absolute value of t-calculated is greater than t-tabulated.

Computations:

$$\bar{x}_1 = \frac{5+7+8+9+6+7+10+8+6}{9} = \frac{66}{9} = 7.3333$$

$$\bar{x}_2 = \frac{9+10+8+6+8+7+9}{7} = \frac{57}{7} = 8.1429$$

$$S_1^2 = \frac{\sum x_i^2 - \frac{(\sum x_i)^2}{n_1}}{n_1 - 1} = \frac{504 - \frac{(66)^2}{9}}{8} = \frac{504 - 448}{8} = \frac{20}{8} = 2.5$$

$$S_2^2 = \frac{\sum x_i^2 - \frac{(\sum x_i)^2}{n}}{n_2 - 1} = \frac{475 - \frac{(57)^2}{7}}{6} = \frac{475 - 464.1429}{6}$$

$$= \frac{10.857}{6} = 1.8095238.$$

$$\text{Pooled variance} = S_p^2 = \frac{(n_1-1)S_1^2 + (n_2-1)S_2^2}{n_1+n_2-2} = \frac{(8)(2.5)+(6)(1.8095)}{14}$$

$$= \frac{20.0+10.85713286}{14} = \frac{30.85714286}{14} = 2.204082$$

$$t^* = \frac{\bar{x}_2 - (\mu_1 - \mu_0)}{\sqrt{S_p^2\left(\frac{1}{9}+\frac{1}{7}\right)}} = \frac{7.3333 - 8.1429}{\sqrt{2.204082(0.11111111 + 0.142857142)}}$$

$$= \frac{-0.8096}{\sqrt{(2.204082)(0.253968253)}}$$

$$= -\frac{0.8096}{\sqrt{0.559766856}} = -\frac{0.8096}{0.748175685} = -1.082$$

Conclusion: Since $|t^*| = 1.08 < t_{0.025\,14} = 2.145$ we accept H_0 and conclude that the two methods of feeding babies are the same at $\alpha = 0.05$.

9.4 TESTING THE EQUALITY OF MEANS WHEN SAMPLE SIZES ARE SMALL (n₁,n₂<30) AND POPULATION VARIANCES ARE UNEQUAL $\sigma_1^2 \neq \sigma_2^2$ AND UNKNOWN

When testing the equality of means of small samples $(n_1, n_2 < 30)$ and the population variances are unequal $(\sigma_1^2 \neq \sigma_1^2)$ and unknown, but can be assumed to be normally distributed populations, we can use the t-distribution as: /=

$$t^* = \frac{x_1 - x_2}{\sqrt{\dfrac{S_1^2}{n_1} + \dfrac{S_2^2}{n_2}}}$$

But in this case, the t* here has approximately a t-distribution with an approximate number of degrees of freedom that must be estimated from the formula:

$$V = \frac{\left(s_1^2/n_1 + s_2^2/n_2\right)^2}{\left[(s_1^2)^2/(n_1 - 1)\right] + \left[(s_2^2)^2/(n_2 - 1)\right]}$$

which may or may not be an integer, but must be approximated to the nearest integer before we can establish the critical region: $t_{\alpha/2}$. However, the hypothesis is tested in the usual way:

$\mathrm{H_0}$: $\mu_1 = \mu_2 = 0$ i.e. the two means are equal.

$\mathrm{H_1}$: $\mu_1 \neq \mu_2 \neq 0$ i.e. the two means are not equal

$\alpha = 0.05 \ or \ 0.01$

$t_{\alpha/2} = t_{0.025}$, the critical region cannot be established until we estimate the approximate degrees of freedom from the formula above.

Example 9.4.1:

Using the data in example 6.7.1, we tested the equality of the two mean rainfalls in Nifor and Udo areas of Edo State and calculated the degrees of freedom (v) as 23.

Therefore, $t_{\frac{\alpha}{2}} = t_{0.025,23} = 20.069$

Critical region: $t_{0.025,23} = 20.069$

Decision Rule: Reject $\mathrm{H_0}$, if $|t^*|$ is greater than 2.069

Test statistic: $t^* = \dfrac{(1.94-1.04)-(0)}{\sqrt{\left(\dfrac{0.2025}{15} + \dfrac{0.06761}{10}\right)}} = \dfrac{0.90}{\sqrt{(0.0135 + 0.00676)}}$

$= \dfrac{0.90}{\sqrt{0.02026}} = \dfrac{0.90}{0.142337626} = 6.323$

143

Conclusion: Since $t^* = 6.323 \ggg t_{0.025,23} = 2.069$, reject H_0 and conclude that the difference between the average rainfalls of Nifor and Udo stations is highly significant.

9.4.1 Pairing of Observations with Unequal Variances

According to Dixon and Massey (1969), if there is evidence that σ_1 and σ_2 are not equal, a more convenient sampling design and analysis is to pair the observations from the two populations. To achieve this, we could take a random sample of size n from each population and pair the individuals in the order they were obtained. Then, the difference of observations in each pair could be obtained. Such differences can be considered as a sample of size n from a population having a mean,

$$\mu_1 - \mu_2, \text{ and a variance, } \sigma_2 = \sigma_2^1 + \sigma_1^2.$$

Extraneous factors cause a significant difference in means, even though there may be no difference in the effects we are trying to measure. But extraneous factors can mask or obscure a real difference. If there are no extraneous effects, we actually lose information by pairing because with only (n-1) degrees of freedom to estimate σ_2^2 we accept larger differences in $(\bar{x}_1 - \bar{x}_2)$ than we might accept, if we had (2n – 2) degrees of freedom.

Of course, if extraneous effects are likely, it is preferable to take the risk of some loss in power arising from an increase in the probability of accepting the hypothesis when it is false. Albeit, the increase is slight, if the number of pairs is moderately large, that is, greater than ten.

Example 9.4.2:

In a study to assess the IQ of SS-1 students in Uniben Demonstration Secondary School (UDSS), 10 boys and 10 girls were randomly selected and paired on the basis of IQ to determine any differences in their learning ability.

The following data were obtained to test the hypothesis that there is no difference in their learning ability, versus the alternative that there is, at 1% level of significance:

Table 9.4.1: The IQ of UDSS Boys and Girls in Uniben

S/N	Boys	Girls	Difference (d_i)	d_i^2
1	28	19	9	81
2	18	38	-20	400
3	22	42	-20	400
4	27	25	2	4
5	25	15	10	100
6	30	31	-1	1
7	21	22	-1	1
8	21	37	-16	256
9	20	30	-10	100
10	27	24	3	0
$\sum d_1$ $\bar{d}=-4.4$ $= -44$			$\sum d_\iota = -44$	1,352

144

$$s_d^2 = \left[\sum d_i^2 - \frac{(\sum d_i)^2}{n}\right] \bigg/ n-1 = \left(1{,}352 - \frac{1936}{10}\right)\bigg/ 9 = \frac{1352 - 193.6}{9} = \frac{1{,}158.4}{9} = 128.71111$$

Solution:

$H_0: \mu_d = 0$

$H_1: \mu_d \neq 0$

$\alpha = 0.01: t_{0.005,9} = 3.250$

Critical Region: $t_{0.005,9} = \pm 3.250$

Decision Rule: Reject H_0, if $|t^*|$ is greater than 3.250

Computations:

$$\mu_d = \bar{d} = \bar{x}_1 - \bar{x}_2 = \frac{\sum d_i}{n} = -\frac{44}{10} = -4.4$$

$$S_d^2 = \frac{\sum d_i^2 - \frac{(\sum d_i)^2}{n}}{n-1} = \frac{1352 - 193.6}{9} = 1{,}158.4 = 128.71111$$

$$|t^*| = \frac{\bar{d} - 0}{\sqrt{\frac{S_d^2}{n}}} = -\frac{4.4}{\sqrt{\frac{128.71111}{10}}} = -\frac{4.4}{\frac{11.34509145}{3.16227766}} = -\frac{4.4}{3.587632924} = 1.2269$$

Conclusion: Since $|t^*| = 1.23 << 3.25$, do not reject H_0 and conclude that there is no difference in the mean learning ability of SS-1 boys and girls in UDSS.

9.4.2 Pairing of Dependent Observations (Paired–Sample T-Test)

Sometimes, two measurements are made on the same individual before and after a treatment to see the change observed. In such a situation, there will be a relationship between the data before and after the treatment. As a result of this relationship, the pairs of data values will not be independent and therefore making our previous methods of hypothesis testing invalid. This calls for the method of

"Paired-Sample t-test".

Example 9.4.3:

A nutritionist in UBTH, Benin-City, developed a vegetable-based food supplement which he claims reduces the weights of patients who use it. In a diet experiment, ten men of a certain age-group, were measured for their weights before and after the experiment and these data were obtained. Is the diet really effective in reducing weight at a 5% level of significance?

S/N	Beginning Weight(X_1)	Ending Weight (X_2)	Difference ($X_2 - X_1$)
1	183	177	-6
2	144	145	+1
3	151	145	-6
4	163	162	-1
5	155	151	-4
6	159	163	+4
7	178	173	-5
8	184	185	+1
9	142	139	-3
10	137	138	+1

$$\sum d_1 = -18$$

$$\bar{d} = \frac{\sum d_i}{10} = -1.8$$

Solution:

$H_0: \mu_d = 0$

$H_1: \mu_d \neq 0$

$\alpha = 0.05: t_{0.025,9} = \pm 2.262$

Critical Region: $t_{0.025,9} = \pm 2.262$

Decision Rule: Reject H_0, if $|t^*|$ is greater than 2.262.

Computations:

$$\bar{d} = \frac{\sum d_1}{10} = -1.8$$

$$S_d^2 \frac{\sum d_i^2 - \frac{(\sum d_i)^2}{n}}{n-1} = \frac{\left(142 - \frac{324}{10}\right)}{9} = 109.6 = 12.177778$$

$$t^* = \frac{\bar{d} - 0}{\sqrt{\frac{S_d^2}{n}}} = \frac{-1.8}{\sqrt{\frac{12.177778}{10}}} = -\frac{1.8}{\sqrt{1.2177778}} = -\frac{1.8}{1.10352969} = -1.63113$$

Conclusion: Since $|t^*| << t_{0.025,9} = 2.262$, do not reject H_0 and conclude that there is no difference between the before and after weights.

9.5 TESTING THE EQUALITY OF TWO PROPORTIONS

Frequently, situations arise that we need to test the hypothesis that two population proportions are equal. When we have two large independent populations whose characteristics are of interest to us, there are also characteristics that we are not interested in. Thus, the "characteristics of interest" and "no-interest" make the two populations binomial populations.

The proportion of the characteristic of interest in the first population is the parameter, P_1, while that of the second population is the parameter, P₂. Suppose we are interested in testing the hypothesis that these two proportions are equal against the alternative that they are not equal:

$H_0: P_1 = P_2$

$H_1: P_1 \neq P_2$

These P_1 and P_2 are the two population proportions of the attribute under investigation, which are unknown and must be estimated before we can test the hypothesis above.

Independent samples of size n_1 and n_2 are randomly selected from the two populations to estimate the proportions of the attributes of interest as X_1 in population 1 and X_2 in population 2:

$$\hat{P}_1 = \frac{X_1}{n_1} \quad \text{as the estimator of } P_1.$$

$$\text{Var}(\hat{P}_1) = \frac{P_1(1-P_1)}{n_1} = \frac{P_1 q_1}{n_1}$$

$$\hat{P}_2 = \frac{X_2}{n_1} \quad \text{as the estimate of } P_2$$

$$\text{Var}(\hat{P}_2) = \frac{P_2(1-P_2)}{n_2} = \frac{P_2 q_2}{n_2}$$

Hence: $E(P_1 - P_2) = E(P_1) - E(P_2) = (P_1) - (P_2)$

$$\text{Var}(\hat{P}_1 - \hat{P}_2) = Var(\hat{P}_1) + Var(\hat{P}_2)$$

$$= \frac{P_1(1-P_1)}{n_1} + \frac{P_2(1-P_2)}{n_2}$$

$$= \frac{P_1 q_1}{n_1} + \frac{P_2 q_2}{n_2}$$

If n_1 and n_2 are reasonably large, then, $\hat{P}_1$ and $\hat{P}_2$ will be approximately normally distributed and so will be their difference, $\hat{P}_1 - \hat{P}_2$. As we saw in section 6.8:

$Z = \dfrac{\hat{P}_1 - \hat{P}_2 - (P_1 - P_2)}{\sqrt{\left(P_1 q_1/n_1\right) + \left(P_2 q_2/n_2\right)}}$ is approximately a standard normal random variable, when n_1 and n_2 are large.

Suppose H_0 is true, so that the two proportions are equal, $P_1 = P_2 = P$, and $P_1 - P_2 = 0$, then:

$$Z = \frac{\hat{P}_1 - \hat{q}_2}{\sqrt{\left(\frac{p_1 q_1}{n_1}\right) + \left(\frac{p_2 q_2}{n_2}\right)}} = \frac{p_1 - p_2}{\sqrt{\left(\frac{pq}{n_1}\right) + \left(\frac{pq}{n_2}\right)}}$$

$$\frac{\hat{P}_1 - \hat{P}_2}{\sqrt{Pq\left(\frac{1}{n_1}\right) + \left(\frac{1}{n_2}\right)}}$$

Thus, we must estimate the common unknown proportion, p, by pooling the data from the two independent samples of size $(n_1 + n_2)$ as:

$X_1 + X_2 = n_1\hat{P}_1 + n_2\hat{P}_2$ -- elements with the characteristic of interest. The estimator of the common unknown proportion, p, is $\hat{P}$ defined as:

$$\hat{P} = \frac{n_1\hat{P}_1 + n_2\hat{P}_2}{n_1 + n_2} = \frac{X_1 + X_2}{n_1 + n_2}$$

Hence:
$$Z = \frac{\hat{P}_1 + \hat{P}_2}{\sqrt{\hat{p}\hat{q}\left(\frac{1}{n_1} + \frac{1}{n_2}\right)}}$$

The procedure for testing the hypothesis is as follows:

H_0: $P_1 = P_2$

H_1: $P_1 \neq P_2$

$\alpha = 0.05$.

Critical Region: $Z_{0.025} = \pm1.960$

Decision Rule: Reject H_0, if $|Z_{cal}| > 1.960$.

Computations:

$$\hat{P}_1 = \frac{X_1}{n_1} : \quad \hat{P}_2 = \frac{X_2}{n_2}$$

$$\hat{P} = \frac{(X_1 + X_2)}{n_1 + n_2}$$

Then find

$$Z = \frac{\hat{p}_1 + \hat{p}_2}{\sqrt{\hat{p}\hat{q}\left(\frac{1}{n_1} + \frac{1}{n_2}\right)}}$$

Conclusion: Reject H_0, if $|Z_{cal}| > 1.960$

But if the sample sizes, n_1 and n_2 are small, the test statistic to use is:

$$t_{cal} = \frac{\hat{p}_1 + \hat{p}_2}{\sqrt{\hat{p}\hat{q}\left(\frac{1}{n_1} + \frac{1}{n_2}\right)}} \; with \; \; v = (n_1 + n_2 - 2)$$

degrees of freedom.

Example 9.5.1:

A Fish Pond in the Department of Fisheries, Uniben, Benin-City, contains 142 Catfish and 72 Tilapia. A random sample of 74 catfish and an independent sample of 61 tilapia were taken from the pond for a fertility test. Do these data indicate that the sampled proportions of the two fish species are equal at 5% level of significance?

Solution: Catfish Tilapia

H_0: $P_1 = P_2.n_1 = 142$:$n_2 = 72$

H_1: $P_1 \neq P_2$: $X_1 = 74$:$X_2 = 61$

$\propto = 0.05$ i.e. 0.025.

Critical Region $Z_{0.025} = \pm 1.960$

Decision Rule: Reject H_0, if $|z_{cal}| > 1.960$

Computations:

$$\hat{P}_1 = \frac{X_1}{n_1} = Catfish = \frac{74}{142} = 0.5211.$$

$$\hat{P}_2 = \frac{X_2}{n_2} = Tilapia = \frac{61}{72} = 0.8472.$$

Pooled estimator, $\hat{P} = \frac{74+61}{142+72} = \frac{153}{214} = 0.63084.\cdot \quad \hat{q} = 0.36916$

Test statistic: $Z_{cal} = \dfrac{\hat{P}_1 - \hat{P}_2}{\sqrt{\hat{P}\hat{q}\left(\frac{1}{142} + \frac{1}{72}\right)}} = \dfrac{0.5211 - 0.8472}{\sqrt{(0.63084)(0.36916)\left(\frac{1}{142} + \frac{1}{72}\right)}}$

$$= \frac{-0.3261}{\sqrt{(0.232880894)(0.00704225352 + 0.013888888)}}$$

$$= \frac{-0.3261}{\sqrt{(0.232880894)(0.020931142)}}$$

$$= \frac{-0.3261}{\sqrt{(0.004874463157)}} = \frac{-0.3261}{0.69817355}$$

$$= -4.6707584$$

Conclusion: Since $|z_{cal}| = 4.671 \gg 1.960$, reject H_0 and conclude that the two proportions are not equal.

Example 9.5.2:

In a palatability test of smoked fish in the Dept of Fisheries, Uniben, Benin-City, 200 Uniben staff were randomly selected and 20% of them favored smoked catfish. Another independent random sample of 300 Uniben staff, 27% favored smoked Tilapia. Can the HOD conclude that there is no difference in staff preference for smoked Catfish fish and Tilapia at 5% level of significance?

Solution:

H_0: $P_1 = P_2$

H_1: $P_1 \neq P_2$

$\alpha = 0.05$: 0.025: $\alpha = 0.01$: 0.005

Critical Region: $z_{0.025} = \pm 1.960$; $z_{0.005} = \pm 2.33$

Decision Rule: Reject H_0, if $|z_{cal}|$ is greater than 1.960.

Computations:

$n_1 = 200 : 20\% \ of \ 200 = 40$

$$\hat{P}_1 = \frac{40}{200} = 0.20$$

$n_2 = 300 : 27\% \ of \ 300 = 81$

$$\hat{P}_2 = \frac{81}{300} = 0.27$$

Pooled Estimator: $\hat{P} = \frac{40+81}{200+300} = \frac{121}{500} = 0.242$

$$\hat{q} = 0.758$$

Test Statistic: $Z^* = \dfrac{\hat{P}_1 - \hat{P}_2 - (P_1 - P_2)}{\sqrt{\hat{P}\hat{q}\left(\frac{1}{n_1} + \frac{1}{n_2}\right)}}$

$$Z = \frac{0.20 - 0.27}{\sqrt{(0.242)(0.758)\left(\frac{1}{200} + \frac{1}{300}\right)}} = Z = \frac{0.07}{\sqrt{(0.183436)(0.005 + 0.003333)}}$$

$$\frac{0.07}{\sqrt{(0.00152863333)}} = \frac{-0.07}{0.0390974} = -1.79038478$$

$$= -1.790$$

Conclusion: Since $|z_{cal}| = 1.790$ is less than $z_{0.025} = 1.960$, accept H_0 and conclude that there is no difference in staff preference for smoked Catfish or Tilapia.

9.6 TESTING THE EQUALITY OF TWO VARIANCES $(\sigma_1^2 = \sigma_2^2)$

In the previous sections, we were interested in comparing two means and proportions under various conditions of whether or not the variances are known or unknown, or whether sample sizes are large or small. But in this section, we will introduce methods for comparing two variances.

In section 8.6.3, we saw that a two-tailed hypothesis may be tested by setting confidence intervals on the difference between two means or proportions. Similarly, the equality of two variances can be tested by setting confidence intervals on the ratio of the two variances.

Suppose we have two independent random samples of size n_1 and n_2 with variances S_1^2 and S_2^2, respectively, from the same population, the ratio of the two variances has a sampling distribution called the F-distribution, with $(n_1 - 1) \ and \ (n_2 - 1)$ degrees of freedom:

$$F = \frac{S_1^2}{S_2^2} \sim F_{(n_1-1),(n_2-1)} = F_{v_1,v_2}$$

where $(n_1 - 1 = v_1)$ = degrees of freedom for estimating S_1^2 -- the numerator;

$(n_2 - 1 = v_2)$ = degrees of freedom for estimating S_2^2 - - the denominator.

Each pair of degrees of freedom determines an F-distribution written as $F_{(n_1-1),(n_2-1)}$. The critical values of $F_{(n_1-1),(n_2-1)}$ can be obtained from the two separate Tables labelled 5% and 1% (Walpole, 1974) which refer to the proportion of the area under the curves to the right of the values given in the Tables.

At the top of the F-Tables are the numbers of degrees of freedom $(n_1 - 1 = v_1)$ in the numerator of the F-ratio, while the numbers of degrees of freedom in the denominator $(n_2 - 1 = v_2)$ are in the left-hand column.

For a two-tailed test, we must know the level of significance, $\alpha = 0.05$ or 0.01 before the critical region can specified:

$$F_{\alpha}, v_1, v_2 \equiv F_{0.05, v_1 v_2} \text{ and}$$

$$F_{1-\alpha}, v_1, v_2 \equiv F_{0.95, v_1 v_2}$$

If $v_1 = 10$ and $v_2 = 12$, $F_{0.05, 10, 12} = 2.75$ and $F_{0.01, 10, 12} = 4.30$

A point estimate of the ratio of two population variances $\frac{\sigma_1^2}{\sigma_2^2}$ is estimated by the ratio $\frac{S_1^2}{S_2^2}$ (Walpole, 1974).

If σ_1^2 and σ_2^2 are the variances of two normal populations, a confidence interval can be established for $\frac{\sigma_1^2}{\sigma_2^2}$ by using the statistic:

$$F = \frac{\sigma_2^2}{\sigma_1^2} \frac{S_1^2}{S_2^2}$$

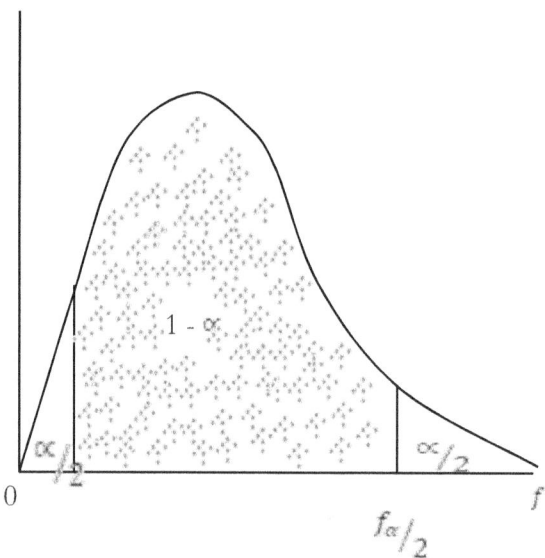

Figure. 9.6.1: $P\left[f_{1-\alpha/_2}(v_1, v_2) < F < f_{\alpha/_2}(v_1 v_2)\right] = 1 - \alpha$

Thus, substituting for F, we have:

$$p\left[F_{1-\frac{\alpha}{2}}(v_1, v_2) < \frac{\sigma_2^2 S_1^2}{\sigma_2^2 S_2^2} < f_{\frac{\alpha}{2}}(v_1, v_2)\right] = 1 - \alpha$$

Manipulating the inequality and multiplying by $\frac{\sigma_2^2}{\sigma_1^2}$ and inverting each term we obtain:

$$p\left[\frac{S_1^2}{S_2^2}\frac{1}{f_{\alpha/2}(v_1,v_2)} < \frac{\sigma_1^2}{\sigma_2^2} < \frac{S_1^2}{S_2^2}\frac{1}{f_{1-\alpha/2}(v_1,v_2)}\right] = 1-\alpha$$

Walpole (1974) asserted that: $f_{1-\alpha(v_1v_2)} = \frac{1}{f_x(v_2v_1)}$.

Hence, $(1-\alpha)100\%$ Confidence Interval for the ratio $\frac{\sigma_1^2}{\sigma_2^2}$ is:

$$\frac{S_1^2}{S_2^2}\frac{1}{f_{\frac{\alpha}{2}}(v_1,v_2)} \leq \frac{\sigma_1^2}{\sigma_2^2} \leq \frac{S_1^2}{S_2^2}f_{\frac{\alpha}{2}}(v_1,v_2)$$

Where: S_1^2 and S_2^2 are the variances of independent samples of size n_1 and n_2 respectively.

Example 9.6.1:

Two independent random samples of Catfish and Tilapia from a fish pond produced the following data:

$$\text{Catfish} \quad \text{Tilapia}$$

$$n_1 = 10 \quad n_2 = 7$$

$$S_1^2 = 16 \quad S_2^2 = 3$$

a) Using a 10% significance, test the hypothesis that the population variances H_0: $\sigma_1^2 = \sigma_2^2$ versus the alternative H_1: $.\sigma_1^2 \neq \sigma_2^2$

b) Then using a 5% level of significance, test the hypothesis: H_0: $\sigma_1^2 = \sigma_2^2$ versus (i) H_1: $\sigma_1^2 > \sigma_2^2$ (ii) H_1: $\sigma_1^2 < \sigma_2^2$

Solution: 0.1:

$$H_0: \sigma_1^2 = \sigma_2^2$$

$$H_1: \sigma_1^2 \neq \sigma_2^2$$

$$\alpha = \frac{\alpha}{2} = \frac{0.1}{2} = 0.05:$$

$$(F_{0.05}(9,6) = 4.10:$$

$$(F_{0.05}(6,9) = 3.37$$

Critical Region: $F_{0.05}(9,6) = 4.10: \frac{1}{F_{\alpha(v_1v_2)}} = \frac{1}{3.37} = 0.2967.$

Decision Rule:

Reject H_0, if $F_{cal} > 4.10$, or $F_{cal} < 0.297$.

Accept H_0: if $0.297 \leq F_{cal} \leq 4.10$

Test Statistic: $F_{cal} = \dfrac{S_1^2}{S_2^2}$

Computations:

$$F_{cal} = \frac{S_1^2}{S_2^2} = \frac{16}{3} = 5.33$$

Conclusion: Since $F_{cal} = 5.33 > 4.10$, reject H_0 and conclude that the variances are not equal.

(bi) $H_0 : \sigma_1^2 = \sigma_2^2$

$H_1 : \sigma_1^2 > \sigma_2^2$

$\alpha = 0.1 \quad i.e \quad \dfrac{0.1}{2} = 0.05 : F_{0.05}(9,6) = 4.10.$

Decision Rule: Reject H_0, if $F_{cal} > 4.10$ as above, computations are the same.

Conclusion: Since $F{cal} = 5.33 > 4.10$, reject H_0 and conclude that the Catfish variance is significantly larger than that of the Tilapia fish.

(bii) $H_0 : \sigma_1^2 = \sigma_2^2$

$H_1 : \sigma_1^2 < \sigma_2^2$

$\alpha = 0.1 \; i.e \dfrac{0.1}{2} = 0.05$

Critical Region: $F_{0.05}(6,9) = 3.37$

Decision Rule: Reject H_0, if $F_{cal} > 3.37$, or Accept H_0, if $F_{cal} \leq 3.37$

Computations:

$$= \frac{S_2^2}{S_1^2} = \frac{3}{16} = 0.1875$$

Conclusion:

Since $F_{cal} = 0.188 < 3.37$, accept H_0 that the variation in Tilapia is much less than that of Catfish.

Example 9.6.2:

A Silviculturist in Uniben raised Moringa seedlings in polypots and another set of Moringa seedlings on nursery beds. The observed data were:

$$\text{Polypot seedlings} \quad \text{Nursery bed seedlings}$$

$$\bar{X}_1 = 94.8mm \qquad \bar{X}_2 = 95.3mm$$

$$S_1^2 = 6.13mm \qquad S_2^2 = 13.7mm$$

$$n_1 = 31 \qquad\qquad n_2 = 31$$

Assuming that the scores are normal, do the two groups of seedlings differ in variability at $\alpha = 0.02$ level of significance?

Solution:

$$H_0 : \sigma_1^2 = \sigma_2^2$$

$$H_1 : \sigma_1^2 \neq \sigma_2^2$$

$$\alpha = 0.02 : \frac{0.02}{2} = 0.01$$

Critical Region: $F_{0.01}, (30,30) = 2.39 : or \frac{1}{2.39} = 0.4184$

Decision Rule: $\begin{bmatrix} \text{Reject } H_0, if \quad F_{cal} > 2.39 : \quad or < 0,418 \\ \text{Accept } H_0, if \quad 0.418 \leq F_{cal} \leq 2.39. \end{bmatrix}$

Computations: $F_{cal} = \frac{S_1^2}{S_2^2} = \frac{6.13}{13.7} = 0.4474.$

Conclusion: Since $F_{cal} = 0.4474$ i.e. $0.418 \leq 0.447 \leq 2.39$ we accept H_0 and conclude that the variability of the two groups is the same.

9.7 ESTIMATION OF VARIANCES USING SEVERAL SAMPLES

According to Dixon and Massey (1969), if two or more samples are from the same population, or from different populations having equal variances, σ^2, then the variances of the several samples, one from each population, can be pooled or averaged to give an estimate of σ^2. The formula for such an estimate of a pooled variance of k-sample is:

$$S_p^2 = \frac{(n_1 - 1)S_1^2 + (n_2 - 1)S_2^2 + (n_3 - 1)S_3^2 + \cdots + (n_k - 1)S_k^2}{n_1 + n_2 + n_3 + \cdots + n_k - k}$$

$$= \frac{\sum_{i=1}^{k}(n_i - 1)s_i^2}{\sum_{i=1}^{k} n_i - k}$$

The pooled variance, S_p^2, is an unbiased estimate of the common variance σ^2.

If the populations have normal distributions but not necessarily the same means, the sampling distribution of $\frac{S_p^2}{\sigma^2}$ is a chi-squared distribution with $n_1 + n_2 + \ldots\ldots + n_k - k$ degrees of freedom.

Thus, 90% confidence interval for σ^2 is:

$$\frac{S_p^2}{\dfrac{\chi_{0.05}^2}{df(\sum n_{i-k})}} < \sigma^2 < \frac{S_p^2}{\dfrac{\chi_{0.95}^2}{df(\sum n_{i-k})}}$$

$$\frac{S_p^2}{\dfrac{\chi_{0.05}^2}{df(\sum n_{i-k})}} < \sigma^2 < \frac{S_p^2}{\dfrac{\chi_{0.95}^2}{df(\sum n_{i-k})}}$$

Another method of estimating a common variance, σ^2 is to group all the observations into one sample and compute the variance of this single set of numbers - - this is valid only if the several populations have the same mean.

9.8 ENUMERATION STATISTICS

In the foregoing sections, we have been concerned with the testing of statistical hypothesis that a single population parameter is equal to a specified value, or that two population parameters are equal. Frequently, however, more than two samples are involved in an experiment so that our previous methods no longer apply. In some cases, the hypotheses apply to attributes or characteristics and we count the number of items, or occurrences of the attributes in a sample.

For example, hypotheses that proportions are equal in several populations from which the samples are obtained can be handled by using a sampling distribution called **Pearson Chi-squared,** χ^2, to compare observed data with expected frequencies or outcomes - - this is called **Goodness-of-Fit Test**.

9.8.1 Goodness-of-Fit Test

For example, the Silviculturist in the Department of Forestry and Wildlife, raised the seedlings of four tree species (Obeche, Mansonia, Irvingia and Garcinia) from two categories of Nurseries (Nursery Beds and Polypots) and obtained the following data:

Table 9.8.1: Observed and Expected Frequencies of Different Seedlings Raised in a Uniben Nursery

	Obeche	Mansonia	Irvingia	Garcinia	Total
Nursery Beds	99	98	98	97	392
Polypots	121	112	92	83	408
Total	220	210	190	180	800

(i)   Each data entry in the Table is called a **Cell**, while the data entries are the observed frequencies in each **Category**– four (k) categories in this case.

(ii) **Expected or Theoretical frequencies** for each cell of each category are estimated from the observed data by using the formula:

$E_i = \dfrac{R \times C}{T}$, where:

E_i = Cell expected frequency.

155

R = Row total containing the cell in which we are interested.

C = Column total containing the cell which we are interested in.

T = Overall Total.

The expected frequency of Obeche seedlings raised in the Nursery Bed is:

$$E_{11} = \frac{(Obeche\ Row\ total)(Obeche\ column\ total)}{Overall\ Total}$$

$$= \frac{(392)(220)}{800} = \frac{86240}{800} = 107.8$$

$$E_{21} = \frac{(Obeche\ Row\ total)(Obeche\ column\ total)}{Overall\ Total}$$

$$= \frac{(408)(220)}{800} = \frac{89,760}{800} = 112.2$$

$$E_{12} = \frac{(Mansonia\ Row\ total)(Mansonia\ Column\ total)}{Overall\ Total}$$

$$= \frac{(392)(210)}{800} = \frac{82,320}{800} = 102.9$$

$$E_{22} = \frac{(Mansonia\ Row\ total)(Mansonia\ Column\ total)}{Overall\ Total}$$

$$= \frac{(408)(210)}{800} = \frac{85,680}{800} = 107.1$$

$$E_{13} = \frac{(Irvingia\ Row\ total)(Irvingia\ Column\ total)}{Overall\ Total}$$

$$= \frac{(392)(190)}{800} = \frac{74,480}{800} = 93.1$$

$$E_{23} = \frac{(Irvingia\ Row\ total)(Irvingia\ Column\ total)}{Overall\ Total}$$

$$= \frac{(408)(190)}{800} = \frac{77,520}{800} = 96.9$$

$$E_{14} = \frac{(Garcinia\ Row\ total)(Garcinia\ Column\ total)}{Overall\ Total}$$

$$= \frac{(392)(180)}{800} = \frac{70,560}{800} = 88.2$$

$$E_{24} = \frac{(Garcinia\ Row\ total)(Garcinia\ Column\ total}{Overall\ Total}$$

156

$$= \frac{(408)(180)}{800} = \frac{73,440}{800} = 91.8$$

Table 9.8.1a is now summarized with the estimated expected frequencies in parenthesis in each cell as:

Table 9.8.1a: Observed and Expected Frequencies of Seedlings Raised in a Uniben Nursery.

	Obeche	Mansonia	Irvingia	Garcinia	Total
Nursery Beds	99(107.8)	98(102.9)	98(93.1)	97(88.2)	392
Polypots	121(112.2)	112(107.1)	92(96.9)	83(91.8)	408
Total	220	210	190	180	800

From these data, we would like to know whether the proportion of seedlings raised in Nursery Beds is the same for each class (or tree species) - - - that is, if the observed differences in proportions are due to chance, or due to samples drawn from populations with different proportions.

To evaluate the differences between the observed and expected values for each class or cell, we use the Chi-squared statistic which is calculated as:

$$\text{Chi-Squared } (\chi^2) = \sum_{i=1}^{k} \frac{(\text{Observed} - \text{Expected})^2}{\text{Expected}}$$

$$i.e \ \chi^2 = \sum_{i=1}^{k} \frac{(O_i - E_i)^2}{E_i}$$

If the observed frequencies are close to the corresponding expected or theoretical frequencies, the χ^2 value will be small, indicating a good fit leading to the acceptance of H_0. But if the observed frequencies differ widely from the expected frequencies, the χ^2 value will be large, indicating a poor fit and thus leading to the rejection of H_0. Generally, a good fit leads to the acceptance of H_0, whereas a poor fit leads to the rejection of H_0 (Walpole, 1974).

Are the observed and expected frequencies in Table 9.9.1a above the same?

Solution:

$H_0 : P_1 = P_2 = P_3 = P_4 = P$ i.e. the true proportions are all equal to a common population proportion.

$H_1 : P_1 \neq P_2 \neq P_3 \neq P_4 = P$ i.e. the true proportions are not all equal.

$\alpha = 0.05$

Critical Region: $\chi^2_{0.05,3} = 7.815$ (From Table A.6 (Walpole, 1974).

Decision Rule: Reject H_0, if $\chi^2_{cal} > 7.815$

Computations:

$$\chi^2 = \sum \frac{(o_i - e_i)^2}{e_i}$$

$$= \frac{(99 - 107.8)^2}{107.8} + \frac{(121 - 112.2)^2}{112.2} + \frac{(98 - 102.9)^2}{102.9} + \frac{(112 - 107.1)^2}{107.1}$$

$$+\frac{(98-107.8)^2}{93.1}+\frac{(92-96.9)^2}{96.9}$$

$$\frac{(97-88.2)^2}{88.2}+\frac{(83-91.8)^2}{91.8}=4.093.$$

Conclusion: Since $\chi^2_{cal} < 7.815$, do not reject H_0 and we conclude that there is no significant difference among the sample proportions.

9.8.2 Single Classification

Dixon and Massey (1969) described a single-classification problem as one in which the theoretical proportion of cases in each category is specified in advance. For example, in tossing a balanced coin, we expect 50% heads and 50% tails. Therefore, our hypothesis would be:

$$H_0 : P_1 = P_2 = 0.5$$

$$H_0 : P_1 \neq P_2 \neq 0.5$$

For example, in genetics, it is reported that when two types of peas are crossed, the Mendelian theory of inheritance states that the frequencies should be in the ratios 9:3:3:1. In an experiment reported by Dixon and Massey (1969), the following data were observed:

Table 9.8.2: Theoretical and Observed Mendelian Ratios

	Observed (O_i)	Theoretical (E_i)	O_i-E_i	$\dfrac{(O_i - E_i)^2}{E_i}$
Round and yellow	315	312.75	2.25	0.01619
Wrinkled and yellow	101	104.25	-3.25	0.10132
Round and green	108	104.25	3.75	0.13489
Winkled and green	32	34.75	-2.75	0.21763
Total	**556**	**556.00**		**0.47003**

The question is: Do the observed data support the Mendelian ratios 9:3:3:1? From the observed data, we will use these ratios to obtain theoretical frequencies as:

$$\text{Round and yellow – probability} = \frac{9}{16} \text{x } 556 = 312.75$$

$$\text{Wrinkled and yellow-probability} = \frac{3}{16} \text{x } 556 = 104.25$$

$$\text{Round and green-probability} = \frac{3}{16} \text{x } 556 = 104.25$$

$$\text{Wrinkled and green-probability} = \frac{1}{16} \times 556 = 34.75$$

H_0: Ratios are 9:3:3:1.

H_1: Ratios are not 9:3:3:1.

$\alpha = 0.05$

Critical Region: $\chi^2_{0.05} = 7.81$

Decision Rule: Reject H_0, if χ^2_{cal} is greater than 7.81.

Conclusion: Since $\chi^2_{cal} = 0.470 < \chi^2_{0.05,3}$ do not reject H_0. We do not have sufficient reason or evidence to reject the Mendelian ratios.

Quite often, the variables of interest are qualitative in nature and the level of measurement is nominal, so that each item observed can be placed in a particular category, hence they are called **Categorical Data**. A set of mutually exclusive and exhaustive qualitative events is often called an **Attribute**.

In a study of educational achievement of Nigerian males of 30 years of age, each subject observed was placed in one and only one of the following categories, according to his maximum formal educational achievement as follows:

1. University graduate (or HND)
2. OND
3. NCE
4. Secondary School (SS3)
5. JSS 3
6. Primary School

It is also known that 10 years ago, in 2005, the distribution of educational achievement, on the scale above, for 25 year-old men was as given in Table 9.9.2. The question we want to answer is whether the present population distribution based on this scale is exactly as it was in 2005.

Table 9.8.3: Distribution of Educational Achievement of 25- year-old Nigerian Men in 2005

Category	Relative Frequency
1. University graduate (or HND).	0.18
2. Ordinary Nigerian Diploma (OND)	0.17
3. National Certificate of Education (NCE)	0.32
4. Secondary School (SS3).	0.13
5. Junior Secondary School (JSS-3)	0.17
6. Primary School	0.03

Using the above relative frequency, the following data were observed in a random sample of 200 (Table 9.9.3) from the present population in 2015 and the expected frequencies estimated:

Table 9.8.4: Frequency Distribution of Educational Achievement in 2015

Category	Observed (O_i)	Expected (E_i)
1. University graduate	35	36
2. OND	40	34
3. NCE	83	64
4. SS-3	16	26
5. JSS-3	26	34
6. Primary School	0	6
Total	**200**	**200**

Solution:

H_0: The observed values are in agreement with the expected values.

H_1: The observed values are not in agreement with the expected values.

$\alpha = 0.05$

Critical Region:

$\chi^2_{0.05,5} = 11.0705$; $\chi^2_{0.01,5} = 15.0863$.

DecisionRule: Reject H_0, if $\chi^2_{cal} > 11.0705$ at $\alpha = 0.05$, or $\chi^2_{cal} > 15.0863$ at $\alpha = 0.01$

Computations:

$$\chi^2 = \sum \frac{(O_i - E_i)^2}{E_i}$$

$$\frac{(35-36)^2}{36} + \frac{(40-34)^2}{34} + \frac{(83-64)^2}{64} + \frac{(16-26)^2}{26} \frac{(26-34)^2}{34} + \frac{(0-6)^2}{6}$$

$$= 0.027777 + 1.058824 + 5.640625 + 3.846154 + 1.882353 + 6.0$$

$$= 18.455733 \approx 18.46$$

Conclusion: Since $\chi^2_{cal} = 18.46 > \chi^2_{0.05} = 15.09$, we reject H_0 and conclude that there is a highly significant difference between the educational achievements of 25 year-old Nigerians in 2005 and 2015.

Note: The real purpose in the comparison of these distributions is to investigate the hypothesis that the expectations are correct and that the current distribution is actually the same as 10 years ago. Therefore, given the hypothesis and the assumption of independent random sampling of individuals (with replacement), the exact probability of a particular sample distribution can be found from the

Multinomial Distribution Rule (Winkler and Hays, 1975):

"If n observations are made independently and at random, then the probability that exactly n_1 will be of event 1, n_2 of event 2, . . . and n_k of event k, where $n_1 + n_2 + n_3 + . . . + n_k = n$, is given by:

$$P = \frac{n!}{n_1!\, n_2!\, n_3!\, \dots\, n_k}(P_1)^{n_1}(P_2)^{n_2}(P_3)^{n_3} \dots (P_k)^{n_k}$$

Thus, for our example above, the exact probability is:

$P(observed \cdot distribution | H_0) =$

$$\frac{200!}{35!\cdot40!\cdot83!\cdot16!\cdot26!\cdot0!}(.18)^{35}(.17)^{40}(.32)^{83}(.13)^{16}(.17)^{26}(.03)^{0}$$

Due to the extremely large amount of calculations, the χ^2 approximation is justified.

9.8.3 Two-way Classification (Contingency Tables)

According to Winkler and Hays (1975), situations often arise where n independent observations are made and each and every observation is classified in two or more qualitative ways – according to sex, hair color, eye color, ear-rings, etc.

In a situation where there are two attributes of interest, A and B, and there are different categories of A = $A_1, A_2 \ldots A_c$ as the columns and different categories of B = $B_2 B_1, \ldots B_R$ as the rows, for attributes A and B respectively, such a table is called a **Contingency Table** which is identified by the number of rows (R) and the number of columns. It is designated as R x C (or R by C) contingency table.

As a result of using two attributes for the classification, such classifications are appropriately called **"Two-Way classification"**. A case of two Rows and two Columns is called a 2 x 2 Table as shown in Table 9.9.3.1:

Table 9.8.5.1: A 2 x 2 Contingency Table.

	A_1	A_2
B_1	(A_1, B_1)	(A_2, B_1)
B_2	(A_1, B_2)	(A_2, B_2)

(i)   **Testing for Independence (Association)**

For two-way classifications, we want to know whether the two attributes are independent . . . that is, the distribution of one attribute should be the same, regardless of the other attribute associated. For example, if eye color and hair color are independent, the proportion of the blue-eyed people having light-colored hair should be the same as the proportion of the brown-eyed people having light-colored hair in the population.

For example, using eye and hair color as attributes for classification, Dixon and Massey (1969) obtained the following data in Table 9.8.5.2:

Table 9.8.5.3: A 3 x 2 Contingency Table

Eye color	Hair Color		
	Light	Dark	Total
Blue	32	12	44
Brown	14	22	36
Other	6	9	15
Total	**52**	**43**	**95**

$$Degrees\ of\ freedom = (R - 1)(C - 1)$$

$$= (3 - 1)(2 - 1) = 2$$

The question we want to answer is: Are Eye color and Hair color independent?

Solution:

H_0: Eye color and Hair color are independent.

H_1: Eye color and Hair color are not independent.

$\alpha = 0.05$

Critical Region: $\chi^2_{0.05,2} = 5.99$.

161

Decision Rule: Reject H$_0$, if $\chi^2_{cal} > \chi^2_{0.05,2} = 5.99$.

Computations:

Record the number of observations that fall into each category (0$_i$).

Look at each attribute separately and record the total number of observations. These are the **"marginal" totals**.

For example, 52 people out of 95 have light hair, 43 have dark hair, while 44 people have blue eyes.

Estimate the expected frequencies (E$_i$) by using the "marginal totals" (Column total and Row total of the cell of interest (R$_i$ x C$_j$) divided by overall total (T): $E_{ij} = \dfrac{R_i \times C_j}{T}$.

For example: E_{11} = *expected No. of people with light hair*

$$= \frac{(Total\ of\ light\ hair)(Total\ of\ Blue\ eyes)}{Overall\ total\ observations}$$

$$E_{11} = \frac{(52)(44)}{95} = \frac{2,288}{95} = 24.08421053 \approx 24.1$$

E_{12} = *Expected no. of people with Dark hair*

$$\frac{(Total\ Dark\ hair)(Total\ Blue\ eyes)}{Overall\ total\ observations}$$

$$E_{12} = \frac{(43)(44)}{95} = \frac{1,892}{95} = 19.91578947 \approx 19.9$$

$$E_{32} = \frac{(43)(15)}{95} = \frac{645}{95} = 6.789473684 \approx 6.8$$

The observed and expected frequencies (within brackets) of Table 9.9.3.2 are as follows:

Table 9.8.5.4: A 3 x 2 Contingency Table With Expected Frequencies in Parenthesis

Eye Color	Light	Dark	Total
Blue	32(24.1)	12(19.9)	44
Brown	14(19.7)	22(16.3)	36
Other	6(8.2)	9 (6.8)	15
Total	**52**	**43**	**95**

$$\chi^2 = \sum \frac{(O_i - E_i)^2}{E_i}$$

$$= \frac{(32-24.1)^2}{24.1} + \frac{(12-19.9)^2}{19.9} + \frac{(14-19.7)^2}{19.7} + \frac{(22-16.3)^2}{16.3} + \frac{(6-8.2)^2}{8.2} + \frac{(9-6.8)^2}{6.8}$$

$$= 2.59 + 59 + 3.14 + 1.65 + 1.99 + 0.59 + 0.71 = 10.67$$

Conclusion: Since $\chi^2_{cal} = 10.67 > \chi^2_{0.05,2}$ reject H$_0$ and conclude that eye-color and hair-color are not independent.

Example 9.8.5.5:

A Sociologist in Uniben studied the relationship between Christian Denominations and Geographical Area of origin of Nigerians living in Eastern and Western areas of the country. The sociologist randomly selected 600 persons from Eastern Nigeria and classified each person as Protestant, Catholic, or Pentecostal; and 400 persons from Western Nigeria and also classified each person according to their denomination. Are the Christian Denominations and Geographical areas of origin independent at $\alpha = 0.05$?

The observed frequencies are presented in Table 9.8.3.2:

Table 9.8.5.5: 2 x 3 Contingency Table: Observed frequencies of Christian Denomination

Area of origin	Protestant	Catholic	Pentecostal	Total
Eastern Nigeria	182	215	203	600
Western Nigeria	154	136	110	400
Total	**336**	**351**	**313**	**1000**

Solution:

H_0: Area of origin and Christian Denominations are independent.

H_1: Area of origin and Christian denominations are not independent.

$\alpha = 0.05$

Critical Region: $\chi^2_{0.05,2} = 5.99$

Decision Rule: Reject H_0, if χ^2_{cal} is greater than 5.99.

Computations:

Estimate the expected frequency of each cell: $\mathbf{E_{ij}} = \dfrac{(R_i)(C_j)}{T}$

$$E_{11} = \frac{(336)(600)}{1000} = \frac{201,600}{1000} = 201.6 \approx 202$$

$$E_{12} = \frac{(351)(600)}{1000} = \frac{210,600}{1000} = 210.6 \approx 211$$

$$E_{13} = \frac{(313)(600)}{1000} = \frac{187,800}{1000} = 187.6 \approx 188$$

$$E_{23} = \frac{(313)(400)}{1000} = \frac{125,200}{1000} = 125.2$$

Table 9.8.5.5a:

Complete Table 9.8 with the Expected frequencies in brackets as:

Observed and Expected Frequencies of Christian Denominations.

Area of origin	Protestant	Catholic	Pentecostal	Total
Eastern Nigeria	182(202)	215(211)	203(187)	600
Western Nigeria	154(134)	136(140)	110(126)	400
Total	**336**	**351**	**313**	**1000**

Table 9.8.5.5: Observed and Expected Frequencies.

$$\chi^2 = \sum \frac{(O_i - E_i)^2}{E_i}$$

$$= \frac{(182 - 202)^2}{202} + \frac{(215 - 211)^2}{211} + \frac{(203 - 187)^2}{187} + \frac{(154 - 134)^2}{134} + \frac{(136 - 140)^2}{140}$$
$$+ \frac{(110 - 126)^2}{126}$$

$=1.98019802+0.075829383+1.368983957+2.985074627+0.114285714+\quad 2.031746032$

$= 8.556117733 \approx 8.556.$

Conclusion: Since $\chi^2_{cal} = 8.556 > 5.99$, reject H_0 and conclude that Christian denomination and geographical area of origin in Nigeria are not independent.

Example 9.8.5.6:

A lecturer who taught a third-year Statistics Course (AGR 311) in the Faculty of Agriculture, Uniben, was interested in studying the relationship between the previous prerequisite courses and performance (or the grade obtained) in AGR 311, to determine, if the classifications are independent. In year 2012, the grades that the students obtained in AGR 311 were distributed as follows, by the category of prior course work:

Table 9.8.5.7: A 4 x 5 Contingency Table of Grades Observed in AGR311. GRADES

S/N	Prior Course Work in	A	B	C	D	E	Total
1	AGR 212	15	20	40	5	0	80
2	MATH 223	10	15	70	20	5	120
3	Both AGR 212 and MATH 233	10	20	25	5	0	6
4	Neither (1) nor (2)	15	15	75	30	55	190

 Is there any relation between prior course work and the grade obtained in AGR 311 at $\alpha = 0.05$ and $\alpha = 0.01$?

Solution:

H_0: The grade scored in Statistics (AGR 311) and previous prerequisite courses in AGR 212 and MATH 223 are independent.

H_1: The grades and prerequisite courses are not independent. $\alpha = 0.05$

$\alpha = 0.01: Df = (R - 1)(C - 1) = (4 - 1)(5 - 1) = 3 \times 4 = 12$

Critical Region: $\chi^2_{0.01,12} = 26.217.$

Decision Rule: Reject H_0 if $\chi^2_{cal} > 26.22$

Computations:

Estimate the Expected frequencies from the Observed frequencies as:

$$E_{11} = \frac{(R_1)(C_1)}{T} = \frac{(80)(50)}{450} = \frac{4000}{450} = 8.8889 \approx 8.9$$

$$E_{12} = \frac{(R_1)(C_2)}{T} = \frac{(80)(70)}{450} = \frac{5600}{450} = 12.4444 \approx 12.4$$

164

$$E_{13} = \frac{(R_1)(C_3)}{T} = \frac{(80)(210)}{450} = \frac{16,800}{450} = 37.3333 \approx 37.3$$

$$E_{14} = \frac{(R_2)(C_4)}{T} = \frac{(80)(60)}{450} = \frac{4,800}{450} = 10.6666 \approx 10.7$$

$$E_{15} = \frac{(R_2)(C_5)}{T} = \frac{(80)(60)}{450} = \frac{4,800}{450} = 10.7$$

$$E_{21} = \frac{(R_2)(C_1)}{T} = \frac{(120)(50)}{450} = \frac{6,000}{450} = 13.3333 \approx 13.3$$

$$E_{45} = \frac{(R_4)(C_5)}{T} = \frac{(190)(60)}{450} = \frac{11,400}{450} = 25.3333 \approx 25.3$$

GRADES

Table 9.8.5.8: Observed and Expected Frequencies (in Parentheses).

S/N	Prior course Work in	A	B	C	D	E
1	AGR 212	15(8.9)	20(12.4)	40(37.3)	5(10.7)	0(10.7)
2	MATH 223	10(13.3)	15(18.7)	7056.0)	20(16.0)	5(16.0)
3	Both AGR 212 and MATH 233	10(6.7)	20(9.3)	25(28.0)	5(8.0)	0(8.0)
4	Neither (1) nor (2)	15(21.1)	15(29.6)	75(88.7)	30(25.3)	55(25.3)
Total		50	70	120	60	60

Compute χ^2 as:

$$\chi^2 = \sum_{i=1}^{k} \frac{(O_i - E_i)^2}{E_i}$$

$$= \frac{(15-8.9)^2}{8.9} + \frac{(20-12.4)^2}{12.4} + \frac{(40-37.3)^2}{37.3} + \frac{(5-10.7)^2}{10.7} + \frac{(0-10.7)^2}{10.7}$$

$$+ \ldots \frac{(15-21.1)^2}{21.1} + \frac{(15-29.6)^2}{29.6} + \frac{(75-88.7)^2}{88.7} + \frac{(30-25.3)^2}{25.3} + \frac{(55-25.3)^2}{25.3} = \underline{106.59}$$

Conclusion: Since $\chi^2_{cal} = 106.59 \gg 26.22$, reject H_0 and conclude that the grade obtained in AGR 311 is not independent of previous prerequisite course work - - in fact, there is a strong relationship.

Note: Two discrete attributes are considered independent if and only if:

$P(A_j, B_k) = P(A_j)P(B_k)$ - - for all possible joint events (A_j, B_k). The marginal population probabilities, $P(A_j)$ and $P(B_k)$ are unknown, but must be estimated from the sample marginal proportions:

$$P(A_j) = \frac{frequency\ of\ A_j}{n}$$

$$P(B_k) = \frac{frequency\ of\ B_k}{n}$$

$$\therefore \text{ Expected frequency of } P(A_j, B_k) = \frac{(\text{frequency of } A_j)(\text{frequency of } B_k)}{n}$$

9.8.6 Two-by-Two Table

Generally, the test for independence can be carried out as above. But in the case where there are only two rows and two columns, Dixon and Massey (1969) and Winkler and Hays (1975) have suggested that the computing forms be simplified by converting the discrete sampling distribution of the χ^2 to the continuous χ^2 distribution by reducing the absolute value of each difference by 0.5 before it is squared. This modification is sometimes called a **Continuity Correction, or Yates' Correction for Continuity**.

For a 2 x 2 Table, we can include this correction and simplify the χ^2 statistic formula. For example:

Table 9.8.6.1: A 2 x 2 Contingency Table - - Observed Frequencies

S/N	I	II	Total
1	A	B	a + b
2	C	D	c + d
Total	**a + c**	**b + d**	**n = a + b + c + d**

Suppose a, b, c, d, are the observed frequencies in a 2 x 2 Table, the χ^2 statistic for the test of independence is given by:

$$X^2 = \frac{\left(|ad - bc| - \frac{n}{2}\right)^2 n}{(a+b)(a+c)(c+d)(b+d)}$$

Example 9.8.6.1:

In a public-opinion survey of students, two questions were asked:

1. Do you drink beer?
2. Are you in support of selling beer on Campus?

	Question 1		
Question 2	Yes	No	Total
Yes	37	20	57
No	15	6	21
Total	52	26	78

$$df = (R - 1)(c - 1) = (2 - 1)(2 - 1) = 1$$

$$\chi^2_{0.05,1} = 3.841$$

Solution:

H_o: Question 1 and Question 2 are independent.

H_1: Question 1 and Question 2 are not independent.

$\alpha = 0.05$.

Critical Region: $\chi^2_{0.05,1} = 3.041$.

166

Decision Rule: Reject H_0, $\chi^2_{cal} >$ if $\chi^2_{0.05.1} >= 3.041$.

Computations:

$$\chi^2 = \frac{\left(|ad - bc| - \frac{n}{2}\right)^2 n}{(a+b)(c+d)(a+c)(b+d)}$$

$$= \frac{(|222 - 300| - 39)^2\, 78}{(52)(26)(57)(21)} = \frac{(78 - 39)^2(78)}{52 \times 26 \times 57 \times 21} = \frac{39^2 \times 78}{1,352 \times 1,197}$$

$$\frac{1,521 \times 78}{1,618,344} = \frac{118,638}{1618,344} = 0.0733$$

Conclusion: Since $\chi^2_{cal} = 0.073 \ll \chi^2_{0.05(1)} = 3.841$, accept H_0 and conclude that Questions (1) and (2) are independent.

10 THE ANALYSIS OF VARIANCE (ANOVA)

Variability or variation is an inherent property of any experimental material (in science, agriculture, engineering or biological sciences) and the measurements of such experimental material, no matter how uniform or homogeneous such material is, nor how carefully the measurements are made, will always vary.

Variability in statistics is also called "noise" (Mendenhall, 1975) which reduces or masks the amount of information obtainable from any experiment. The greater the variability, or noise, the less information will be obtained from the sample.

The true population mean (called the parametric mean), μ, and its variance (σ^2) for any defined, or finite population, are called parameters which are fixed constants, but they are usually unknown in real life; hence they must be estimated from the information contained in a sample mean($\bar{X}$) and a sample variance (S^2). Therefore, μ and σ^2, the population mean and variance, respectively, are parameters which are constant for any given population, but may vary from one population to another.

For example, the mean height, or the mean diameter (μ) and its variance (σ^2) for all **Triplochiton** trees in Edo State (Nigeria) are fixed or constant values which are unknown, and must be estimated from a sample of **Triplochiton** trees in Edo State.

The Analysis-of-variance, generally called **Anova**, was developed by a British statistician, Sir Ronald A. Fisher, fondly called the **"Father of statistics",** as an arithmetic process for partitioning a total sum of squares deviations into components associated with recognized sources of variation. Steel and Torrie (1981) observed that Anova has been used to great advantage in all fields of research where quantitative data are measured. Therefore, Anova is fundamental to much of the applications of statistics in agriculture, biological sciences, social sciences and engineering, etc., and is the most powerful weapon for the researcher's statistical arsenal.

The Analysis-of-variance provides any researcher the tool to test whether two or several sample means could have been obtained from populations with the same parametric mean, with respect to a given variable like tree height, or diameter. Alternatively, the researcher would conclude that the sample means differ from each other to such an extent that we must say they are samples from different populations.

The knowledge of Anova is indispensable to any modern scientist for testing scientific hypotheses. Once understood, Anova provides an insight into the nature of variation of natural events. The elegance and beauty of statistical methodology is found in Anova.

10.1 TERMINOLOGIES AND SYMBOLISM OF ANOVA

Before we discuss the methodologies of Anova, it is only fair to understand the terminologies and symbolism that are frequently used in analysis of variance. According to Mendenhall (1975), the analysis-of-variance procedure attempts to analyze the variation of a response and to assign portions of this variation to each of a set of independent variables.

The variability of a set of n measurements (Mendenhall, 1975) is proportional to the sum of squares of deviations (Total SS) $=\sum_{j=1}^{k} \sum_{i=1}^{n} \left(X_{ij} - \bar{X} \right)^2$, which is used to calculate the sample variance. The analysis-of-variance partitions the total sum of squares of deviations into portions each of which is attributed to one of the independent variables in the experiment, plus a remainder that is associated with random error from unknown and uncontrolled sources of variation.

10.1.1 Total Sum of Squares Deviation (Total SS)

The total sum of squares deviation (Total SS) $= \sum_{j=1}^{k} \sum_{i=1}^{n} \left(X_{ij} - \bar{X} \right)^2$ can be partitioned into two portions -- one corresponding to deviation of the treatment mean from the grand mean $\left(\bar{X}_{.j} - \bar{X}_{..} \right)$ and the other, the deviation of an observation (X_{ij}) from its own treatment means $(\bar{X}_{.j})$:

$$\sum_{j=1}^{k}\sum_{i=1}^{nj}(X_{ij} - \bar{X}_{..})^2 = \sum_{j=1}^{k}\sum_{i=1}^{nj}(\bar{X}_{.j} - \bar{X}_{..})^2 + \sum_{j=1}^{k}\sum_{i=1}^{nj}(X_{ij} - \bar{X}_{.j})^2 \ldots\ldots\ldots(10.1)$$

Total SS=   SS Treatment + SSError.

When these components of SS deviations are divided by their corresponding degrees of freedom, they are called **Mean Squares (MS)** which are used to designate variances in Analysis-of-variance procedure.

Thus, if the components of equation 10.1 are divided by their respective degrees of freedom, they become Mean Squares, or variances as;

$$\frac{\sum_{j=1}^{k}\sum_{i=1}^{nj}(X_{ij} - \bar{X}_{..})^2}{N-1} = \frac{\sum_{j=1}^{k}\sum_{i=1}^{n}(\bar{X}_{.j} - \bar{X}_{..})^2}{k-1} + \frac{\sum_{j=1}^{k}\sum_{i=1}^{n}(X_{ij} - \bar{X}_{.j})^2}{N-k}$$

Total MS = Treatment MS + Error MS

Unbiased Population variance = Unbiased Estimate of Treatment variance + Unbiased Estimate of Error variance.

Hence there are three independent unbiased estimates of variance (σ^2) from the data in a one-way Anova, if the hypothesis is true.

10.1.2 Treatment

A statistical, or experimental treatment (Steel and Torrie, 1981; Wahua, 1999) is any process or procedure whose effects are to be measured and compared with others. For example, different amounts of fertilizer may be applied to a crop and their yields are recorded to determine the best concentration. Such quantitative or qualitative components of a treatment are called **Levels.**

When we have two or more types of treatments in an experiment, each type is referred to as a **FACTOR**. A Factor is a type of treatment with levels. According to Sokal and Rohlf (1969), samples in Anova are variously called Groups, Classes, Treatments or Factors.

10.1.3 Experimental Material

Any object, or element of an environment on which treatments are applied and observations made, is an **experimental material**. Thus, the plant, the air, the soil and the field where the experiment is located constitute the experimental material which can be further partitioned into smaller units for different treatments to be applied.

10.1.4 Sampling Unit

The sampling Unit is that portion of Experimental Unit that is taken for chemical or statistical analysis and subsequently used for decision making. The sampling unit may involve the entire experimental plot. A sampling unit may be an animal on a treatment ration; a tree on some fertilizer trial, or a random sample of leaves from a sprayed tree.

10.2.1 Principles of Scientific Experimentation

An Experiment is a planned investigation to furnish information necessary for increasing knowledge and improving human welfare. The results of an experiment may help to obtain new facts, or to confirm, or deny, the results of previous experiments to aid decision-making.

Every experiment is set up to provide answers to one or more questions. These questions are generally expressed as statements of hypotheses that have to be verified or disproved through experimentation or observation. Therefore, ask questions of importance or relevance in the field of research.

Once a hypothesis is framed, the next step is to design a procedure for its verification. The experiment is the set of rules used to draw the sample from the population. The set of rules is the **experimental procedure or experimental design**. The experimental procedure usually consists of four phases:

   i.   Selecting appropriate materials to test e.g. native maize and hybrid maize yields, disease resistance and other characters to measure.
   ii.   Specifying the characters to measure.
   iii. Selecting the procedure to measure these characters.
   iv. Specifying the procedure to determine whether the measurements made support the hypothesis.

The purpose of any experimental design is for:

   a) Estimation of Error.
   b) Control of Error.
   c) Proper interpretation of Results.

10.2.2 Objectives of an Experiment

It is of paramount importance (Steel and Torrie, 1981) to define the population for which inferences are to be drawn and to sample that population randomly.

State the objectives clearly as questions to be answered, hypotheses to be tested, and effects to be estimated. It should be noted that as the variability (σ^2) of the experimental material increases, the amount of information decreases, and as the sample size (n) increases the amount of information increases.

10.2.3 An Experimental Unit (or an Observation)

An Experimental Unit, or Observation, or an experimental plot, is the unit of material to which one application of a treatment is applied. An experimental unit may be an animal, a leaf, a fertilizer level, or ten birds in a pen, etc. Observations are the raw materials with which research workers deal. Statistics cannot be applied to observations, unless they are in form of numbers. Such numbers constitute **"data"** and their common characteristic is variability or variation. In general, observations are denoted by a letter in the alphabet, like $x_i = x_1, x_2, x_3, \ldots, x_n$ - - where x_1, and x_n refer to the first and last term, respectively.

10.2.4 Variables

Characteristics which show variability, or variation, are called **variables, chance variables, or random variables.** A variable may be either quantitative or qualitative. A quantitative variable is one for which the resulting observations can be measured, e.g. age, weight, height, GPA, etc. Quantitative variables may be further classified as discrete or continuous.

Discrete variables (or observations) are integers, or count data, because they arise from counting, like the number of students in AGR 311 class, number of trees in the **ARBORETUM**, number of insects on a Khaya tree, etc.

Continuous variables are those for which all values in a range are possible.

Qualitative variables are those for which numerical measurements are not possible- - they are for classification purposes only. For example, the sex of a baby can be a boy or girl, eye color, skin color, chicken color, etc., are qualitative variables.

10.2.5 Experimental Error

A common characteristic of all experimental material (Steel and Torrie, 1981; Wahua, 1999) is variation, or variability. Experimental Error is a measure of the variation which exists among observations or experimental units treated alike.

Variation comes from three main sources:

(a) Inherent variability

This is the variability that exists in the experimental material to which treatments are applied - - like genetic constitution, non-uniformity in soil qualities, differences in heat, light intensity in cages, etc.

(b) Lack of Uniformity in the Application of Treatments

This variation results from lack of uniformity in the physical conduct of the experiment - - like any lack of uniformity either in the quantity, or distribution, or improper handling of experimental material can cause experimental error.

(c) Uncontrolled or Inadequately Controlled External Factors

Pests attack plants indiscriminately and we must take adequate care so that plants, or species, or cultivars, that are more susceptible to pests would contribute to the variation among materials originally treated like. Therefore, we need to handle experimental material so that the effects of inherent variability are reduced.

10.2.6 Why Control Experimental Error?

The greater the variability, the more difficult it is to obtain a reliable estimate of the average yields and estimate the effect of a given treatment accurately and also compare the effects of different treatments. Experimental error will mask true treatment effects, if it is not adequately controlled.

10.2.7 How to Control and Reduce Experimental Error

There are several methods for reducing experimental error. Mendenhall (1975) described such methods as **"noise-reducing"** experimental designs which reduce background noise, or variation, caused by uncontrolled nuisance variables. Reduction of noise can also be accomplished through the following methods:

10.2.7.1. Choice of Appropriate Experimental Design

The use of experimental design as a means of controlling experimental error has been widely investigated (Steel and Torrie, 1981). The control of experimental error consists of designing an experiment so that some of the natural variation among the set of experimental units is physically handled so as to contribute nothing to differences among treatment means.

Once the experimenter recognizes the sources of variation, choosing an appropriate design to remove those sources of variation becomes a matter of routine. Thus, the use of **Randomized Complete Block Design** reduces variation between blocks, while units within blocks are more homogeneous and therefore reduces experimental error.

Designs are available where the Complete Block is sub-divided into a number of Incomplete Blocks such that each incomplete block contains only a portion of the treatments- - so that the experimental error can be estimated among the units within the incomplete blocks. Such designs, called **Incomplete Block Designs** include the **Split-Plot Design.**

10.2.7.2. Choice of Appropriate Sizes and Shapes of Experimental Units (Plots)

Large experimental units, as a general rule, show less variation than small units. However, adequate replication of small plots is easier to obtain than large plots. The size and shape of the experimental unit or plot, as well as that of complete or incomplete block, are important to the precision of field experiments.

Relatively long and narrow individual plots in several studies have been shown (Steel and Torrie, 1981) to give the greatest precision, but extremely oblong plots increase variability within plots (Wahua, 1999). The block should be approximately square to maximize variation among blocks, while minimizing variation among plots within blocks.

For Greenhouse or Nursery experiments, the containers (or polypots) should be the same and appropriate size (usually 5kg of soil and above). The crop species, types of treatments and data to be collected should be carefully considered before choosing plot sizes and shapes.

Wahua (1999) suggested the following field plot sizes for various crops:

Rice	= 2m x 1.5m
Cassava	= 8m x 6m
Maize	= 6m x 5m
Cotton	= 12m x 8m
Water melon	= 10m x 10m
Cowpeas	= 3m x 3m

10.2.7.3. Proper Handling of Experimental Materials

Experimental materials should be properly handled in order to reduce inherent variability. Seeds for planting must be of the same physical condition, planted at the same depth and uniformly treated with pesticides or fungicides. Seed yams and other vegetatively propagated plants like cassava and sweet potatoes should be from the same portion and length of stem and be as uniform as possible. Animals should be of same sex, age, body weight and physiological condition to reduce variability.

10.2.7.4. The Use of Concomitant Information, or Observations

There are many experiments where precision may be increased by the use of accessory or concomitant observations, which are not controllable enough to be allocated an experimental unit. For example, it may be necessary to record the initial weight of goats before being fed different forage grasses because their initial body weights may affect their response to the diet. The use of Covariance Analysis can separate variability due to initial body weights from the treatment effects.

10.2.7.5. The Use of Refined Experimental Techniques

The experimenter must ensure that everything is done to improve accuracy and precision of an experiment because no statistical analysis can improve the data obtained from poorly performed experiments. Variation resulting from careless techniques is not random and therefore not subject to the laws of chance on which statistical inference is based.

Refinement of techniques includes:

- Uniformity in the application of treatments like spreading of fertilizer, filling of test-tubes or polypots, spraying of pesticides, etc.
- Control of external influences (biotic and abiotic factors)- - if a complete experiment cannot be set out, or harvested in one day, it is desirable to do complete blocks in one day.
- Data should be recorded in duplicates to avoid recopying of data for input.
- Suitable and unbiased measures of the effects of several treatments and using the right instrument for each measurement.
- Adequate supervision of assistants and close scrutiny of data will prevent gross errors which can occur in experimentation.

10.2.8 Replication and its Functions

Replication is the application of a treatment more than once in an experiment. The treatment appearing more than once in an experiment is said to be replicated and its functions are:

(i)To provide an estimate of experimental error.

(ii)To improve the precision of an experiment by reducing the standard deviation of a treatment mean.

(iii)To increase the scope of inference of the experiment by selection and appropriate use of more variable experimental units.

(iv)To effect control of the error variance.

10.2.9 Factors Affecting the Number of Replicates

The number of replications for an experiment depends on several factors:

(i)   The Degree of precision required. The smaller the precision required, the more the number of replicates required.
(ii) The more variable the experimental units are, the more replicates are needed and the more uniform the experimental units, the fewer the number of replicates required.
(iii) The number of treatments affects the precision of an experiment and hence the number of replications.
(iv) The experimental design affects precision and the required number of replications. Unfortunately, the number of replicates is very often determined by the funds and time available for the experiment.

10.2.10 Randomization

Randomization involves the use of some chance device like using a **Table of Random Numbers**, or flipping a coin, to ensure that each experimental unit has an equal chance of being included in the experiment.

The function of randomization is to ensure that we have a valid or unbiased estimate of experimental error, and of treatment means and differences among them. Systematic designs often result in either under-estimation, or over-estimation of experimental error. Randomization tends to destroy the correlation among errors and make valid the usual tests of significance.

10.2.11 Statistical Inference

The object of experiments is to draw conclusions and determine if there are real differences among our treatment means and to estimate the magnitude of such differences, if they exist. A statistical inference about such differences involves the assignment of a measure of probability to the inference.

10.3.1 Classification of Experimental Designs

Experimental Designs vary from the simple one-way model to the complex multi-way models. The classifications of experimental designs are many and varied. Experimental Designs are just methods of arranging treatments in order that their effects may be meaningfully tested.

(i)   **Single-Classification Anova** are also called **"One-way Anova", or "Cell-Means Model"**. This group of Anova evaluates the effect of one factor on the mean response.
(ii) **Two-way Classification Anova** evaluates the simultaneous effects of two factors on the mean response.
(iii) **Multi-way Classification Anova** evaluates the simultaneous effects of several factors on the mean response.

Other classifications also used in Anova include:

1. Model I Anova, or "Cell-Means Models" are used to denote "Fixed Effects Models", where the conclusions will pertain to just the factor levels included in the study.
2. Model II Anova, or "Variance Components Models", or "Random Effects Models", are those experimental designs whose conclusions will apply to a population of factor levels of which the levels in the study are a sample.
3. Model III Anova are also called "Mixed Models" which are a combination of both Models I and II, where some of the factors are fixed, while others are random.

10.3.2 Fundamental Assumptions of Analysis of Variance

Before applying the methodologies of Anova, a review of the fundamental assumptions on which Anova is based is necessary for valid statistical tests.

1. **Randomness**

   All Anovas require that the sampling of individuals be random, that is, each individual or treatment has an equal chance of being selected. Random selection of individuals is a necessary condition for the application of statistical tests. Therefore, randomization may overcome the problem of non-independence of errors.

2. **Independence**

   The error terms, ε_{ij}, are independent and identically distributed $(i.i.d.)$ i.e. $\varepsilon_{ij} \sim (0, \sigma^2)$. If $\varepsilon_{ij}'s$ are not independent, the validity of the usual F-tests of significance can be seriously impaired.

3. **Homogeneity of Variances.**

   Homogeneity of variances, or homoscedasticity which simply means equality of variances in a group of samples, is an important pre-condition for several statistical tests. The inequality of variances among samples is called heteroscedasticity. To test whether two samples have equal variances, or are homoscedastic, we use an F-test for the hypothesis:

   $$H_0: \sigma_1^2 = \sigma_2^2$$

   $$H_1: \sigma_1^2 \neq \sigma_2^2$$

   When there are more than two groups, Bartlett's Test for homogeneity of variances is frequently suggested.

4. Normality of Error Terms (ε_{ij}).

   The Error terms are normally distributed with mean zero and variance $\sigma^2 - (\varepsilon_{ij}) \sim N(0, \sigma^2)$.

5. **Additivity.**

   For the one-way classification, an observation is described as a linear model which is a sum of three components - - an overall population mean, μ, common to all treatments, a treatment effect, α_i, peculiar to each treatment and a random component from unknown sources of variation as:

   $$X_{ij} = \mu + \alpha_i + \varepsilon_{ij} \quad \begin{matrix} i = 1,2,3,...,k. \\ j = 1,2,3,...,n. \end{matrix}$$

   An observation = fixed overall population mean + Treatment effect + Random component. These

174

components of an observation are additive. There are two basic statistical requirements for a good experiment, namely Replication and Randomization:

6. **Replication** is the number of times that a treatment is applied to an experimental unit. Replication provides an estimate of variability; the more the number of units used for each treatment, the lower will be the standard errors of the estimates of the treatment effects and therefore the more accurate the experiment - - thus ensuring the validity of the conclusions drawn from the results of the experiment.
7. **Randomization** is the process of allocating treatments to experimental units so that each treatment has the same, or equal chance, or probability, of being allocated to a particular unit. Randomization produces a statistical analysis which is mathematically valid.

11 THE ONE-WAY (SINGLE-FACTOR) CLASSIFICATION: COMPLETELY RANDOMIZED DESIGN (CRD)

The Completely Randomized Design (CRD) is useful when the experimental units are considered to be approximately homogeneous (Hicks, 1975 and Steel and Torrie, 1981) where the variation among them is small.

The CRD is flexible because the number of treatments and replicates is only limited by the number of experimental units available, or by cost considerations. It is desirable to have equal number of observations, or experimental units, or replicates per treatment. However, statistical analysis is still simple, even if the number of replicates varies, or are missing, or of unequal variances.

Each experimental unit, X_{ij} (where (i) denotes the row and (j) the column of each observation) is made up of three components called the mathematical or structural model as:

$X_{ij} = \mu + \tau_j + \varepsilon_{ij}$, where:

X_{ij} = each random experimental unit (or observation) estimates:

μ = Constant (or fixed) unknown overall population mean.

τ_j = Fixed Treatment Effect of the jth population, subject to the restriction that:

$$\sum_{j=1}^{k} \tau_j$$

ε_{ij} = Random variation, or Experimental Error.

n_j = Equal for all the k-treatments.

A Sample Layout for One-Way Anova is shown in Table 11.1:

Table 11.1: Sample Layout for One-Way Anova

			Treatments			
	1	2	3 . .	J	. . k	
	X_{11}	X_{12}	X_{13} . .	X_{1j}	. . X_{1k}	
	X_{21}	X_{22}	X_{23} . .	X_{2j}	. . X_{2k}	
	:	:	: : :	:	: : :	
	X_{i1}	X_{i2}	X_{i3} . .	X_{ij}	. . X_{ik}	
	:	:	: : :	:	: : :	
	X_{n1}	X_{n2}	X_{n3}	X_{nj}	X_{nk}	
Totals	$T_{.1}$	$T_{.2}$	$T_{.3}$ . .	$T_{.j}$	. . $T_{.k}$	T.. = Grand Total;
Number	n_1	n_2	n_3 . .	n_j	. . n_k	N=Total observations.
Means	$\bar{X}_{.1}$	$\bar{X}_{.2}$	$\bar{X}_{.3}$	$\bar{X}_{.j}$	$\bar{X}_{.k}$	$\bar{X}_{..}$ = Grand Mean.

The use of **"dot notation"** indicates a summing over all observations in the population:

176

$T._j$ – represents the total of the observations taken under treatment j.

n_j – represents the number of observations taken for treatment j.

$\bar{X}._j$ – is the observed mean for treatment j.

$T..$ – represents the grand total of all observations taken, where:

$$T.. = \sum_{j=1}^{k} \sum_{i=1}^{nj} X_{ij} = \sum_{j=1}^{k} T._j$$

$$N = \sum_{j=1}^{k} n_j$$

$\bar{X}.. =$ the mean of all the N observations. $= \left[\sum_{j=1}^{k} n_j \bar{X}._j\right]/N$.

Random samples of size n are selected from each of the k-populations (see Layout). If it is assumed that such populations are independent and normally distributed with means $\mu_1, \mu_2, ..., \mu_k$ and the structural or mathematical model of the CRD is given by:

$X_{ij,} = \mu + \tau_j + \varepsilon_{ij}$, subject to the restriction that $\sum \tau_j = 0$

with a common variance, σ^2, and we wish to test the hypothesis that:

$H_0: \mu_1 = \mu_2 = ... = \mu_k = \mu$ i.e. the k-treatment population means are all equal and also equal to the overall population mean.

$H_1: \mu_1 \neq \mu_2 \neq ... \neq \mu_k \neq \mu$ i.e. the k-treatment population means are not all equal and also not equal to the overall population mean.

OR

H_1 : At least two of the k-treatment population means are not equal (or not all the means are equal).

An alternative way of testing the same hypothesis is to use the treatment effect of the jth population as:

$H_0: \tau_1 = \tau_2.. = ... = \tau_k = 0$ i.e. the treatment has no effect (Treatment Effect is zero).

$H_1: \tau_1 \neq \tau_2.. \neq ... \neq \tau_k \neq 0$ i.e. All the treatment effects are not equal to zero.

OR

H_1 : At least one of the τj is not equal to zero.

The test of this hypothesis will be based on three unbiased, independently estimated common population variance,. σ^2 These estimates will be obtained by splitting the total variability of our data into two components referred to as "the fundamental equation of analysis of variance":

$$\sum_{j=1}^{k}\sum_{i=1}^{nj}(X_{ij}-\bar{X}..)^2 = \sum_{j=1}^{k}\sum_{i=1}^{nj}(X_{j}-\bar{X}..)^2 + \sum_{j=1}^{k}\sum_{i=1}^{nj}(X_{ij}-\bar{X}_{j})^2$$

$$\text{Total SS Deviations} = \text{SS Deviations Between Treatment Means} + \text{SS Deviations within Treatments}$$

OR

$$SS_{Total} = SS_{Treatment} + SS_{Error}$$

Dividing each component by its corresponding degrees of freedom, we obtain the three unbiased independent estimates of the population variance, σ^2, if the null hypothesis is true:

$$\frac{SS_{Total}}{kn-1} = \frac{SS_{Treatment}}{k-1} + \frac{SS_{Error}}{k(n-1)} = MS_{Total} = MS_{Treatment} + MS_{Error}$$

The unbiased unknown population variance, σ^2, of all the observations grouped as a single sample of size nk is given by:

$$S_1^2 = \frac{\sum_{j=1}^{k}\sum_{i=1}^{nj}(X_{ij}-\bar{X}_{..})^2}{N-1} = \frac{SS_{Total}}{kn-1} = MS_{Total,}$$ which is based on kn-1 degrees of freedom.

The second independent unbiased estimate of the unknown population variance, σ2, based on k-1 degrees of freedom, is given by:

$$S_2^2 = \frac{\sum_{j=1}^{k}\sum_{i=1}^{nj}(\bar{X}_{j}-\bar{X}_{..})^2}{k-1} = \frac{SS_{Treatment}}{k-1} = MS_{Treatment}$$

If H0 is true, S_2^2 is an unbiased estimate of the unknown population variance, σ^2. However, if H1 is true, $MS_{Treatment}$ will have a larger numerical value and S_2^2 will over-estimate σ^2.

The third unbiased independent estimate of the unknown population variance, σ^2, based on k(n-1) degrees of freedom is given by:

$$S_3^2 = \frac{\sum_{j=1}^{k}\sum_{j=1}^{nj}(X_{ij}-\bar{X}_{j})^2}{k(n-1)} = \frac{SS_{Error}}{k(n-1)} = MS_{Treatment}$$

The estimate, S_3^2, is unbiased, regardless of the truth or falsity of the null hypothesis.

These computations are summarized in Anova Table 11.1:

Table 11.1: Anova Table for One-Way Classification with Equal Replications

Source of Variation	Degrees of freedom	Sum of squares Definition Formula	Sum of squares Working Formula	Mean Square	F-Ratio
Treatments (between Treatments)	k-1	$n_j \sum (\bar{X}_{.j} - \bar{X}_{..})^2$	$\sum_j^k \dfrac{T_j^2}{n_j} - \dfrac{T^2}{nk}$	$MS_T = \dfrac{SS_{Trt}}{k-1}$	$F = \dfrac{MS_T}{MSE}$
Error (Within Treatments)	k(n-1)	$\sum_{i=1}^{k} \sum_{i=1}^{n_i} (X_{ij} - \bar{X}_{.j})^2$	$\begin{aligned} & SS_{total} - SS_{Trt} \\ & = SS_{Error} \end{aligned}$	$MSE = \dfrac{SSE}{k(n-1)}$	
Total	kn-1	$\sum_{j=1}^{k} \sum_{i=1}^{nj} (X_{ij} - \bar{X}_{..})^2$	$\sum_{j=1}^{k} \sum_{i=1}^{nj} X_{ij}^2 - \dfrac{T_{..}^2}{nk}$	$\begin{aligned} & \dfrac{SS_{TOTAL}}{kn-1} = \\ & MS_{TOTAL} \end{aligned}$	

11.2.1 One-Way Classification (CRD) with Equal Replications (Equal Sample Sizes)

The 400-level students of the Department of Forestry and Wildlife, University of Benin, Benin-City, raised the tree-seedlings of *Tectona grandis, Gmelina arborea, Terminalia ivorensis and Khara grandifoliola* each in six plots, in order to compare the growth in the height of the seedlings over a period of six months and obtained the data in Table 11.2:

Table 11.2: Growth in Height of Four Tree Species in Six Months

Tree Species' Height Growth (mm)			
Tectona	**Gmelina**	**Terminalia**	**Khaya**
25.12	40.25	18.30	28.05
17.25	35.26	22.60	28.05
26.42	31.98	25.90	33.20
16.80	36.52	15.05	31.68
22.15	43.32	11.42	30.32
15.92	37.10	23.68	27.58
$\sum X_{ij} = 123.66$	224.43	116.95	178.88
$\bar{T}_{.j} = 20.61$	37.41	19.49	29.81
$\sum X_{ij}^2 = 2{,}652.9022$	8,473.5984	2,434.1213	5,359.4262

a) State the structural model and explain the terms in your model.
b) Test the hypothesis at $\alpha = 0.05$ that there is no difference in the early height growth rate of the four tree-species.
c) Test the hypothesis that the variances are equal at $\alpha = 0.05$, using Bartlett's test.

Assume that $b = 2.3026\dfrac{q}{h}$, where:

$$q = (N - k)\log S_p^2 - \sum_{j=1}^{k}(n_i - 1)\log S_i^2;$$

$$h = 1 + \frac{1}{3(k-1)}\left[\sum \frac{1}{n_{i-1}} - \frac{1}{N-K}\right]$$

d) Estimate the coefficient of variation and comment on it.

e) Using LSD, compare the mean height growth of the four tree-species to identify the species that are really significantly different. Summarize your results.

Before we can answer the questions above, we must understand the **"Procedure for Anova"**, treat additional topics like **"Comparison of Treatment Means"**, etc., as we solve the problem.

Procedure for a CRD Anova:

1. The very first step in Anova is to state the structural (or mathematical) model and explain the terms in the model. For a CRD, the structural model is:

a. $X_{IJ} = \mu + \alpha_j + \mathcal{E}_{ij}$ subject to the restriction that

$$\sum_{j=1}^{k}\alpha_j = 0$$

where:

X_{ij} = Random experimental unit which is independent and normally distributed with mean, μ, and variance, $\sigma^2, i.e. X_{ij} \sim N(\mu, \sigma^2)$

μ = Constant, or Fixed overall population mean which is independent and normally distributed.

α_j = Fixed treatment effect which is normally distributed with mean zero and variance, σ^2 subject to the restriction that $\sum \alpha_j = 0$

$\mathcal{E}_{ij}$ = Random component which is independent, depending upon uncontrolled sources of variation and normally distributed with mean zero and variance, σ^2 i.e $\varepsilon_{ij} \sim N(0, \sigma^2)$ for all treatments;

$\left(\mu + \alpha_j + \varepsilon_{ij}\right)$ implies that the components of the model are additive

2. State the Null and Alternative Hypotheses as:

A. $H_0: \mu_1 = \mu_2 = \ldots = \mu_k = \mu$. i.e. the treatment means are all equal to the common population mean.

$H_1: \mu_1 \neq \mu_2 \neq \ldots \neq \mu_k \neq \mu$. the treatment means are not all equal. Or, at least one is different.

B. An Equivalent hypothesis is:

$H_0: \alpha_1 = \alpha_2 = \ldots = \alpha_k = 0$ i.e. Treatment Effect is zero; or the Treatment has no effect on tree height growth.

$H_1: \alpha_1 \neq \alpha_2 \neq \ldots \neq \alpha_k \neq 0$ i.e. Treatment has an effect; or Treatment has some effect on tree height growth.

Note: The Null hypothesis can only be stated as A or B, not both; because A and B test the same thing.

3. State the significance level: $\alpha = 0.05 \ or \ 0.01$
4. Decision Rule: (a) if F-calculated is greater than F-tabulated, reject H_0 and conclude that F is significant. (b) If F-calculated is not greater than F-tabulated, accept H_0.
5. Computations:
   i. Grand Total:

$$\sum_{j=1}^{k}\sum_{i=1}^{ni} X_{ij} = 25.12 + 17.25 + \cdots 30.32 + 27.58 = 643.92$$

   ii. Correction Factor (CF):

$$CF = \left(\sum_{j=1}^{k}\sum_{i=1}^{ni} X_{ij}\right)^2 / \sum nj = (643.92)^2/24 = 44{,}632.9664/24 = 17{,}276.3736$$

   iii. Sum of Squares Total (SSTotal) $=. \sum_{j=1}^{k}\sum_{i=1}^{ni} X_{ij}^2 - CF \ \sum_{j=1}^{k}\sum_{i=1}^{nj} X_{ij}^2 - CF$

$$= (25.12)2 + (17.25)2 + \ldots + (30.32)2 + (27.58)2 - 17.276.3736.$$

$$= 18{,}191.243 - 17{,}276.3736 = 1{,}642.8694$$

   iv. Sum of Squares Treatment Total divided the Treatment Sample size:

$$SS_{TREATMENT} = \sum_{J=1}^{k} \frac{T_{.j}^2}{n_j}$$

$$= \left[\frac{(213.66)^2}{6} + \frac{(224.43)^2}{6} + \frac{(116.95)^2}{6} + \frac{(178.88)^2}{6}\right] - CF$$

$$= \frac{111{,}335.9774}{6} - 17{,}276.3736 = 18{,}555.9962 - 17{,}276.3736$$

$$= 1{,}279.6226.$$

   v. Sum of Square Error $= SS_{Total} - SS_{Treatment}$

$$SS_{Error} = 1{,}642.8694 - 1{,}279.6226 = 363.24677$$

   Note: (iii) > (iv) > (v) always.

a) Vi Table 11.2.1: Anova Table:

Source of variation	DF	SS	MS	F_{cal}	$F_{0.05}; F_{0.01}$
Treatments (Between Treatments)	$k - 1 = 3$	1,279.62 26	1,279.622/3	$\dfrac{426.5409}{18.1623} =$ 23.485***	3.10 4.94
Error (Within Treatments)	$k(n - 1) = 20$	363.2468	363.2468/20		
Total	$kn - 1 = 23$	1,642.86 94			

Vii Conclusion (Decision):

Since, $F_{cal} = 23.49 >>> F_{0.5,3,20} = 3.10$ and $F_{0.01,3,20} = 4.49$, reject Ho and conclude that the early mean height growth rate among the four tree-species are highly significantly different.

(c) To test the hypothesis that the variances are equal, we must discuss the topic: **"Test for the Homogeneity (Equality) of several variances using Bartlett's Test"**.

11.2.1.1 Test for the Homogeneity (Equality) of Several Variances Using Bartlett's Test)

One of the assumptions of Anova is that the variances among treatments are equal or homogeneous. To confirm whether the variances are equal or not, Bartlett's Test must be carried out (Snedecor and Cochran, 1967; Dixon and Massey, 1969; Sokal and Rohlf, 1969; Alika, 1997; Walpole, 1974).

The Anova procedure is insensitive to departures from the assumption of equal variances for the k-populations, when the sample sizes are equal. This is not the case for unequal sample sizes. Consequently, when an experiment results in unequal numbers of observations in the various samples, one may wish to test the hypothesis:

$H_0: \sigma_1^2 = \sigma_2^2 = \ldots = \sigma_k^2 = \sigma^2$. i.e. The Treatment variances are all equal to the population variance.

$H_1: \sigma_1^2 \neq \sigma_2^2 \neq \ldots \neq \sigma_k^2 \neq \sigma^2$ .i.e. The variances are not all equal.

Bartlett's Test:

This test is based on a statistic whose sampling distribution is very closely approximated by the Chi-Squared distribution when the random samples are drawn from independent normal populations.

Procedure for Bartlett's Test:

1. First compute the k-sample variances from the data in Table 11.2 as:

$$S_j^2 = \left[\sum_{j=i}^{nj} X_{ij}^2 - \frac{(\sum X_j)^2}{n_j} \right] / (n_j - 1), j = 1,2,3, \ldots k$$

$$S_1^2 = \left[2,652.9022 - \frac{(123.66)^2}{6} \right] / 5 = \frac{2,652.9022 - 2,548.6326}{5}$$

$$S_1^2 = \frac{104,2696}{5} = 20.85392$$

$$S_2^2 = \left[8,473.5984 - \frac{(224.43)^2}{6} \right] / 5 = \frac{8,473.598 - 8,394.8072}{5}$$

182

$$= \frac{78.79425}{5} = 15.75885$$

$$S_3^2 = \frac{\left[2{,}434.1213 - \frac{(166.95)^2}{6}\right]}{5} = \frac{2{,}434.1213 - 2279.5504}{5} = \frac{154.5708}{5} = 30.91418$$

$$S_4^2 = \frac{\left[5{,}539.4262 - \frac{(178.88)^2}{6}\right]}{5} = \frac{5359.4262 - 5{,}333.009067}{5}$$

$$= \frac{26.4171333}{5} = 5.283427$$

2. **Estimate the pooled variance $\left(S_p^2\right)$ as:**

   Combine the sample variances to give the pooled variance, $\left(S_p^2\right)$ as:

   $$\left(S_p^2\right) = \frac{\sum(n_j - 1)\left(S_j^2\right)}{N - k}$$

   $$= \frac{5(20.85392) + 5(15.75885) + 5(30.91418) + 5(5.283427)}{24 - 4}$$

   $$\frac{5(72.81037367)}{20} = \frac{364.0518683}{20} = 18.20259342$$

   *Instead of going through all these computations, just use the value of MSE from the Anova Table.

3. **Use the Bartlett's Equation:**

   $$b = 2.3026\frac{q}{h}, where:$$

   $$q = (N - k)\log S_p^2 - \sum(n_j - 1)\log S_j^2$$

   $$h = 1 + \frac{1}{3(k - 1)}\left[\sum\frac{1}{(n_j - 1)} - \frac{1}{(N - k)}\right]$$

   q = (24 - 4)log18.20259342 − (5log20.85392 + 5log15.75885 + 5log30.91418 + 5log5.283427)

   = 20(1.260133267) − 5(1.319187703 + 1.197524522 + 1.490157685 + 0.722915684)

   = 25.20266534 − 5(4.729785594) = 25.20266534 − 23.64892797

   = 1.55373737

   $$h = 1 + \frac{1}{3(3)}\left[\frac{1}{5} + \frac{1}{5} + \frac{1}{5} + \frac{1}{5} - \frac{1}{20}\right] = 1 + \frac{1}{9}\left[\frac{4}{5} - \frac{1}{20}\right] = 1 + 0.111111111\,(0.8 - 0.05)$$

   = 1 + (0.111111111)(0.75) = 1 + 0.083333333

$$= 1.083333333$$

$$b = 2.3026 \left(\frac{1.55373737}{1.083333333} \right) = 2.3026 \,(1.43421911)$$

$$= 3.302432925 \approx 3.302$$

$$\chi^2_{.05,3} = 7.81$$

Conclusion:

Since $\chi^2_{cal} = 3.302 < \chi^2_{0.05,3} = 7.81$, accept H_0 and conclude that the treatment variances are equal or homogeneous. Therefore, no assumption of Anova was violated.

d) Coefficient of variation (CV) $= \dfrac{S}{\bar{x}} \times \dfrac{100}{1} \%$

where: s2=MSE;

$$\frac{\sqrt{MSE}}{T} \times \frac{100}{1} \%$$

$$\bar{T}.. = \frac{Grand\ Total}{N} = \frac{643.92}{24} = 26.83$$

$$CV = \frac{\sqrt{MSE}}{26.83} = \frac{\sqrt{18.1623}}{26.83} = \frac{4.261728}{26.83} \times \frac{100}{1} \%$$

$$= 0.15884 \times 100\% = 15.88\%$$

Comment:

A CV of 15.88% implies that variability is moderate and the data are fairly stable. Ideally, CV should be less than 10%. The higher the CV is, the more unreliable and unstable the data set is.

e) Before we can compare treatment means, we need to discuss some other topics in Anova

like— **"After Anova Tests on Means".**

11.2.2 One-Way Classification (CRD) with Unequal Replications (Unequal Sample Sizes)

A numerical example:

A silviculturist in Uniben, Benin-City, compared the effectiveness of three organic fertilizers – **Compost**, **Cow** Dung and **Chicken Droppings**- - on the height growth of six-month old Tectona grandis seedlings. The data in Table 11.2.2 (in millimeters) were obtained:

Table 11.2.2: Height Growth of 6-month old Teak Seedlings (in mm)

Compost	Cow Dung	Chicken Droppings
48.6	68.0	67.5
49.4	67.0	62.5
50.1	70.1	64.2
49.8	64.5	62.5
50.6	68.0	63.9
50.8	68.3	64.8
47.1	71.9	62.3
52.5	71.5	61.4
49.0	69.9	67.4
46.7	68.9	65.4
-	67.8	63.2
-	68.9	61.2
-	-	60.5
$T_{.j} = \sum X_{ij} = 494.6$	824.8	826.8
$n_1 = 10$	$n_2 = 12$	$n_3 = 13$

i.   State the mathematical model of this design.
ii. Test the hypothesis that the fertilizers have no effect on the height growth of Teak seedlings. Use $\alpha = 0.05$ and $\alpha = 0.01$ and state your conclusions clearly.
iii. Using LSD at $\alpha = 0.05$, identify which fertilizers are really significantly different.
iv. Calculate the coefficient of variation (CV).

   i.   Mathematical model:

$$X_{ij} = \mu + \alpha_j + \epsilon_{ij}$$ where:

X_{ij} = Random, independent and normally distributed experimental unit with mean, μ, and variance, σ^2 i.e. $X_{ij} \sim N(\mu, \sigma^2)$.

μ = Fixed overall population mean, which is independent and normally distributed.

σ_j = Fixed treatment effect which is normally distributed with mean zero and variance, σ^2, subject to the restriction, $\sum \alpha j = 0$

ε_{ij} = Random component which is independent, depending upon uncontrolled sources of variation, and normally distributed with mean zero and variance, σ^2. i.e. $\varepsilon_{ij} \sim N(0, \sigma^2)$.

$(\mu + \alpha j + \varepsilon_{ij})$ implies that all the components of the model are additive.

   ii. State the Null and Alternative hypotheses:

$H_0 : \alpha_1 = \alpha_2 = \alpha_3 = 0.$ i.e. The Treatments have no effect on seedling height growth.

$H_1 : \alpha_1 \neq \alpha_2 \neq \alpha_3 \neq 0$ i.e. The Treatments have some effect on seedling height growth, or Not all Treatment Effects are zero.

Significance level $\alpha = 0.05$ and 0.01

Decision Rule:

If F-calculated is greater than F-tabulated, reject H0 and conclude that the fertilizers have a significant effect on the height growth of Teak seedlings.

If F-calculated is less, or smaller than F-tabulated, accept H0 that the treatments have no effect on the height growth of Teak seedlings.

Computations:

Grand Total (G) $= \sum_{j-1}^{k} \sum_{i-1}^{nj} X_{ij} = 48.6 + 49.4 + \cdots + 61.2 + 60.5 = 2146.2$

$N = n_1 + n_2 + n_3 = 10 + 12 + 13 = 35$

Correction Factor (CF) $= \dfrac{G^2}{N} = \dfrac{(2146.2)^2}{35} = 4{,}606{,}174.44$

$= 131{,}604.984$

$$SS_{Total} = \sum_{j=i}^{k} \sum_{i=1}^{nj} X_{ij}^2 - CF$$

$= (48.6)^2 + (49.4)^2 + \cdots (61.2)^2 (48.6)^2 + (60.5)^2 - 131{,}604.984$

$= 133{,}868.94 - 131{,}604.984 = 2{,}263.956$

$$SS_{Treatment} = \dfrac{\sum_{j=i}^{k} T_j^2}{n_j} - CF = \dfrac{(494.6)^2}{10} + \dfrac{(824.6)^2}{12} + \cdots \dfrac{(826.8)^2}{13} - 131{,}604.984$$

$= 24{,}462.916 + 56{,}691.253 + 52{,}584.48 - 131{,}604.984.$

$= 133{,}738.64 - 131{,}604.98 = 2{,}133.66$

e. $SSError = SS_{Total} - SS_{Treatment}$

$= 2{,}263.956 - 2{,}133.66 = 130.296$

V Anova Table

Sources of variation	DF	SS	MS	F_{cal}	$F_{0.05}$	$F_{0.01}$
Treatments	k-1 = 2	2,133.66	1,066.83	262.01***	3.32	5.39
Error	N − k = 32	130.296	4.07175			
Total	**N − 1 = 34**	**2,263.956**				

Conclusion: Since F-cal = 262.01 >>>Ftab = 3.32 and 5.39 at $\alpha = 0.05$ and 0.01 respectively, we reject H0 and conclude that the fertilizers have a highly significant effect on the height growth of Teak seedlings.

vi) Coefficient of Variation (CV) $= \frac{S}{G} \times \frac{100}{1} \% = \frac{\sqrt{MSE}}{G} \times \frac{100}{1} \%$

$$\bar{G} = \frac{Grand\ total}{N} = \frac{2146.2}{35}$$

$$= 61.32$$

$$CV = \frac{\sqrt{4.07175}}{61.32} \times \frac{100}{1} = \frac{2.017857775}{61.32} \times \frac{100}{1} = 0.032907 \times \frac{100}{1} \%$$

$$= 3.29\% \approx 3.3\%$$

Note: It is customary to place one asterisk (*) on the calculated F-value, if it is only significant at $\alpha = 0.05$; if the calculated F-value is also significant at $\alpha = 0.01$, then two asterisks (**) are placed in front of the calculated F-value. However, if the F-calculated is not significant at any level, it is customary to place ns (not significant) in front of the calculated F-value.

11.3.0 The Linear Additive Model

In Section 10.3.1., we observed that Experimental Designs are just methods of arranging treatments so that their effects may be meaningfully tested. We now want to take a closer look at an experimental unit, or an observation, in the one-way classification also known as Model I.

11.3.1 Model I Anova (Fixed Effects Model, or Fixed Constants Model, or Cell-Means Model)

In Model I Anova, the structural, or mathematical model is given by:

$X_{ij} = \mu + \tau_j + \varepsilon_{ij}$, where:

X_{ij} = An experimental unit, or observation, measured on a randomly selected element i in treatment population j.

μ = Grand mean of treatment populations.

τ_j = Effect of treatment j, subject to the restriction, $\sum \tau_j = 0$.

ε_{ij} = Experimental error.

The term, μ, is constant for all measurements in all treatment populations.

The effect, τ_j, is constant for all measurements within population j; however, a different value, say τ_j' is associated with population j', where j', represents some treatment other than j.

The experimental error, ε_{ij}, represents all uncontrolled sources of variance affecting individual measurements; this effect is unique for each of the elements i in the basic population. This effect is further assumed to be independent of τ_j .

Since both μ and τ_j are constant for all measurements within population j, the only source of variance for these measurements is that due to experimental error.

Assume the population of available treatments is K and a sample of k is selected. If k = K, then the observed data include all available treatments in the domain to which inferences are to be made. Then, we say that the treatments are fixed because, if the experiment is repeated, all the treatments will still be included in the sample i.e. k = K.

When k = K, the grand mean, μ is:

$$\mu = \frac{\sum_j^k \mu_j}{k}$$

The effect of treatment j, designated as τ_j, is the difference between the mean for treatment j and the grand mean of the population of means:

$$\tau_j = \mu_j - \mu$$

Thus, τ_j is a parameter which measures the degree to which the mean for treatment j differs from the means of all other relevant population means.

$$\sigma_\tau^2 = \frac{\sum_{j=1}^k \tau_j^2}{k-1} = \frac{\sum(\mu_j - \mu)^2}{k-1}$$

An equivalent definition in terms of differences between treatment effects is:

$$\sigma_\tau^2 = \frac{\sum(\tau_j - \tau_{j\prime})^2}{k(k-1)} = \frac{\sum(\mu_j - \mu_{j\prime})^2}{k(k-1)},$$

when the treatment effects are equal, i.e. when $\tau 1 = \tau 2 \ldots = \tau k$, $\sigma_\tau^2 = 0.$ The larger the differences between the τ's, the larger will be σ_τ^2. Thus, the hypothesis specifying that $\sigma_\tau^2 = 0$ is equivalent to the hypothesis that specifies:

$$\tau_1 = \tau_2 = \cdots = \tau_k, \text{ or } \mu_1 = \mu_2 = \cdots = \mu_k.$$

The εij's within population j is assumed to be approximately normal in form with $E(\mu_{\varepsilon j}) = 0$ and variance, $\sigma_{\varepsilon j}^2$.

If sources of experimental error are comparable in each of the treatment populations, then:

$$\sigma_{\varepsilon 1}^2 = \sigma_{\varepsilon 2}^2 = \ldots = \sigma_{\varepsilon_k}^2 = \sigma^2$$

where σ_ε^2 is the variance due to experimental error within any of the treatment populations.

11.3.2 Assumptions of Model I (or Fixed-Effects Model, or Fixed-Constants Model, or Cell-Means Model)

$$X_{ij} = \mu + \tau_j + \varepsilon_{ij}, \textbf{subject to } \sum \tau_j = 0.$$

A measurement, X_{ij}, is expressed as the sum of these components:

1. A component μ which is constant for all treatments and all elements.
2. A component τ_j which is constant, systematic, or fixed for all elements within a treatment population, but may differ for different treatment populations.
3. A component ε_{ij}, independent of τj, and distributed as $N(0, \sigma_\varepsilon^2)$ within each treatment population – i.e. a random component depending upon uncontrolled sources of variances or variation.

$$\sigma_\varepsilon^2$$

4. The population variances are all equal to   and the best estimate of this parameter is the pooled within-class sample variance:

$$S^2_{pooled} = \frac{\sum_{j=1}^{k} S_j^2}{k} = \text{Mean Square Error, } \sigma_\varepsilon^2 .$$

5. All the treatments about which inferences are to be made are included in the experiment.
6. If the experiment were to be replicated, the same set of treatments would be included in each of the replications.

When $\tau_{1_2} = \tau_{2_1} = \cdots = \tau_k$, $\sigma_\tau^2 = 0$ and hence the expected value of

$MS_{Treatment}$ $i.e. E(MS_{TRT}) = n\sigma_\tau^2 + \sigma_\varepsilon^2$

This implies that MSTRT is an unbiased estimate of the variance due to experimental error when there are no differences among treatment affects.

Therefore:

$$F = \frac{E(MS_{TRT})}{(MS_{ERROR})} = \frac{n\sigma_\tau^2 + \sigma_\varepsilon^2}{\sigma_\varepsilon^2} \text{ estimate of } \sigma_\varepsilon^2 \text{ Hence :} \frac{E(MS_{TRT})}{(MS_{ERROR})} = \frac{0 + \sigma_\varepsilon^2}{\sigma_\varepsilon^2} = 1.0$$

When $\sigma = 0$, the $E(MS_{TRT}) = \sigma_\varepsilon^2$. Thus the numerator and denominator are unbiased, independent

$$F = \frac{E(MS_{TRT})}{E(MS_{ERROR})} = \frac{n\sigma_\tau^2 + \sigma_\varepsilon^2}{\sigma_s^2} \text{ estimates of } \sigma_\varepsilon^2 \text{ Hence } \frac{E(MS_{TRT})}{E(MS_{ERROR})} = \frac{0 + \sigma_\varepsilon^2}{\sigma_\varepsilon^2} = 1.0.$$

When $\sigma_\tau^2 = 0$, $E(MS_{TRT}) = \sigma_\varepsilon^2$. thus the numerator and denominator are unbiased, independent

When $\sigma_\tau^2 \neq 0$, the expected value of the F-statistic will be greater than 1.0 by an amount which depends, in part, upon the magnitude of σ_τ^2. Thus, if the F-ratio is larger than 1.0 by an amount having a low probability, when $\sigma^2 = 0$, the inference is that $\sigma_\tau^2 \neq 0$. So that:

$$E(MS_{TRT}) = \sigma_\varepsilon^2 + \frac{n \sum \tau_j^2}{k - 1}.$$

11.3.3 Model II Anova for Single-Factor Experiments (Random Effects Model, or Variance-Components Model)

The Model II Anova for the Single-factor experiments is also known as the **"Random Effects Model"**, or the **Variance-Components Model"**.

Models serve as guides in formalizing statistical bases of data analysis and they are also useful tools in guiding test procedures (Winer, 1971).

The structural model for Model II has the same form as that of Model I:

$$X_{ij} = \mu + \tau_i + \varepsilon_{ij} \text{ or: } X_{ij} = \mu + \alpha_j + \varepsilon_{ij}. + \text{ where } \tau_j \text{ and } \alpha_j \text{ are fixed. However,}$$

189

Snedecor and Cochran (1967) used a slightly different structural model to differentiate Model II from Model I as:

Model II: $X_{ij} = \mu + A_j + \varepsilon_{ij}$, where: τ_1, α_j are fixed while A_j and ε_{ij} are random:

$$A_j = \hat{N}(0, \ \sigma_A^2); = \varepsilon_{ij} = N(0, \sigma^2) \ \ and \ \ i = 1,2, \ldots, n_j; j = 1,2, \ldots, k.$$

Model I is usually the most appropriate for a single-factor experiment. If, however, the k treatments that are included in a given experiment constitute a random sample from a collection of K treatments, where k is small relative to K, then upon replication, a different random sample of k-treatments will be included in the experiment. This constitutes a Model II.

In Variance-Components Model (Model II), the primary objective is to estimate σ_τ^2 or σ_A^2, rather than the individual τ_j treatment effects. The computational procedures and test of the hypothesis, $\sigma_\tau^2 = 0$ are identical with that of Model I.

11.3.3.1 Assumptions of Model II

1. μ is still assumed to be constant for all observations.
2. The term, ε_{ij}, is still assumed to have the distribution $N(0, \sigma_\varepsilon^2)$ for all treatments, and is a random component, depending upon uncontrolled sources of variance.
3. The component, τ_j (or A_j), is now considered to be a random variable. The distribution of τ_j is now assumed to be $N(0, \sigma_\tau^2)$. In variance components model, the primary objective is to estimate (σ_τ^2), rather than the individual τ_j treatment effects.
4. The error terms, ε_{ij}, are independent and normally distributed $N(0, \sigma_\varepsilon^2)$.
5. The treatment means, μ_j, are independent and normally distributed i.e. $\mu_j \sim N(\mu, \sigma_\varepsilon^2)$.

11.3.4 Model III Anova (or Mixed Model)

A Model III Anova is called a Mixed Model which must have at least two factors - - one of which is a fixed factor, while the other factor is random.

The structural model is given as:

$$X_{ijk} = \mu + \alpha_i + \beta_j + \alpha\beta_{ij} + \varepsilon_{ijk}.$$

Assumptions of Model III

1. μ is constant.
2. α_i is constant
3. β_j is random and $N(0, \sigma_\beta^2)$.
4. $\alpha\beta_{ij}$ is the interaction which is random and $N(0, \sigma_{\times\beta}^2)$.
5. ε_{ijk} is random and $N(0, \sigma_\varepsilon^2)$

Model III combines the assumptions of Models I and II.

11.4.0 MULTIPLE COMPARISONS

The Anova is only the first step in studying results. If the null hypothesis is rejected in Anova, the investigator still does not know which of the treatment means are really different. Therefore, the next step in the analysis is to examine the treatment (or class, or group) means and the sizes of the differences among them. This leads us to the subject of comparisons among pairs and groups of means or totals called multiple comparisons.

"Pair Comparison" is the simplest and most commonly used comparison in agricultural research. There are two types of pair comparisons:

11.4.1 Planned Pair Comparison ("A-Priori Tests")

An important point about planned or "a-priori tests" is that they are designed and chosen independently of the results of the Anova of the experiment – i.e. whether Anova test is significant or not. They should be planned before the experiment has been actually carried out and the results obtained. Such comparisons are called **"Planned", or "A- Priori" Comparisons.**

For k-treatments, the sum of the separate "a-priori" tests should not exceed k-1 i.e., the number of treatments minus one. In addition, it is desirable to structure the tests in such a way that each test tests an independent relationship among the means.

One concern when making many comparisons in a single experiment is whether significant differences obtained are due to real differences in the functions being compared, or simply due to the very large number of comparisons being made, which increases the chance of finding differences that appear significant. For example, in conducting 25 independent tests in an experiment, and finding one significant difference at 0.05 level, one should not put too much faith in the result because we should expect to find (0.05)(25) = 1.25 differences just by chance alone.

The most commonly used "Planned" or "A-Priori" test procedure for pair comparison is **the Least Significant Difference (LSD) Test:**

The Least Significant Difference Test (LSD)

The LSD test, according to Cochran and Cox (1957), is basically a two-sample t-test which provides a single LSD value, at a prescribed level of significance, which serves as the boundary between significant and non-significant differences between any pair of treatment means, i.e. two treatment means or totals are declared to be significantly different at a prescribed level of significance, if their difference exceeds the computed LSD value; otherwise, the two means or totals are not significantly different.

The LSD test is most appropriate for making "planned pair comparisons", but not valid for comparing all possible pairs of means, especially when the number of treatments is large (less than six treatments are preferred (Gomez and Gomez, 1984).

a) Equal Sample Sizes LSDα:

$$LSD_\alpha = t^*_{\frac{\alpha}{2}}(v) \times S_{\bar{Y}_i - \bar{Y}_j}$$

*Note: t-value $df\ (v)\ is\ the\ same\ for\ MSE$

$$LSD\alpha = t^*_{\alpha/2}(v) \times S\sqrt{\frac{2}{r}}$$ — for equal r, or equal sample sizes:

where:

$$S = \sqrt{MSE}$$

r = number of replications within the two treatments.

If the number of replications within the two treatments is different $i.e.\ (r_1 \neq r_2)$, or $(n_1 \neq n_2)$, then:

Unequal Sample Sizes LSDα:

$$LSD\alpha = t_{\alpha/2}(v) \times S \sqrt{\frac{1}{r_i} + \frac{1}{r_j}} = t_{\alpha/2}(v) \sqrt{MSE\left(\frac{1}{r_i} + \frac{1}{r_j}\right)}$$

$$LSD_{0.05} = t_{0.025}(v) \times \sqrt{MSE\left(\frac{1}{r_i} + \frac{1}{r_j}\right)}$$

$$LSD_{0.01} = t_{0.025}(v) \times \sqrt{MSE\left(\frac{1}{r_i} + \frac{1}{r_j}\right)}$$

To avoid multiple and tedious calculations because of different sample sizes, it is advisable to use the harmonic mean $(\bar{n})$ of the different sample sizes.

Estimate Harmonic Mean as: $\bar{n} = r \left(\frac{1}{n_1} + \frac{1}{n_2} + \dots + \frac{1}{n_r}\right)^{-1}$

where: r = t = number of treatments or replications.

11.4.1.1 Procedure for Applying the LSD to Compare two Treatments $(i^{th}$ and $j^{th})$:

1. Compute the absolute mean differences between the ith and jth treatments as:

   $d_{ij} = |\bar{T}_i - \bar{T}_j|$, where $\bar{T}_i$ and $\bar{T}_j$ are the means of the ith and jth treatments. Using the data in examples 11.2.1 and 11.2.2:

Example 11.2 for Equal Sample Sizes:

Treatment Means:

$$\bar{T}_1 = 20.61: \bar{T}_2 = 37.41; \bar{T}_3 = 19.49; \bar{T}_4 = 29.81: MSE = 18.1623. n_j = 6; df = 20.$$

2. Compute LSD value at α - level of significance from Tables 11.2.1 and 11.2.2 as:

$$LSD_{0.05} = t_{0.025}(20)\sqrt{\frac{2MSE}{6}}$$

$$= (2.086)\sqrt{\frac{2(18.1623)}{6}} = (2.086)\sqrt{\frac{36.3246}{6}} = (2.086)\sqrt{\frac{6.0541}{6}}$$

$$(2.086)(2.460508078) = 5.1326$$

$$= 5.13.$$

Compute absolute mean differences (/dij/) as:

$d_{ij} = |\bar{T}_1 - \bar{T}_2| = 20.61 - 37.41| = 16.8 > LSD = 5.13$ - - Significant.

$|\bar{T}_1 - \bar{T}_3| = |20.61 - 19.49| = 1.12 < SD$ - -not significant.

$|\bar{T}_1 - \bar{T}_4| = |20.61 - 29.81| = 9.2 > LSD$ - - significant.

$|\bar{T}_2 - \bar{T}_3| = |37.41 - 19.49| = 17.92 > LSD$ - - significant.

$|\bar{T}_2 - \bar{T}_4| = |37.41 - 29.81| = 7.6 > LSD$ - - significant.

- $|\bar{T}_2 - \bar{T}_4| = |19.49 - 19.49| = 10.32 > LSD$ - - significant.

Compare the absolute mean differences in (2) above with the LSD value computed in (2) and declare the ith and jth treatments to be significantly different at the α-level of significance, if the absolute value of $|d_{ij}|$ is greater than the LSD value, otherwise, it is not significantly different as showed in (2).

Summarize the results of the comparisons:

Arrange the Treatment Means in increasing order of magnitude:

$$\bar{T}_3 \quad \bar{T}_1 \quad \bar{T}_4 \quad \bar{T}_2$$

$$19.5 \quad 20.6 \quad 29.8 \quad 37.4$$

Underscore any pair of Treatment Means that are not significantly different. Here, only treatments 1 and 2 are not different.

i. Example 11.2.2 unequal sample sizes:
   a) Since the replications or sample sizes of the treatments are unequal, estimate the Harmonic Mean $(\bar{n})$ as:

$$\bar{n} = t\left(\frac{1}{n_1} + \frac{1}{n_2} + \frac{1}{n_3}\right)^{-1}$$

where t = no. of treatments.

nj = treatment sample sizes $(n_1 = 10; n_2 \neq 12; n_3 \neq 13)$.

$$= 3\left(\frac{1}{10} + \frac{1}{12} + \frac{1}{13}\right)^{-1}$$

$$= 3(0.1 + 0.08333 + 0.7692)^{-1} = 3(0.26025641)^{-1}$$

$$= 3(3.842364532) = 11.5270936$$

$$\approx 11.5$$

   b) Compute the LSD as: $LSD_{0.05} = t_{0.025}(32)\sqrt{\frac{2MSE}{\bar{n}}}$

where: MSE = 4.07175; df = 32; $\bar{n} = 11.5$; $t_{0.025,(32)} = 1.960$.

$$LSD_{0.05} = (1.960)\left(\sqrt{\frac{8.1435}{11.5}}\right) = (1.960)(\sqrt{0.78130434})$$

$$= (1.960)(0.841544863) = 1.649349532 = 1.649 \approx 1.6.$$

Compute the absolute mean differences between the ith and jth treatments as:

$d_{ij} = |\bar{T}_i - \bar{T}_j|$. Using the data in Example 11.2.2:

$\bar{T}_1 = 49.5; \bar{T}_2 = 68.7; \bar{T}_3 = 63.6$

$d_{ij} = |\bar{T}_1 - \bar{T}_2| = |49.5 - 68.7| = 19.2$

$|\bar{T}_1 - \bar{T}_3| = |49.5 - 63.6| = 14.1$

$$|\bar{T}_2 - \bar{T}_3| = |68.7 - 63.6| = 5.1$$

Comparing d_{ij} with LSD:

$$\bar{T}_1 - \bar{T}_2 = 19.2 > 1.6 \text{ -- significant}$$

$$\bar{T}_2 - \bar{T}_3 = 141.1 > 1.6 \text{ -- significant}$$

$$\bar{T}_2 - \bar{T}_3 = 5.1 > 1.6 \text{ -- significant}$$

Arrange the Treatment Means in ascending order of magnitude:

$\bar{T}_1 \quad \bar{T}_2 \quad \bar{T}_3$ -- the three means are all significantly different from one another.

49.5 63.6 68.7

Since all the means are significantly different, we do not underscore any pair of means.

11.4.2 Contrasts

A contrast, or a comparison, between two treatment totals or means is by definition, **the difference between the two totals or means, with appropriate algebraic sign.** For example:

$\bar{T}_1 - \bar{T}_2$ defines a comparison or contrast between the Totals of treatment 1 and treatment 2.

$\bar{T}_1 - \bar{T}_2$ defines a contrast between the Means of treatment 1 and treatment 2.

Comparisons among three treatment means can be made in different ways:

$\bar{T}_1 - \bar{T}_2 \, ; \, \bar{T}_1 - \bar{T}_3; \, \bar{T}_2 - \bar{T}_3;$ are contrasts.

$\frac{T_1 + T_2}{2} - \bar{T}_3 \; or \; \frac{T_2 + T_3}{2} - \bar{T}_1$ are also contrasts.

In general, a contrast or comparison among k-treatment means is an expression of the linear combination of the treatment means of the form:

$$L = c_1 \bar{T}_1 + c_2 \bar{T}_2 + c_3 \bar{T}_3 + \cdots + c_k \bar{T}_k,$$

$$where \sum_{j-1}^{k} c_j = 0: cj's \; are \; called \; coefficients$$

$$L = c_1 \bar{T}_1 + c_2 \bar{T}_2 + c_3 \bar{T}_3 + \cdots + c_k \bar{T}_{k'}$$

$$where \sum_{j=1}^{k} c_j = 0: cj's \; are \; called \; coefficients.$$

Therefore:

$$L = \sum_{j=1}^{k} c_j \bar{T}_j.$$

For comparisons among three treatment means, the c_j's must be carefully chosen so that $\sum c_j = 0$ will hold, otherwise, the linear combination is not a contrast.

If we set c1 = 1, c2 = -1 and c3 = 0, we find that the sum of c_j's is equal to zero ($\sum c_j = 0$). Thus, the linear combination of the coefficients is a contrast:

$$c_1 + c_2 + c_3 = 1 - 1 + 0 = 0$$

$L = c_1 \bar{T}_1 + c_2 \bar{T}_2 + c_3 \bar{T}_3 = (1)\bar{T}_1 + (-1)\bar{T}_2 + (0)\bar{T}_3 = \bar{T}_1 - \bar{T}_2$ is a contrast between the Means of Treatments 1 and 2.

Thus, a contrast between Treatments 2 and 3 is obtained by setting

$$c_1 = 0, c_2 = 1 \ and \ c_3 = -1: \sum cj = 0$$

$$L = c_1 \bar{T}_1 + c_2 \bar{T}_2 + c_3 \bar{T}_3$$

$$= (0)\bar{T}_1 + (1)\bar{T}_2 + (-1)\bar{T}_3 = \bar{T}_2 - \bar{T}_3$$

Note: Instead of using Treatment Means, Treatment Totals may also be used:

$$L = \frac{T_1 + T_2}{2} - T_3, \ is \ a \ contrast \ because: c_1 = \frac{1}{2}, c_2 = \frac{1}{2} \ and \ c_3 = -1$$

$$\sum cj = \frac{1}{2} + \frac{1}{2} - 1 = 0$$

$$L = c_1 T_1 + c_2 T_2 + c_3 T_3 = \left(\frac{1}{2}\right)T_1 + \left(\frac{1}{2}\right)T_2 + (-1)T_3 = \frac{1}{2}T_1 + \frac{1}{2}T_2 - T_3.$$

11.4.1.3 Orthogonal Contrasts (or Comparisons)

Two comparisons are orthogonal, if the sum of the products of the corresponding coefficients is equal to zero:

$$\sum c_{1i} c_{2i} = 0$$

For example, if we define three linear combinations as:

$$L_1 = c_1 T_1 + c_2 T_2 + c_3 T_3 + c_4 T_4 \text{, where } c_1 = -3, \ c_2 = -1; c_3 = 1; \ c_4 = 3: \sum ci = 0$$

$$L_2 = c_1 T_2 + c_2 T_2 + c_3 T_3 + c_4 T_4 \text{, where } c_1 = 1, \ c_2 = -1; \ c_3 = 1; \ c_4 = 1: \sum_{ci} = 0$$

$$L_3 = c_1 T_2 + c_2 T_2 + c_3 T_3 + c_4 T_4 \text{,where } c_1 = 1, \ c_2 = 1; \ c_3 = -1; \ c_4 = 1: \sum_{ci} = 0$$

For comparisons L1 and L2, the sum of the products of the corresponding coefficients is:

$$L_1 L_2 = c_1 c_1 + c_2 c_2 + c_3 c_3 + c_4 c_4$$

$$= (-3)(1) + (-1)(-1) + (1)(-1) + (3)(1)$$

$$= -3 + 1 - 1 + 3 = 0$$

Therefore, the comparisons L1 and L3 are orthogonal.

But the contrast L1 and L3 which is:

$$L_1 L_2 = c_1 c_1 + c_2 c_2 + c_3 c_3 + c_4 c_4$$
$$= (-3)(1) + (-1)(-1) + (1)(-1) + (3)(1)$$
$$= -3 + 1 - 1 + 3 = 0$$

The concept of orthogonality in this context is analogous to the concept of non-overlapping, or uncorrelated sources of variation. In practice, the comparisons that are constructed are those having some meaning in terms of the experimental variables, whether these comparisons are orthogonal or not, makes little or no difference.

In general, k-treatments have k-1 degrees of freedom and the treatment sum of squares can be partitioned into k-1 single-degree-of-freedom components which add to this sum of squares. The k-1 components must be derived from a set of k-1 orthogonal contrasts. Such a set of contrasts may be considered to include all the information available in the data.

11.4.2 Unplanned Pair Comparisons ("A Posteriori Comparisons", or "Post-Mortex Comparisons").

"A Posteriori Comparisons" called "Post-Mortex Comparisons", or "Unplanned Comparisons", are comparisons which were not specified before the data were obtained and analyzed. But they are "meaningful" or "interesting" comparisons which are made after the inspection of the experimental data. These comparisons are carried out only if the preliminary overall Anova is significant.

Such tests are made on all possible pairs of means which are "meaningful", or "interesting" comparisons to identify pairs of treatments that are significantly different. When there are k-means, there can, of course, be $\frac{k(k-1)}{2}$ possible comparisons between the means.

When a large number of comparisons are made, following a significant overall F, some of the decisions which reject H0, may be due to Type I error. For example, if 5 independent tests are each made at the 0.05 level, the probability of a Type I error in one or more of the five decisions is 1-(0.95)5 = 0.23.

Thus, when the number of comparisons is large, the number of decisions that can potentially be wrong owing to Type I error, can be relatively large. This is described as "Experimentwise Error Rate". **Experimentwise Error Rate** (if H0 is true) = (Number of Experiments with at least one Erroneous inference)/Number of experiments conducted.

11.4.2.1 The Use of the Studentized Range Statistic (qr)

Another method of testing the hypothesis: $\mu_1 = \mu_2 = \ldots = \mu_k$, is through the use of the Studentized Range Statistic, defined by:

$$q_r = \frac{\bar{T}_{largest} - \bar{T}_{smallest}}{\sqrt{\frac{MS_{ERROR}}{n}}}$$

;

$$q_r = \frac{\bar{T}_{largest} - \bar{T}_{smallest}}{\sqrt{MS_{ERROR}/n}}$$

,where n is the number of observations in each treatment mean $\bar{T}$, if they are equal;

r = number of steps the two means are apart on an ordered scale.

The Studentized Range Statistic (also called the q-statistic) is defined as the difference between the largest and smallest treatment means (i.e. the range of treatment means) divided by the square root of the quantity Mean-Square Experimental Error over n.

An equivalent form of the q-statistic for testing treatment totals is given by:

$$q_r = \frac{T_{largest} - T_{smallest}}{\sqrt{n\, MS_{ERROR}}} \; ,$$

where a T represents a treatment total with n observations.

For unequal number of observations in each treatment (unequal sample sizes, nj's), the harmonic mean $(\bar{n})$ of the nj's should be used in place of n in the above expression.

The harmonic mean $(\bar{n})$ is defined as:

$$(\bar{n}) = \frac{k}{\left(\frac{1}{n_1}\right) + \left(\frac{1}{n_2}\right) + \cdots \left(\frac{1}{n_k}\right)}$$

, where a k is the number of treatments; nj = sample size of each treatment.

$$q_r = \frac{T_{largest} - T_{smallest}}{\sqrt{MS_{ERROR}/n}} \qquad \text{for Treatment Means of equal sample sizes.}$$

$$q_r = \frac{T_{largest} - T_{smallest}}{\sqrt{n\, MS_{ERROR}}} \qquad \text{-- for Treatment Totals of equal sample sizes.}$$

We see that:
$$q_r \sqrt{\frac{MS_{ERROR}}{n}} = \bar{T}_{largest} - \bar{T}_{smallest} \; .$$

Therefore, Winer (1971) suggested the use of the critical value for the difference between two means which is:

$$q_{1-\alpha}(r, df) \sqrt{\frac{MS_{ERROR}}{n}}.$$

Thus, any pair of means whose difference exceeds the critical value is significantly different. The critical values of the studentized range statistic for $\alpha = 0.05 \text{ and } 0.01$ are listed at the back of most statistical texts.

Numerical Example 11.4.2.1

The partial data and analysis below were provided for us to use the studentized range statistic.

Anova Table:

Treatments	1	2	3	4	5	
Total (T$_j$)	10	12	18	16	14	k = 5
Means ($\bar{T}_j$)	2.50	3.00	4.50	4.00	3.50	n = 4

Source of variation	DF	SS	MS	F	$F_{0.01, 4, 15}$
Treatments	k-1 = 4	10	2.50	$\frac{2.50}{0.50} = 5.00^{**}$	4.89
Error	k(n-1) = 15	7.50	0.50		
Total	Kn-1=19	17.50			

$**$significant at $\alpha = 0.01$

$$q_r = \frac{T_{largest} - \bar{T}_{smallest}}{\sqrt{\frac{MS_{ERROR}}{n}}} = \frac{4.50 - 2.50}{\sqrt{\frac{0.50}{4}}} = \frac{2.00}{\sqrt{0.125}}$$

$$= \frac{2.00}{0.35355339} = 5.65685$$

$$\approx 5.66$$

$q_{1-\alpha}$: (5,15)Winer (1971) Table C.4 $= \frac{(5.63+5.49)}{2} = \frac{11.12}{2} = 5.56$

Steel and Torrie (1981) Table A.8 = 5.56.

Since $q_r > q_{1-\alpha}(5,15) = 5.56$, the difference between the largest and smallest means is significant. Other differences between other pairs of means were similarly tested and compared with the critical value but were not significant.

11.4.2.2 The Least Significant Range (LSR)

The Least Significant Range (LSR) is the threshold, or critical value, that the difference between any pair of means must exceed before the pair of means can be declared to be significantly different, in a posteriori comparisons.

A related way of carrying out **"a posteriori tests"** is the use of the largest difference found among a set of means (their range) as a statistic in place of their Sum of Squares. A sample range is then compared with the Least Significant Range obtained by multiplying the critical values

q_α (r, v) of **the significant studentized ranges with the Standard Error (MSE) from the Anova Table** $\left(\frac{S}{\sqrt{n}}\right)$: where:

LSR $= \frac{q_\alpha(r,v)\sqrt{\frac{MS_{Error}}{n}}}{}$, where: r = no of treatments in the set; v = df of MSE and $q_\alpha(r,v)$ is the critical value of a special statistic for significant testing called the **"Significant studentized Ranges (SSRs)"** obtained from a statistical table, depending on the number of treatments and $\alpha = 0.05$ or 0.01.

Any set of means with ranges greater than LSR are significantly heterogeneous. If we wish to compare a pair of means that do not have the same sample size, the LSR must be modified slightly to take this into account by replacing n by $\frac{2n_1 n_2}{n_1} + n_2$

; or simply by the harmonic mean;

For unequal sample sizes:

$$\text{LSR} = q_\alpha(r,v)\sqrt{(MS_{Error})/(2n_1 n_2/n_1 + n_2)}$$

Or, we simply replace n by the harmonic mean$(\bar{n})$:

$$LSR = q_\alpha(r,v)\sqrt{\frac{(MS_{Error})}{\bar{n}}}$$

Note: There are several procedures for carrying out "A Posteriori comparisons". All of these procedures use the "Significant Studentized Ranges (SSRs) which depend on the number of treatments involved and the α level. Such procedures include:

1. Duncan's Multiple Range Test (DMRT).
2. Tukey's Honestly Significant Difference (HSD).
3. Student-Newman-Keuls Test (S-N-K).
4. Scheffe's Method.
5. Dunnett's Test (For comparing all means with a control).

Generally, Significant Studentized Range (SSR) = $q_\alpha(p, f_e)$ also denoted as qa (r,v), are obtained from a Statistical Table, where:

p (or r) = number of treatments to be tested; f_e (or v) = df for MSE.

$$LSR = q_\alpha(p, f_e)\left(\sqrt{\frac{MSE}{n}}\right)$$ and is used in DMRT, S-N-K and Tukey's Test, under different names:

  i.   In DMRT, SSR = $q_\alpha(p, f_e)$ is denoted as rp and LSR is denoted as Rp..
  ii. In S-N-K, LSR is known as Wp.
  iii. In Tukey's Test (HSD) LSR is known as "W".

LSR should only be used in a posteriori, or unplanned comparisons, when the overall F-test in Anova is significant.

Let us use the various procedures above to compare the multiple treatment means in the following numerical example:

Numerical Example 11:4:2:2:

In a CRD data analysis, the following results were obtained:

Treatment Means: A = 3.5; B = 4.5; C = 8.5; D = 1.8; E = 10.5.

MSE = 1.52; Error df = 15:r = no. of observations per treatment;

p = no. of treatments involved.

11.4.2.2.1 The Least Significant Difference (LSD - - "A-Priori Tests")

Procedure:

1. Compute *LSD* value as:

$$LSD_{0.05} = t_{\alpha/2}(v)\sqrt{\frac{2S^2}{r}},$$

where: $S^2 = MSE \; from \; Anova \; Table.$

$r = no. of \; observations \; per \; treatment.$

$t = t - value\ from\ T - table.$

MSE = 1.52; Error df = 15; $r = 4$

$$LSD_{0.05} = t_{0.025}(15)\sqrt{\frac{2MSE}{r}} = 2.131\sqrt{\frac{2(1.52)}{4}} = 2.131(\sqrt{0.76})$$

$$= (2.131)(0.871779788) = 1.85776273$$

2. Array the treatment means in ascending order of magnitude:

D	A	B	C	E
1.8	3.5	4.5	8.5	10.5

3. Compute the absolute mean differences between treatments means ith and jth as:

$$d_{ij} = |\bar{T}_i - \bar{T}_j|:$$

E-D = 10.5 − 1.8 = 8.7:

C-D = 8.5 − 1.8 = 6.7

E-A = 10.5 − 3.5 = 7.0:

C-A = 8.5 − 3.5 = 5.0

E-B = 10.5 − 4.5 = 6.0:

C-B = 8.5 − 4.5 = 4.0

E-C = 10.5 − 8.5 = 2.0:

B-D = 4.5 − 1.8 = 2.7

B-A = 4.5 − 3.5 = 1.0

A-D = 3.5 − 1.8 = 1.7

4. Compare the LSD with each. $|d_{ij}|$ If $|d_{ij}|$ > LSD, then the two means are significantly different; if $|d_{ij}|$ < LSD, then the two means are not significantly different.

Thus:

E − D = 8.7>1.86 − significant.

C − D = 6.7>1.86 − significant.

E − A = 7.0>1.86 − significant.

C − A = 5.0>1.86 − significant.

E − B = 6.0>1.86 − significant.

C − B = 4.0>1.86 − significant.

E − C = 2.0>1.86 − significant.

B – D = 2.7>1.86 – significant.

B – A = 1.0<1.86 – Not significant.

A – D = 1.7<1.86 - - Not Significant.

5. Summarize the comparisons by under-scoring or underlining the means that are not significantly different:

D   A   B   C   E

1.8   3.5   4.5   8.5   10.5

Any pair of means that are not joined by an underscore are significantly different.

(B) A more compact and faster way of computing d_{ij} as suggested by Sokal and Rholf (1969) and Wahua (1999) is:

1. Arrange the means in a decreasing order, omitting the lowest mean horizontally and in increasing order vertically, omitting the highest mean as:

	E	C	B	A
	10.5	8.5	4.5	3.5
D =1.8	8.7	6.7	2.7	1.7
A =3.5	7.0	5.0	1.0	-
B =4.5	6.0	4.0	-	
C =8.5	2.0	-		

2. Subtract the "vertical" from the "horizontal" figures (means) to obtain a matrix of mean differences as shown above.
3. In summary, array the means in a decreasing order of magnitude. Compare the mean differences with the LSD (1.86) and summarize the result as shown. The Means that are

- not under-scored by the same line are significantly different; if two or more means have
- the same underscore, they are not significantly different:

E   C   B   A   D

10.5 8.5 4. 5 3.5   1.8

11.4.2.2.2 Duncan's Multiple Range Test (DMRT)

Duncan purposely developed this test (Wahua, 1999) to reduce the Experimentwise Error Rate (if H0 is true) defined as:

Experimentwise Error Rate (H0 true) =

$$\frac{No.\,of\ experiments\ with\ at\ least\ one\ erroneous\ inference}{Total\ number\ of\ experiments\ conducted.}$$

If there is no true difference among a set of means, DMRT is less likely to declare a difference than LSD.

Given the data in 11.4.2.2.1:

The means are: A B   C   D E :   MSE = 1.52

        3.5 4.5 8.5 1.8 10.5:   df = 15

        r = 4.

Procedure for DMRT:

Step 1: Arrange the Means in increasing order of magnitude:

    D   A   B   C   E

    1.8   3.5   4.5   8.5   10.5

Step 2: Write down the standard error as: $S_{\bar{x}} = \sqrt{\dfrac{MSE}{r}} = \sqrt{\dfrac{1.52}{4}}$

$= 0.6164414 = 0.62$

Step 3: Look up the Significant Studentized Ranges (SSRs) called rp from a Statistical Table (A.7- - from Steel and Torrie, 1981), using $\alpha = 0.05$ and the Error df = 15.

	P			
$(\alpha, v) = (0.05,15)$	2	3	4	5
SSR(r_p) (0.05,15)	3.01	3.16	3.25	3.31

Note: p is the number of steps between the ordered means or totals:

    p = j − i + 1.

Step 4: Calculate the Least Significant Ranges (LSRs) (Rp) by multiplying each SSR(rp) by the standard error from Anova Table = 0.62 as:

$$LSR = SSR \sqrt{\dfrac{S^2}{n}}$$

$$= SSR \sqrt{\dfrac{MSE}{n}}$$

. For example, at P=2, LSR(RP)=

$$\left(SSR(r_r) \times \sqrt{\dfrac{MSE}{n}} \right) = 3.01 \times \sqrt{\dfrac{1.52}{4}} = 3.01(0.5154414) = 3.01 \times 0.62 = 1.8662 \approx 1.87$$

P	2	3	4	5
SSR(r_p)	3.01	3.16	3.25	3.31
LSR(R_p)	1.87	1.96	2.02	2.05

Step 5: To compare the Means, write the LSRs with their corresponding ordered means:

	D	A	B	C	E
Ordered Means	1.8	3.5	4.5	8.5	10.5
LSRs (R_p)		1.87	1.96	2.02	2.05

Step 6: Obtain a matrix of differences between the means as we did in LSD above.

	E	C	B	A
	10.5	8.5	4.5	3.5
D = 1.8	8.7	6.7	2.7	1.7
A = 3.5	7.0	5.0	1.0	-
B = 4.5	6.0	4.0	-	
C = 8.5	2.0	-		

- To compare E with any of the means, the LSR we expect is 2.05 under column E, all the mean differences, except C, are bigger than 2.05. All the means, except C, are significantly different from E.
- For C, compare the differences in column C with LSR, which is 2.02, meaning that C is bigger than the other means, D, A and B.
- For B, the LSR is 1.96 and the differences in column B are 2.7 and 1.0. Thus, 2.7 is bigger than 1.96 and so B and D are significantly different. But B and A are not.
- For A, the LSR is 1.87 which is larger than the difference between A and D, which is 1.7 and so, they are not significantly different.

4. Summarize the comparisons by underscoring the means that are not different.

D   A   B   C   E

1.8 3.5 4.5 8.5 10.5

b   b   a   a

c   c

The Means with the same alphabetic letter are not significantly different.

11.4.2.2.3 Student-Newman-Keuls'Test (S-N-K Test, or Keuls' Test)

Still using the data in 11.4.2.2.1.above:

The means: A B C D E: MSE = 1.52.

               3.5 4.5 8.5 1.8 10.5:   df = 15.

               r = 4.

               p = t = no. of treatments.

S-N-K Procedure:

1. Order the set of Treatment Means in ascending magnitude:

$$\bar{T}_{(1)}, \bar{T}_{(2)}, \bar{T}_{(3)}, ... \bar{T}_{(t)},\text{ where, } \bar{\bar{T}}_{(1)} \text{ is the smallest mean and } \bar{\bar{T}}_{(t)} \text{ is the largest mean.}$$

2. Obtain the Significant Studentized Ranges (SSRs) from Table A.8 (Steel and Torrie, 1981) as: $q_{0.05}(p, f_e)$ with $p = 5$; $f_e = 15$

P	2	3	4	5
$q_{0.05}(5,15)$	3.01	3.67	4.08	4.37

3. Compute a set of LSRs by multiplying the SSRs $(q_\alpha(p, f_e))$ by the standard error:

$$S_{\bar{x}} = \sqrt{\frac{S^2}{r}} = \sqrt{\frac{MSE}{r}} = \sqrt{\frac{1.52}{4}} = 0.6164414.$$

$$LSR = W_p = q_{0.05}(5,15)S_{\bar{x}} = q_{0.05}(5,15)(0.62)$$

p	2	3	4	5
SSR$q_{0.05}(5,15)$	3.01	3.67	4.08	4.37
LSR= (W$_p$)	1.87	2.28	2.53	2.71

4. To compare the Means, write the LSRs with their corresponding ordered Means:

	D	A	B	C	E
Ordered means	1.8	3.5	4.5	8.5	10.5
LSR= W$_p$		1.87	2.28	2.53	2.71

- Compare the maximum and minimum means, $\bar{T}_{(t)} - \bar{T}_{(1)}$. If the range is not significant, no further testing is done, and the set of means is declared homogeneous.

- If this maximum difference is declared significant, it is concluded that $\mu_{(1)} \neq \mu_{(t)}$ and testing continues.

- The next stage tests: $\bar{T}_{(1)}$ versus $\bar{T}_{(t-1)}$ and $\bar{T}_{(2)}$ versus $\bar{T}_{(s)}$ using a test criterion for t-1 means. At any stage where a difference is not significant, testing stops and the set is declared homogenous. Otherwise testing continues.

5. Obtain a matrix of mean differences as we did in DMRT above and compare the means:

	E	C	B	A
	10.5	8.5	4.5	3.5
D = 1.8	8.7	6.7	2.7	1.7
A = 3.5	7.0	5.0	1.0	-
B = 4.5	6.0	4.0	-	
C = 8.5	2.0	-		

6. Summarize the comparisons by underscoring the arrayed means that are not significantly different:

D A   B   C E

1.8 3.5 4.5 8.5 10.5

b b   a   a

c c

The Means with same underscore, or same letter, are not significantly different.

11.4.2.2.4 Tukey's "W" Test Procedure, or Tukey's Honestly Significant Difference (HSD)

Tukey's test procedure (Steel and Torrie, 1981) makes use of the Studentized Range (SSR) and is applicable to pairwise comparisons of means. It requires a single value (like the LSD) for judging the significance of all differences and is thus quick and easy to use. Since only pairwise comparisons are made, the critical value is smaller than that required by Scheffe's method. All pairs of means constitute a family and error rate is familywise, as is the confidence coefficient when constructing interval estimates of differences.

The family of interest (Winer, 1971) is the set of all pairwise comparisons of factor level means; in other words, the family consists of estimates of all pairs:

$$D = \mu_j - \mu_{j'}, \text{ is estimated by the range:}$$

W = max (Xi) – min (Xi). Then, the ratio W/S is called the Studentized Range (where s = std. dev.) denoted as:

$$q\,(r,v) = {}^W\!/\!_S.$$

For unequal sample sizes: $S^2(\bar{D}) = MSE\left(\frac{1}{nj} + \frac{1}{nj'}\right)$.

This procedure is called Tukey's W-procedure, or Tukey's Honestly Significant Difference (HSD).

When all sample sizes are equal, the family confidence coefficients for the Tukey method is exactly $1 - \alpha$ - - that is, the Tukey method is conservative.

Tukey's "W"- Procedure:

1. Order the set of Treatment Means in increasing order of magnitude: $\bar{T}_{(1)}, \bar{T}_{(2)}, \bar{T}_{(3)}, \ldots, \bar{T}_{(p)}$

   ,where $\bar{T}_{(1)}$ is the smallest mean, while $\bar{T}_{(p)}$ is the largest mean in the set.

2. Obtain the Significant Studentized Ranges (SSRs) from a Statistical Table like A.8 of Steel and Torrie (1981) as: $q_{(0.05)}(p, f_e)$ with p = 5; fe = 15.

P	2	3	4	5
SSR$q_{0.05}$ (5,15)	3.01	3.16	3.25	3.31

   p = number of treatment means.

   $q_{0.05}$ (5,15) = SSRs from a Statistical Table.

3. Compute only one critical LSR value using the maximum $q_{0.05}(p, f_e)$ as:

$$W = q_\alpha(p, f_e)S_{\bar{x}}$$

$$= q_{0.05}(5,15)\sqrt{\frac{MSE}{n}}$$

   From the Table above, the largest SSR = 3.31 which is substituted for $q_{0.05}$ (5,15)

$$W = q_{0.05}(5,15)\sqrt{\frac{MSE}{n}}$$

$$= (3.31)(0.6164414) = 2.0404 \approx 2.04$$

4. Obtain a matrix of mean differences as we did in DMRT and other procedures above:

	E	C	B	A
	10.5	8.5	4.5	3.5
D =1.8	8.7	6.7	2.7	1.7
A =3.5	7.0	5.0	1.0	-
B =4.5	6.0	4.0	-	
C =8.5	2.0	-		

5. Compare these mean differences with the single LSR value = 2.04 and summarize the results by underscoring the means that are not significantly different:

```
D    A    B    C    E

1.8  3.5  4.5  8.5  10.5

          b    b    a    a

c    c
```

The means having the same underscore, or same letter, are not significantly different.

11.4.2.2.5 Scheffe's Test

Scheffe's method is very general (Steel and Torrie, 1981) in that all possible contrasts can be tested for significance, or confidence intervals constructed for the corresponding linear functions of parameters. Therefore, infinitely many simultaneous tests are permitted, although a finite number must be made, resulting in an error rate no larger than planned; the set of confidence intervals will have confidence coefficient at least as large as stated.

The critical value for a contrast, Q, requires the computation of S as:

$$S = \sqrt{f_t F_\alpha (f_t f_e)},$$

where: f_t = treatment degree of freedom;

f_e = error degree of freedom.

F_α = the tabulated F-value.

To compare all possible pairs of means, Scheffe's critical value is computed as:

$$Ss_{\bar{x}_i - \bar{x}_{i'}} = \sqrt{f_t F_\alpha (f_t f_e) S^2_{\bar{x}_i - \bar{x}_i}}$$

$$= \sqrt{f_t F_\alpha (f_t, f_e) \frac{2S^2}{r}} = \sqrt{f_t F_\alpha (f_t, f_e) \frac{2MSE}{r}}$$

All tests use this critical value.

According to Winer (1971), the Scheffe's method applied to testing differences between all possible pairs is even more conservative with respect to Type I errors than the Tukey method. The Scheffe' approach has this optimum property that Type I error is at most α for any of the possible comparisons.

Using the data for other tests above and the Scheffe's critical value:

1. $$Ss_{\bar{x}} = \sqrt{f_t F_\alpha(f_t, f_e)^2 \frac{MSE}{n}}, where: f_t = 4 = k - 1, \ f_e = 15;$$

$$\sqrt{\frac{MSE}{n}} = 0.6164414$$

$$f_{0.05}(4,15) = 3.06$$

$$Ss_{\bar{x}} = \sqrt{(4)(3.06)2(0.62)} = \sqrt{15.1776} = 3.8958$$

$$= 3.8958 \approx 3.90$$

2. Obtain a matrix of mean differences as in Tukey's method:

	E	C	B	A
	10.5	8.5	4.5	3.5
D =1.8	8.7	6.7	2.7	1.7
A =3.5	7.0	5.0	1.0	-
B =4.5	6.0	4.0	-	
C =8.5	2.0	-		

3. Compare these mean differences with the single Scheffe's critical value of 3.90 and summarize the results by underscoring the means that are not significantly different:

D	A	B	C	E
1.8	3.5	4.5	8.5	10.5
b	b	b	a	a

Means having the same underscore, or same letter, are not significantly different.

11.4.2.2.6 Comparing all Means with a Control – The Dunnett Procedure

If one of the k-treatments in an experiment represents a control condition, the experimenter is generally interested in comparing each treatment with the control condition, regardless of the outcome of the overall F-value (Winer, 1971). Rather than setting a level of significance equal to α for the collection of k-1 decisions, it is considered as a single decision for summarizing the outcomes.

The objective of an experiment (Steel and Torrie, 1981) is sometimes to locate treatments which are different or better than some standard, but not to compare them. Such a family of comparisons of control against each treatment is not an independent set.

Like Tukey, Scheffe's and LSD, Dunnett's procedure requires a single value for judging the significance of observed differences between each treatment and control. A Table (Table C.6 in Winer(1971)) is available for comparisons against one-tailed and two-tailed alternatives.

Dunnett critical value is given by:

$$d' = t(Dunnett)S_{\bar{y}_i - y_i'}$$

$$t_D S_{\bar{d}} = t_D \sqrt{\frac{2MSE}{n}}$$

$$t = \frac{\bar{T}_j - \bar{T}_o}{\sqrt{\frac{2MSE}{n}}} = t \sqrt{\frac{2MSE}{n}} = \bar{T}_j - \bar{T}_o$$

Simultaneous confidence intervals are included simultaneously under a single confidence coefficient for true differences $(\mu_i - \mu_o)$ are given by:

$$CI = (\bar{Y}_i - \bar{Y}_o) \pm t_s \sqrt{\frac{2}{r}}$$ — Dunnett's two-sided comparison Table.

$$CI = (\bar{Y}_i - \bar{Y}_o) - t_s \sqrt{\frac{2}{r}}, \infty$$ — Dunnetts' one-sided comparison Table. ∞ - implies no finite end-point on the right.

11.5.0 Two-Way Classification: The Randomized Complete Block Design (RCBD)

Randomized Complete Block Designs are called (Snedecor and Cochran, 1967) the Two-Way Classification because each experimental unit (observation) is classified by the treatment which it received and the replication (block) to which it belonged. The two criteria of classification are treatments and replications.

In RCBD, blocks are laid in such a way that each block represents as homogeneous an environment as possible. Each block is then sub-divided into the number of plots and the treatments are numbered and allocated at random to these plots by using a table of Random Numbers. A new randomization is carried out for each block. These practices (Steel and Torrie, 1981) help to control variation within blocks, and hence experimental error. The random arrangement of the treatments among the plots is necessary so that any given treatment would not always maintain the identical relative position in the field.

It is anticipated that most of the heterogeneity is among the blocks than within blocks. Mendenhall (1975) described RCBD as "Noise-reducing experimental designs" because they decrease the background noise (variation) caused by uncontrolled nuisance variables.

According to Winer (1971), RCBD is a restricted randomization design in which the experimental units are first sorted into homogeneous groups called Blocks, and the treatments are then assigned at random within the blocks. The order of treatments is subject to the restriction that each treatment occurs once in each block. Each block is a replication of the treatment combinations.

A distinguishing feature of RCBD is that Blocks are of equal size, each of which contains all the treatments. Blocks are not necessarily lands, but could be days during which treatments are carried out – each day representing a block; or in different laboratories, each serving as a block; or different age classes as blocks.

Blocking Technique

The primary purpose of blocking, according to Gomez and Gomez (1984), is to reduce experimental error by eliminating the contribution of known source of variation among experimental units:

- Plot shape and block orientation can be chosen so that much of the variation is accounted for by the difference among blocks, and experimental plots within the same block are kept as uniform as possible.
- An ideal source of variation to use as the basis for blocking is one that is large and highly predictable like:
  - ✓ Soil heterogeneity in fertilizer trial;
  - ✓ Direction of insect migration in insecticide trial;
  - ✓ Slope of the field.

- When the gradient is unidirectional, use long and narrow blocks and orient them so that their length is perpendicular to the direction of the gradient.
- When the fertility gradient occurs in two directions with one gradient much stronger than the other, ignore the weaker gradient and follow the preceding guideline for unidirectional gradient.
- When fertility gradient occurs in two directions, with both gradients equally strong and perpendicular to each other, choose one of these alternatives:
  - ✓ Use blocks that are as square as possible.
  - ✓ Use long and narrow blocks with their lengths perpendicular to the direction of one gradient and use the covariance technique to take care of the other gradient.
  - ✓ Use the Latin Square Design with two-way blockings, one for each gradient.
- When the pattern of variability is not predictable, blocks should be as square as possible.

Efficiency of Blocking

Snedecor and Cochran (1967) suggested the examination of the "Efficiency of Blocking" to know how effective the blocking was in increasing the precision of the comparisons and whether the criterion used in constructing the replications is a good one:

1. Determine the level of significance of the replication variation by computing the F-value for replication as:

   $$F(\text{Replication}) = \frac{Replication\ MS}{Error\ MS}, \text{ and test its significance by comparing it to the tabular F-values.}$$

   Blocking is considered effective in reducing the experimental error, if F-Replication is significant.

2. Determine the magnitude of the reduction in experimental error due to blocking by computing the Relative Efficiency as:

$$R.E. = \frac{s^2_{CR}}{s^2_{RE}}\ Estimated\ \frac{MSE(CRD)}{(S^2_{RE})} = \frac{F_b MSB + (f_t + f_e)MSE}{f_b + f_t + f_e}$$

f_b = df for block.

f_t = df for treatment.

f_e = df for error.

If MB and ME are the mean squares for blocks and error in the analysis of variance of randomized blocks experiment, then:

$$R.E. = \frac{S^2_{CR}}{S^2_{RB}} = \frac{(b-1)M_B + b(a-1)M_E}{(ab-1)M_E}$$

where: MB = Mean Square Block.

ME = Mean Square Error.

If the error degree of freedom is less than 20, Snedecor and Cochran (1967) suggested that the ratio S^2_{CR}/S^2_{RB}

be replaced by:

Relative amount of information $= \frac{(f_{RB}+1)(f_{CR}+3)\ S^2_{CR}}{(f_{RB}+3)(f_{CR}+1)\ S^2_{RB}}$.

where: f_{RB} = blocks degrees of freedom.

209

f_{RC} = CRD Error degrees of freedom.

However, Gomez and Gomez (1984) used the formula for Relative Efficiency (R.E) as:

$$R.E = \frac{(r-1)E_b + r(t-1)E_r}{(rt-1)E_r}$$

where: E_b = Replication Mean square in RCBD Anova.

E_e = Error Mean Square in RCBD Anova.

If the error df is less than 20, the R.E., value should be multiplied by an Adjustment Factor, k, defined as:

$$\text{Adjustment Factor, k} = \frac{[(r-1)(t-1)+1][t(r-1)+3]}{[(r-1)(t-1)+3][t(r-1)+1]}$$

For the RCB design, the numerator of the R.E. formula, is the comparable error had the CRD been used; the difference in the magnitude of experimental error between a CRD and an RCB design is essentially due to blocking; the value of the relative efficiency is indicative of the gain in precision due to blocking.

For example, if R.E. is computed as 1.63 and because the error df is only 15, the adjustment factor, k, is computed as k = 0.982, the adjusted R.E. value is computed as:

Adjusted R.E. = (k)(R.E.)

= (0.982)(1.63) = 1.60.

This result indicates that the use of the RCB design, instead of a CRD, increased experimental precision by 60%.

11.5.1 Two-Way Classification with a Single Observation per Cell

In RCBD, a set of observations (Walpole, 1974) may be classified according to two criteria at once, by means of a rectangular array in which the columns represent one criterion (blocks) of classification and the rows represent a second criterion of classification (treatments). Each treatment combination defines a cell in our array as shown the sample layout below:

Table 11.5.1: Sample Layout for a Two-Way Classification with one Observation per cell

Treatments	Replications (Blocks) $_{j=1,3...k}$				Treatment Total	Treatment Mean
	1	2	3 . . .	K		
i = 1,2...n 1.	X_{11}	X_{12}	X_{13} . . .	X_{1k}	$T_{1.}$	$\bar{T}_{1.}$
2.	X_{21}	X_{22}	X_{23} . . .	X_{2k}	$T_{2.}$	$\bar{T}_{2.}$
.	.	.	.	.	.	.
.	.	.	.	.	.	.
r.	X_{r1}	X_{r2}	X_{r3} . . .	X_{rk}	$T_{r.}$	$\bar{T}_{r.}$
Block-Totals	$T_{.1}$	$T_{.2}$	$T_{.3}$ . . .	$\bar{T}_{.k}$	$T_{..}$	$\bar{T}_{..}$
Mean	$\bar{T}_{.1}$	$\bar{T}_{.2}$	$\bar{T}_{.3}$ . . .	$\bar{T}_{.k}$		

The symbols:

X_{ij} = the measurement obtained for the unit that is in the ith row (treatment) and jth column

(replication).

210

$T_{i\cdot}$ = the total of the observations taken under Treatment i.

$T_{\cdot j}$ = the total of the observations taken under Block (Replication) j.

$\bar{T}_{\cdot j}$ = the mean observed under Block j.

$\bar{T}_{i\cdot}$ = the mean observed under Treatment i.

$T_{\cdot\cdot}$ = the grand total of all observations taken.

$\bar{T}_{\cdot\cdot}$ = the grand mean of all observations.

N = the total number of all the observations, where: $N = n1 + n2 + \ldots + nk$.

Procedure for RCBD Anova:

1. The Structural Model of RCBD

The structural (or mathematical) model of RCBD with one observation per cell, is given by:

$$X_{ij} = \mu + \alpha_i + \beta_j + \varepsilon_{ij},$$ where:

X_{ij} = Random experimental unit which is independent and normally distributed with mean, μ and variance, σ^2 i.e. $(X_{ij} \sim N(\mu, \sigma^2))$

μ = Constant overall population mean which is independent and normally distributed, and common to all treatments and replications.

α_i = Constant (Fixed) Treatment Effect which is normally distributed with mean zero and variance σ^2, subject to the restriction that: $\sum \alpha_i = 0$

β_j = Constant (Fixed) Block Effect which is normally distributed with mean zero and variance, σ^2, subject to the restriction that: $\sum \beta_j = 0$

ε_{ij} = Random Experimental Error which is independent, depending upon uncontrolled sources of variation and normally distributed with mean zero and variance, σ^2. $\left(\varepsilon_{ij} \sim N(o, \sigma^2)\right)$.

$\left(\mu + \alpha_i + \beta_j + \varepsilon_{ij}\right)$ implies that the components of the model are additive.

2. State the Null and Alternative Hypotheses:

In section 11.0, we stated that the k-treatment means, $\mu_{\cdot j}$, are equal and therefore equal to the population mean, μ, is equivalent to testing the hypothesis:

i. $H_0 : \alpha_1 = \alpha_2 = \ldots = \alpha_r = 0$ i.e. the Treatment has no effect.

$H_1 : \alpha_1 \neq \alpha_2 = \ldots \neq \alpha_r = 0$ i.e. the Treatment has some effect;

or At least one of the effects α_i is not equal to zero.

ii. Similarly, the null hypothesis that the Block means, $\mu_{.j}(\bar{T}_{.j})$ are equal to the population mean, μ, is equivalent to testing the hypothesis:

$H_0: \beta_1 = \beta_2 = \cdots = \beta_k = 0$. i.e. the Block Effect is zero, or Block has no effect.

$H_1: \beta_1 \neq \beta_2 \neq \cdots \neq \beta_k = 0$. i.e. the Block Effect is not equal to zero, or Block has some effect, or at least one of the β_j is not equal to zero.

The test of these two hypotheses will be based on four unbiased, independently estimated common population variance, σ^2. These estimates will be obtained by splitting the total variability of our data into three components represented symbolically as:

$$SS_{TOTAL} = SS_{BLOCKS} + SS_{TREATMENTS} + SS_{ERROR}$$

$$\sum_{k=1}^{r}\sum_{j=1}^{k}(X_{ij} - \bar{T}_{..})^2 = \sum_{i=1}^{r}\sum_{j=1}^{k}(\bar{T}_{.j} - \bar{T}_{..})^2 + \sum_{i=1}^{r}\sum_{j=1}^{k}(\bar{T}_{i.} - \bar{T}_{..})^2 + \sum_{i=1}^{r}\sum_{j=1}^{k}(X_{ij} - \bar{T}_{i.} - \bar{T}_{.j} - \bar{T}_{..})^2$$

Because these SS formulae are usually difficult to apply, they are rewritten to give the formulae which are easier to apply:

$$SS_{TOTAL} = \sum_{i=1}^{r}\sum_{j=1}^{k}X_{ij}^2 - \frac{T_{..}^2}{rk}$$

$$SS_{BLOCKS} = \frac{\sum_{j=1}^{k}T_{.j}^2}{r} - \frac{T_{..}^2}{rk}$$

$$SS_{TREATMENTS} = \frac{\sum_{i=1}^{r}T_{i.}^2}{k} - \frac{T_{..}^2}{rk}$$

$$SS_{ERROR} \sum_{i=1}^{r}\sum_{j=1}^{k}X_{ij}^2 - \frac{\sum_{i=1}^{r}T_{i.}^2}{k} - \frac{\sum_{j=1}^{k}T_{.j}^2}{r} + \frac{T_{..}^2}{rk}$$

Source of Variation	DF	SS	MS	F_{CAL}	$F_{0.05}:$ $F_{0.01}$
Between Blocks (β_j)	$r-1$	$\frac{\sum_{i=1}^{r}T_{.j}}{r} - \frac{T^2}{rk} = SS_B$	$\frac{SS_B}{r-1} = MS_B$	$F_1 = \frac{MS_B}{MSE}$	
Between Treatments (α_i)	$k-1$	$\frac{\sum_{j=1}^{k}T_{i.}}{r} - \frac{T^2}{rk} = SS_B$	$\frac{SS_{TRT}}{k-1} = MS_{TRT}$	$F_2 = \frac{MS_{TRT}}{MSE}$	
Error (ε_{ij})	$(r-1)$ $(k-1)$	$\sum X_{ij}^2 - \frac{\sum^r T_{i.}^2}{k} - \frac{\sum^k T_{.j}^2}{r} + \frac{T^2}{rk} = SS_{ERROR}$	$\frac{SS_{ERROR}}{(r-1)(k-1)} = MSE$		
Total	$rk-1$	$\sum_{i}^{r}\sum_{j}^{k}X_{ij}^2 - \frac{T_{..}^2}{rk} = SS_{TOTAL}$			

Table 11.5.1.1: Anova Table for RCBD with One Observation per Cell

Note: The degrees of freedom for Blocks is simply the number blocks or replications minus one (b-1).

The degrees of freedom for Treatment is simply the number of treatments minus one (t-1).

11.5.2 Numerical Example of RCBD with one Observation per Cell

A Silviculturist in the Department of Forestry and Wildlife, Uniben, treated 12 seeds of Garcinia cola in different substrates as shown in Table 11.5.1 in RCBD to see which treatments gave the best germination rate:

Table 11.5.1: The Germination Rate of Garcinia cola under various treatments

Treatments	Blocks				Treatment Total
	1	2	3	4	
Control	2	3	4	5	14
Hot water	3	6	4	5	18
Sulphuric Acid	7	8	9	10	34
Hydrochloric acid	1	2	1	3	7
Nitric Acid	10	9	11	12	42
Block Total	23	28	29	35	115

1. Test the hypotheses at $\alpha = 0.05 \; and \; 0.01$ that the treatments and the blocks were effective.
2. Estimate the efficiency of blocking in this experiment.

The procedure for the analysis of CRD also applies in RCBD:

1. State the mathematical model and explain the meaning of the terms in the model:

$$X_{ij} = \mu + \alpha_i + \beta_j + \varepsilon_{ij}$$ (See the meaning of the terms under the procedure for RCBD Anova).

2. State the Null and Alternative Hypotheses:

i. $H_0 : \alpha_1 = \alpha_2 = \ldots = \alpha_r = 0.$ i.e. Treatment has no effect on germination.

$H_1 : \alpha_1 \neq \alpha_2 \neq \ldots \neq \alpha_r = 0$ i.e. Treatment has some effect on the germination. At least one of the α_i is not equal to zero.

ii. $H_0 : \beta_1 = \beta_2 = \ldots = \beta_k = 0.$ i.e. Blocks have no effect on the germination.

$H_1 : \beta_1 \neq \beta_2 \neq \ldots \neq \beta_k = 0$ i.e. The effect of blocks is not zero; or At least one of the β_j is not equal to zero.

3. State the significance level $= \alpha = 0.05 \; and \; 0.01.$
4. **Decision Rule**: If the F-calculated is greater than F-tabulated, reject H0 and conclude that Fcal is significant; otherwise, accept H0.
5. **Computations**:
   i. **Compute the Grand total (T..):**

$$T_{..} = \sum_{i=1}^{r} \sum_{j=1}^{k} X_{ij} = 2 + 3 + 4 + \cdots + 9 + 11 + 12 = 115$$

ii. Compute the Correction Factor (CF):

$$CF = \frac{(\Sigma X_{ij})^2}{rk} = \frac{(115)^2}{5x4} = \frac{13.225}{20} = 661.25$$

iii. Compute Sum of Squares Total:

$$SS_{TOTAL} = \sum_{i=1}^{r}\sum_{j=1}^{k} X_{ij}^2 - CF$$

$$= 22 + 32 + 42 + \ldots + 92 + 112 + 122 - 661.25$$

$$= 895 - 661.25 = 233.75.$$

iv. Compute Blocks SS:

$$SS_{BLOCKS} = \frac{\Sigma_i^r T_{.j}^2}{r}$$

$$= \frac{23^2 + 28^2 + 29^2 + 35^2}{5} - 661.25$$

$$= 675.80 - 661.25$$

$$= 14.55$$

v. Compute Treatment SS:

$$SS_{TREATMENT} = \frac{\Sigma T_{i.}^2}{k} - CF.$$

$$= \frac{214^2 + 18^2 + 34^2 + 7^2 + 42^2}{4} - 661.25 \quad = \quad \frac{14^2 + 18^2 + 34^2 + 7^2 + 42^2}{4} - 661.25$$

$$= 872.25 - 661.25 = 211.0$$

vi. Compute Error SS:

$$SS_{ERROR} = SS_{TOTAL} - SS_{BLOCKS} - SS_{TREATMENTS}.$$

$$= 233.75 - 14.55 - 211.0$$

$$= 8.20$$

Table 11.5.1.2a: Anova Table for RCB with one observation for Cell

Source of Variation	DF	SS	MS	F-CAL	$F_{0.05}$ $F_{0.01}$
Blocks	$k-1=3$	14.55	4.85	7.13**	1.49 : 5.95
Treatments	$t-1=4$	211.00	52.75	77.57**	3.26 : 5.41
Error	$(k-1)(t-1)=12$	8.20	0.68		
Total	$kt-1=19$	233.75			

1. (a).Conclusion:

   Since Fcal= 7.13>> F0.05, 3, 12 =3.49 and F0.01, 3, 12 = 5.95, reject H0 and conclude that the Blocks are highly significantly different from zero.

   (b) Conclusion:

Since Fcal = 77.55>>F0.05, 4,12 = 3.26 and F0.01, 4, 12 = 5.41, reject H0 and conclude that the treatments are highly significantly different from zero.

2. Efficiency of Blocking:

$$R.E = \frac{(b-1)M_B + b(a-1)ME}{(ab-1)M_E}$$

$$= \frac{(3)(4.85) + (4)(4)(0.68)}{((5)(4)-1)(0.68)} = \frac{14.55 + 10.88}{12.92} = \frac{25.43}{12.92}$$

$$= 1.968266254.$$

Since the Error df is less than 20, the R.E. value should be multiplied by the Adjustment Factor, k, defined as:

$$k = \frac{[(r-1)(t-1)+1][t(r-1)+3]}{[(r-1)(t-1)+3][t(r-1)+1]}$$

$$= \frac{[(3)(4)+1][5(3)+3]}{[(3)(4)+3][5(3)+1]}$$

$$= \frac{(13)(18)}{(15)(16)} = \frac{234}{240} = 0.975$$

$$R.E = (1.968)(0.975) = 1.9188 \approx 1.92$$

Therefore, using RCBD increased efficiency by 92%

Note: If Blocks, or Replications, are not statistically significant in the Anova Table, we lost power because there are fewer degrees of freedom left to estimate the experimental error. It therefore means that the data should have been better analyzed by using CRD which would have provided higher degrees of freedom for estimating the experimental error.

11.5.3 TWO-WAY CLASSIFICATION WITH SEVERAL OBSERVATIONS PER CELL

The major difference between the two-way classification with one observation per cell and that with several observations per cell is that the observations within each cell are repeated several (n) times. These repeated measurements are now said to interact and thus making the model non-additive without it.

The rectangular array of data consists of "r" rows (treatments) and "c" columns (blocks) with "rc" cells, each containing "n" observations. Hence, each kth observation in the ith row and jth column is now denoted as Xijk.

Table11.5.2: The Sample Layout for a Two-Way Classification with several observations per cell

Treatments	Replications (Blocks)				Treatment Total	Treatment Mean
	1	**2**	**3 . . .**	**K**		
1	X_{111}	X_{121}	X_{131} . . .	X_{1k1}	$T_1..$	$\bar{T}_{1}..$
.	X_{112}	X_{122}		X_{1k2}		
.	:	:	:	:		
.	X_{11n}	X_{12n}		X_{1Kn}		
2	X_{211}	X_{221}	X_{231}...	X_{2k1}	$T_2..$	$\bar{T}_{2}..$
.	X_{212}	X_{222}	X_{232}...	X_{2k2}		
.	:	:				
.	X_{21n}	X_{22n}	X_{23n}...	X_{2kn}		
.	:	:	: ...	:		
R	X_{r1n}	X_{r2n}	X_{r3n}	X_{rkn}	$T_r..$	$\bar{T}_{r}..$
Total	$T._1.$	$T._2.$	$T._3....$	$T.._k.$	$T...$	$\bar{T}\ ...$

The symbols:

X_{ijk} = Each observation per cell.

$T_{ij}.$ = Sum of the observations in the ijth cell.

$T_i..$ = Sum of the observations in the ith row.

$T.j.$ = Sum of the observations in the jth column.

$T...$ = Sum of all the rcn observations.

N= The total number of all observations, where $N = rcn = n1+n2+n3 +.......+ nk$.

Procedure for RCBD Anova with Several Observations per Cell:

1. The structural (or mathematical) model is given by:

$$X_{ijk} = \mu + \alpha_i + \beta_j + (\alpha\beta)ij + \varepsilon_{ijk},\ \text{where:}$$

X_{ijk} = Random experimental unit which is independent and normally distributed with mean, μ, and variance, σ^2, i.e. $X_{ijk} \sim N(\mu, \sigma^2)$

μ = Constant (or fixed) overall population mean which is independent and normally distributed, and common to all treatments and replications.

α_i = Constant Treatment Effect which is normally distributed with mean, zero and variance, σ^2,
$$\sum \sigma_i = 0$$

216

subject to the restriction that: .

β_j = Constant Block Effect which is normally distributed, with mean, zero and variance, σ^2 and subject to the restriction that: $\sum \beta_j = 0$.

$(\sigma\beta)_{ij}$ = Constant interaction Effect which is normally distributed with mean zero and variance, σ^2, and subject to the restriction that: $\Sigma(\sigma\beta)_{ij} = 0$.

ε_{ijk} = Random Experimental Error which is independent, depending up $\left(\varepsilon_{ijk} \sim N(0,\sigma^2)\right)$ urces of variation and normally distributed, with mean, zero and variance, σ^2 i.e

$(\mu + \sigma_i + (\alpha\beta)_{ij} + \varepsilon_{ijk}$

implies additivity of model components.

2. State all the Null and Alternative Hypotheses:

i. $H_0: \alpha_1 = \alpha_2 = \ldots = \alpha_r = 0$. i.e. The Treatment Effect is zero; or Treatment has no effect.

$H_1: \alpha_1 \neq \alpha_2 \neq \ldots \neq \alpha_r \neq 0$. i.e. The Treatment Effect is not zero; Treatment has some effect. At least one α_i is not equal to zero.

ii. $H_0: \beta_1 = \beta_2 = \cdots = \beta_c H_0: \beta_1 = \alpha_2 = \ldots = \beta_c = 0$. i.e. The Block Effect is equal to zero; Block has no Effect.

$H_1: \beta_1 \neq \beta_2 \neq \ldots \neq \beta_c \neq 0$ i.e. The Block Effect is not equal to zero; No Block effect. At least one β_j is not equal to zero.

iii. $H_0: (\alpha\beta)_{11} = (\alpha\beta)_{12} = \cdots = (\alpha\beta)_{rc} = 0$ i.e. Interaction Effect is equal to zero.

$H_1: (\alpha\beta)_{11} \neq (\alpha\beta)_{12} \neq \cdots \neq (\alpha\beta)_{rc} \neq 0$ i.e. Interaction Effect is not equal to zero. At least one $(\alpha\beta)_{ij}$ is not equal to zero.

The test of these three hypotheses will be based on the estimates of the effects which are obtained by partitioning the total variability of our data into four components represented symbolically as:

$$SS_{TOTAL} = SS_{BLOCKS} + SS_{TREATMENTS} + SS_{ERROR}$$

These SS can be obtained from these formulas:

$$SS_{TOTAL} = \sum_{i=1}^{r}\sum_{j=1}^{c}\sum_{k=1}^{n} X_{ijk}^2 - \frac{T^2}{rcn}$$

$$SS_{COLUMNS} = \frac{\sum_{j=1}^{c} T_{.j.}^2}{rn} - \frac{T^2}{rcn}$$

$$SS_{ROWS} = \frac{\sum_{i=1}^{i} T_{i..}^2}{cn} - \frac{T_{...}^2}{rcn}$$

$$SS_{R \times C} = \frac{\sum_{i=1}^{r} \sum_{j}^{c} T_{ij.}^2}{n} - \frac{\sum_{j=1}^{c} T_{.j.}^2}{rn} + \frac{\sum_{i=1}^{r} T_{i..}^2}{cn} + \frac{T^2}{rcn}$$

$$SS_{ERRORS} = SS_{TOTAL} - SS_{COLUMNS} - SS_{ROWS} - SS_{RC}$$

Table 11.5.2a: Anova Table with several Observations per Cell

Source of variation	DF	SS	MS	Fcal
Blocks (Column)	c-1	$SS_{COLUMNS} = \frac{\sum_{j=1}^{c} T_{.j.}^2}{rn} - \frac{T_{...}^2}{rcn}$	$\frac{SS_B}{c-1} = MS_B$	$F_1 = $ MSB/MSE
Rows (Trt)	r-1	$SS_{TRT} = \frac{\sum_{i=1}^{r} T_{i..}^2}{cn} - \frac{T_{...}^2}{rcn}$	$\frac{SS_{TRT}}{r-1} = MS_{TR}$	$F_2 = $ MS$_{TR}$/MSE
Interaction	(c-1)(r-1)	$SS_{RC} = \frac{\sum_{i=1}^{r} \sum_{j=1}^{c} T_{ij.}^2}{n} - \frac{\sum_{k}^{c} T_{.j.}^2}{rn} - \frac{\sum^{r} T_{i..}^2}{cn} + \frac{T_{...}^2}{rcn}$	$\frac{SS_{RC}}{n} = MS_{RC}$	$F_3 = $ MS$_{RC}$/MSE
Error	rc(n-1)	$SSE = SS_{TOT} - SS_B - SS_{TRT} - SS_{RC}$	$\frac{SSE}{rc(n-1)} = MSE$	
Total	rcn-1	$\sum_{i=1}^{r} \sum_{j=1}^{c} \sum_{k=1}^{n} X_{ijk}^2 - \frac{T_{...}^2}{rcn}$		

11.5.3.1 Numerical Example of RCBD with Several Observations per Cell

A Mensurationist in the Department of Forestry and Wildlife, Uniben, investigated the effects of four different Organic Fertilizers on the height (mm) growth of six-month old tree species (Teak, Gmelina and Terminalia) raised from the Nursery. The tree species were randomly assigned to each of the four organic fertilizers, as shown in Table 11.5.3.1:

Table 11.5.3.1: Effects of Organic Fertilizers on Height Growth of Three Tree Species

Organic Fertilizer	Tree Species			
	Teak	Gmelina	Terminalia	Total
Cow Dung	64	72	74	
	66	81	51	607
	70	64	65	
Compost + Cow Dung	65	43	47	
	63	52	58	510
	58	57	67	
Chicken Droppings	59	66	58	
	68	71	39	527
	65	59	42	
Compost	58	57	53	
	41	61	59	466
	46	53	38	
Total	723	736	651	2,110

k = 1, 2, 3.

i = 1, 2, 3, 4.

$$j = 1, 2, 3.$$

The three hypotheses that can be tested in this example include:

i. Test the hypothesis that there is no difference in the height growth of the tree species under the different organic fertilizers.
ii. Test the hypothesis that the tree species are not different in their growth rates.
iii. Test the hypothesis that there is no interaction between the fertilizers and the tree species.

1. Mathematical Model: $X_{ijk} = \mu + \alpha_i + \beta_j + (\alpha\beta)_{ij} + \varepsilon_{ijk}$, (See last section).

(i) H_0 : $\alpha_i = \alpha_2 = \alpha_3 = \alpha_4 = 0$.i.e. The fertilizers have no effect on species height growth.

H_1 : $\alpha_i \neq \alpha_2 \neq \alpha_3 \neq \alpha_4 \neq 0$.i.e. The fertilizers have some effect on species height growth.

(ii) H_0: $\beta_i = \beta_2 = \beta_3 = 0$ .i.e. The tree species are not different in their response to the organic fertilizers.

H_1: $\beta_i \neq \beta_2 \neq \beta_3 \neq 0$ .i.e. The tree species are different in their response to the organic fertilizers.

(iii) H_0: $\alpha\beta_{11} = \alpha\beta_{12} = \alpha\beta_{13} = \alpha\beta_{21} = \cdots \alpha\beta_{42} = \alpha\beta_{43} \neq 0$. i.e. There is no Fertilizer by Species Interaction.

H_1: $\alpha\beta_{11} \neq \alpha\beta_{12} \neq \alpha\beta_{13} \neq \alpha\beta_{21} \neq \cdots \alpha\beta_{42} \neq \alpha\beta_{43} \neq 0$. i.e. There is Fertilizer by Species Interaction.

2. State Significance level: $\alpha = 0.05 \ and \ 0.01$.
3. **Decision Rule**: If the F-calculated is greater than F-tabulated, reject H0 and conclude that F is significant; otherwise, accept H0.
4. **Computations**: The first computation is to construct a Table of Cell Totals (Tij) as:

Table 11.5.3.1: R x C Summary Table(Tij.).

	Teak	Gmelina	Terminalia	Total
Cow Dung	200	217	190	607
Compost/Cow Dung	186	152	172	510
Chicken Droppings	192	196	139	527
Compost	145	171	150	466
Total	723	736	651	2110

$$Grand\ Total\ (T_{...}) = \sum_{i=1}^{r} \sum_{j=1}^{c} \sum_{k=1}^{n} X_{ijk} = 64 + 66 + 70 + \cdots + 53 + 59 + 38 = 2{,}110$$

$$Correction\ Factor\ (CF) = \frac{T_{...}^2}{rcn} = \frac{(\sum_{i=1}^{r} \sum_{j=1}^{c} \sum_{k=1}^{n} X_{ijk})^2}{r \times c \times n} = \frac{(2{,}110)^2}{4 \times 3 \times 3}$$

$$= \frac{4{,}452{,}100}{36} = \underline{123{,}669{,}4444}.$$

$$SS_{TOTAL} = \sum_{i=1}^{r}\sum_{j=1}^{c}\sum_{k=1}^{n} X_{ijk}^2 - CF$$
$$= 64^2 + 66^2 + 70^2 + \ldots + 53^2 + 59^2 + 38^2 - 123{,}669.4444.$$

$$= 127{,}448 - 123{,}669.4444 = \underline{3{,}778.5556}.$$

$$SS_{SPECIES} = \sum_{k}^{c} T_{.j.} - CF = \sum_{i=1}^{c} \frac{\left(\sum_{j=1}^{r}\sum_{k=1}^{n} X_{ijk}\right)^2}{rn} - CF$$

$$= \frac{723^2 + 736^2 + 651^2}{4 \times 3} - 123{,}669.4444$$

$$= \frac{1{,}488{,}226}{12} - 123{,}669.4444 = 124{,}018.8333 - 123{,}669.4444.$$

$$= \underline{349.3889}.$$

$$SS_{FERTILIZERS} = \frac{\sum\, ^{r}T_{i..}^2}{cn} - CF = \frac{\sum\, _{i=1}^{r}\left(\sum\, _{j=1}^{c}\sum\, _{k=1}^{n} X_{ijk}\right)^2}{cn} - CF.$$

$$= \frac{607^2 + 510^2 + 527^2 + 466^2}{3 \times 3} = 123{,}669.4444$$

$$= \frac{1{,}123{,}434}{9} - 123{,}669.444 = 124{,}826 - 123{,}669.444$$

$$= \underline{1{,}156.5556}.$$

$$SS_{INTERACTION} = \frac{\sum_{i=1}^{r}\sum_{j=1}^{c} T_{ij.}^2}{n} - CF = \frac{\sum_{i=1}^{r}\sum_{j=1}^{c}\left(\sum_{k}^{n} X_{ijk}\right)^2}{n} - \frac{\sum_{j}^{c} T_{.j.}^2}{rn} - \frac{\sum_{i}^{r} T_{i..}^2}{cn} - CF.$$

$$= \frac{200^2 + 156^2 + \cdots 1 \;\; 139^2 + 150^2}{3} - 349.3889 - 1.156.5556 - CF$$

$$= \frac{377{,}840}{3} - 123{,}669.4444 - 125{,}946.6669 - 123{,}6694444$$

$$= 2{,}277.2223 - 349.3889 - 1{,}156.5556 = \underline{771.2777}.$$

$$SS_{ERROR} = SS_{TOTAL} - SS_{SPECIES} - SS_{FERT} - SS_{INTERATION}.$$

$$= 2{,}887.5556 - 349.3889 - 1{,}156 - 771.2777$$

$$= \underline{1{,}501.3333}.$$

Table 11.5.3.1a: Anova Table of Effects of Organic Fertilizers On Height Growth of Three Tree Species.

Source of variation	DF	SS	MS	F-CAL	$F_{0.05}$	$F_{0.01}$
Species	c-1 = 2	349.3889	174.69447	$F_1 = 2.7926^{ns}$	3.40	5.61
Fertilizers	r-1 = 3	1,156.5556	385.51853	$F_2 = 6.1628^{**}$	3.01	4.72
Interaction	(c-1) (r-1) = 6	771.2777	128.54629	$F_3 = 2.055^{ns}$	2.51	3.67
Error	rc (n-1) = 24	1,501.3333	62.55555			
Total	**rcn-1 = 35**	**3,778.5556**				

i. **Conclusion** : Since F1-cal = 2.79 < F0.05 = 3.40, accept H0 and conclude that the Tree species are not different in their response to the fertilizers.

ii. **Conclusion** : Since F2-cal = 6.16 >> F0.05 3.01 and F0.01 = 4.72 respectively, reject H0 and conclude that the fertilizers have a highly significant effect on the tree species height growth.

iii. **Conclusion** : Since F3-cal = 2.06 < F0.05 = 2.51, accept H0 and conclude that the interaction effect is zero. It means the interplay between the species and the fertilizers is not really important in this study.

However, we can estimate the cell-means and use them for Multiple Means Comparisons as in section 11.4.2.

Table 11.5.3.1b: Cell Means $\dfrac{(T_{ij})}{3}$ of Tree Species and Organic Fertilizers.

Fertilizers	Tree species			
	Teak	Gmelina	Terminalia	Fertilizer Mean
F_1	66.7	72.3	63.3	67.4
F_2	62.0	50.7	57.3	56.7
F_3	64.0	65.3	46.3	58.6
F_4	48.3	57.0	50.0	51.8
Species Mean	60.3	61.3	54.3	58.6

Note: These Cell Means can be used to plot the graphs of Geometric Interpretations of Interactions which we shall discuss under **"Factorials"**.

It is extremely important to observe that since interaction is not significant, it were better to analyze the data as RCBD with one observation per cell. All we need to do to achieve this is to adjust the Mathematical Model and Anova Table as:

$$X_{ijk} = \mu + \alpha_i + \beta_j + \varepsilon_{ijk}$$

The $SS_{Interaction}$ is added to SS_{ERROR} and the degrees of freedom for interaction (r-1)(c-1) is added to Error df (rc(n-1)) and their Mean Squares recalculated as shown in the new Anova Table below:

Thus Error SS = 1501.3333 + 771.2777 = 2,272.611

Error df = rc(n-1) + (r-1)(c-1) = 24 + 6 = 30.

Table 11.5.3.1 c: New Anova Table of Effect of Organic Fertilizers on Height Growth of Three Tree Species.

Source of variation	DF	SS	MS	F-CAL	$F_{0.05}$	$F_{0.01}$
Species	$c - 1 = 2$	349.3889	174.69447	$F_1 = 2.306^{ns}$	3.32	5.39
Fertilizers	$r - 1 = 3$	1,156.5556	385.51853	$F_2 = 5.89^{**}$	2.92	4.51
Error	$rc(n-1) + (c-1)$ $(n-1) = 30$	771.2777 + 1,501.3333 = 2,272.511	75.7537			
Total	$rc\ n - 1 = 35$	3,778.5556				

Finally, when interaction is significant, the main effects must be interpreted with caution because the main effects take on a different meaning in the presence of interaction. The presence of interaction "masks" the main effects.

11.6.0 THE LATIN SQUARE DESIGN (LSD)

The major feature of the Latin Square Design (LSD) is its capacity to simultaneously handle two known sources of variation among experimental units. It treats the sources as two independent blocking criteria, instead of only one as in RCBD. The two-directional blocking in a LS design, commonly referred to as row-blocking and column-blocking is accomplished by ensuring that every treatment occurs only once in each row-block and once in each column-block. This procedure makes it possible to estimate variation among row-blocks, as well as among column-blocks and to remove them from experimental error.

The presence of row-blocking and column-blocking in a Latin Square Design, while useful in taking care of two independent sources of variation, also becomes a major restriction in the use of the design. This is so because the requirement that all treatments appear in each row-block and in each column-block can be satisfied only if the number of replications is equal to the number of treatments.

In practice, the Latin Square Design is applicable only for experiments in which the number of treatments is not less than four and not more than eight. Because of such limitation, the Latin Square Design has not been widely used in agricultural experiments despite its great potential for controlling experimental error. Thus, in LSD, the number of rows, columns and treatments are equal.

Procedure for LSD Anova:

1. The Mathematical Model is given by:

   $X_{ijk} = \mu + \alpha_i + \beta_j + Y_k + \varepsilon_{ijk}$, where:

   X_{ijk} = Random experimental unit which is independent and normally distributed with mean, μ and variance, $\sigma^2 i.e\ \left(X_{ijk} \approx N(\mu, \sigma^2)\right)$. σ^2. $\left(X_{ijk} \sim N(\mu, \sigma^2)\right)$.

   μ = Constant overall population mean which is independent and normally distributed, and common to all treatments and replications.

   α_i = Constant Row Effect, subject to the restriction that: $\sum \alpha_i = 0$

   β_j = Constant Column Effect, subject to the restriction that: $\sum \beta_j = 0$

   Y_k = Constant Treatment Effect, subject to the restriction that: $\sum Y_k = 0$

ε_{ijk} = Random, or uncontrolled sources of variation which is normally distributed with mean, zero and variance, σ^2 i.e. $\varepsilon_{ijk} \sim N(0, \sigma^2)$.

$(\mu + \alpha_i + \beta_j + \Upsilon_k + \varepsilon_{ijk})$ implies additivity of all model components.

11.6.1 A Numerical Example (1) of a Latin Square Design (LSD)

A Crop Scientist investigated the yields of three hybrids of maize and a control in a Latin Square Design to know which hybrid is the highest yielding among the hybrids. The data are presented in Table 11.6.0.

Table 11.6.0: Grain Yield of three Hybrids and a Control.

Rows	COLUMNS				Row Total
Rows	1	2	3	4	Row Total
1	1.640 (B)	1.210(D)	1.425(C)	1.345(A)	5.620
2	1.475(C)	1.185(A)	1.400(D)	1.290(B)	5.350
3	1,670(A)	0.710©	1.665(B)	1.180(D)	5.225
4	1.565(D)	1.290(B)	1.655(A)	0.660(C)	5.170
Column Total	6.350	4.395	6.145	4.475	21.365

1. The Mathematical Model for the Latin Square Design:

   $X_{ijk} = \mu + \alpha_i + \beta_j + \Upsilon_k + \varepsilon_{ijk}$, where: the meanings of the terms in the model are as described above.

2. The Hypotheses that can be tested include:

(i) $H_0: \alpha_1 = \alpha_2 = \alpha_3 = \alpha_4 = 0$. i.e. Row-blocking has no effect on grain yields.

   $H_1: \alpha_1 \neq \alpha_2 \neq \alpha_3 \neq \alpha_4 \neq 0$. i.e. Row-blocking has some effect on grain yields.

(ii) $H_0: \beta_1 = \beta_2 = \beta_3 = \beta_4 = 0$. i.e. Column-blocking has no effect on grain yields.

   $H_1: \beta_1 \neq \beta_2 \neq \beta_3 \neq \beta_4 \neq 0$. Column-blocking has some effect on grain yields.

(iii) $H_0: \Upsilon_1 = \Upsilon_2 = \Upsilon_3 = \Upsilon_4 = 0$. i.e. The Yields of the hybrid maize are equal, or there is no difference in the hybrid yields of maize.

   $H_1: \Upsilon_1 \neq \Upsilon_2 \neq \Upsilon_3 \neq \Upsilon_4 \neq 0$ .i.e. The yields of the hybrid maize are not equal; There is a difference in the hybrid yields of maize.

3. State Significance level: $\alpha = 0.05$ or α 0.01
4. **Decision Rule**: If F-calculated is greater than F-tabulated, reject H0 and conclude that F is significant, otherwise, accept H0 as specified.
5. **Computations:**

   Since Rows and Columns totals are known, we need to construct the Treatment totals from the data before we compute the SS.

Treatments

	A	B	C	D
Col 1 =	1.670	1.640	1.475	1.565
Col 2 =	1.185	1.290	0.710	1.210
Col 3 =	1.655	1.665	1.425	1.400
Col 4 =	1.345	1.290	0.660	1.180
Total =	5.855	5.885	4.270	5.355

$$Grand\ Total\ (T...) = \sum_{i=1}^{r}\sum_{j=1}^{c}\sum_{k=1}^{n} X_{ijk} = 1.640 + 1.475 + \cdots + 1.180 + 0.660$$

$$= 21.365$$

$$Correction\ Factor\ (CF) = \frac{\left(\sum_{i}^{r}\sum_{j}^{c}\sum_{k}^{n} X_{ijk}\right)^2}{r^2} = \frac{(21.365)^2}{4^2} = \frac{456.463225}{16}$$

$$= 28.52895156.$$

$$SS_{TOTAL} = \sum_{i=1}^{r}\sum_{j=1}^{c}\sum_{k=1}^{n} X_{ijk}^2 - CF$$

$$= (1.210)^2 + (1.475)^2 + 1.670)^2 + \cdots + (1.180)^2 + (0.660)^2 - CF$$

$$= 29.942975 - 28.52895256 = 1.41392344$$

$$SS_{ROWS} = \frac{\sum_{i=1}^{r} T_i^2}{r} - CF$$

$$= \frac{(5.620)^2 + (5.350)^2 + (5.225)^2 + (5.170)^2}{4} - 28.52895156$$

$$= \frac{114.236425}{4} - 28.52895156$$

$$= 28.55910625 - 28.52895156 = \underline{0.03015469}$$

$$SS_{COLUMNS} = \frac{\sum_{j=1}^{c} T_{.j.}^2}{r} - CF$$

$$= \frac{(6.350)^2 + (4.395)^2 + (6.145)^2 + (4.475)^2}{4} - 28.52895156$$

$$= \frac{117.425175}{4} - 28.52895156 = 29.35629375 - 2852895156$$

$$= \underline{0.82734219}.$$

$$SS_{TREATMENT} = \frac{\sum_{i=1}^{n} T_t^2}{r} - CF$$

$$= \frac{(5.855)^2 + (5.885)^2 + (4.270)^2 + (5.355)^2}{4} - 28.52895156$$

$$= \frac{115.823175}{4} - 28.52895156 = -28.95579375 - 28.52895156$$

$$= \underline{0.42684219}.$$

$$SS_{ERROR} = SS_{TOTAL} - SS_{ROWS} - SScolumns - SS_{TRTs}.$$

$$= 1.41392344 - 0.03015469 - 0.82734219 - 0.42684219.$$

$$= \underline{0.12958437}.$$

Table 11.6.1 AnovaTable for a Latin Square Design on Hybrid Yields Maize and a Control.

Source of variation	DF	SS	MS	F-CAL	F$_{0.05}$	F$_{0.01}$
Row	r-1 = 3	0.03015459	0.010051563	F$_1$ = 0.4654^{ns}	4.76	9.78
Column	c-1 = 3	0.82734219	0.27578073	F$_2$= 12.7692**	4.76	9.78
Treatment	t-1 = 3	0.42684219	0.14228073	F$_3$ =6.5879*	4.76	9.78
Error	(t-1)(t-2) = 6	0.12958437	0.021597393			
Total	**t²-1 = 15**	**1.41392344**				

1. **Conclusions**:

   Since FRows = 0.47 << F0.05 = 4.76 and F0.01 9.78, accept H0 and conclude that Row Effect is zero.

   Since FColumns = 12.77 >> F0.05 = 4.76 and F0.01= 9.78, reject H0 and conclude the Column Effect to be highly significantly different and so significantly affect grain yields.

   Since FTrt = 6.59 > F0.05 = 4.76 but less than F0.01 = 9.78, reject H0 at F0.05 and accept H0 at F0.01.

2. **Efficiencies of Row-and Column-Blockings**

   The efficiencies of both Row-and Column-blockings in a Latin Square Design indicate the gain in precision relative to either the CRD, or the RCBD.

3. The Relative Efficiency of a LSD as compared to a CRD is given by:

$$R.E.(CRD) = \frac{E_r + E_c + (t-1)E_e}{(t+1)E_e}$$

   where: Er = Row MS;

   Ec = Column MS;

   Ee = Error MS in the Latin Square Anova and

   t = the number of Treatments.

   From the Anova Table above:

$$R.E.\,(LSD\;Compared\;to\;CRD)$$
$$= \frac{0.010051563 + 0.27578073 + (3)(0.021597395)}{(4+1)(0.021597395)}$$

$$= \frac{0.285832293 + +0.064792185}{0.107986975}$$

$$\frac{0.350624478}{0.107986975} = 3.2469 \approx 3.25$$

This means that the use of LSD increased the experimental precision by 325%. This implies that, if the CRD had been used, an estimated 3.25 times more replications would have been required to detect the treatment difference of the same magnitude as that detected with the LSD.

4. The Relative Efficiency of a LSD as compared to RCBD can be computed in two ways:

when Rows are considered as blocks and (ii) when Columns are considered as blocks of the RCBD:

$$R.E.\,(RCBD, ROW) = \frac{E_r + (t-1)E_e}{(t)(E_e)}$$

$$R.E.\,(RCBD, COLUMN) = \frac{E_c + (t-1)E_e}{(t)(E_e)}$$

In our example:

$$R.E.\,(RCBD, ROW) = \frac{0.010051563 + (3)(0.021597395)}{(4)(0.021597395)}$$

$$\frac{0.010051563 + 0.064792185}{0.08638958} = \frac{0.074843748}{0.08638958}$$

$$= 0.866351566.$$

$$R.E.\,(RCBD, COLUMN) = \frac{0.27578075 + (3)(0.021597395)}{(4)(0.021597395)} = \frac{0.340572935}{0.08638958}$$

$$= 3.9423 \approx 3.94.$$

Because the error df in the LSD is only 6, and thus less than 20, the estimated R.E. values must be adjusted by multiplying each by an Adjustment Factor, k, defined as:

$$k = \frac{[(t-1)(t-2)+1][(t-1)^2+3]}{[(t-1)(t-2)+3][(t-1)^2+1]} = \frac{[(4-1)(4-2)+1][(4-1)^2+3]}{[(4-1)(4-2)+3][(u-1)^2+1]}$$

$$\frac{[(3)(2)+1][9+3]}{[(3)(2)+3][9+1]} = \frac{(7)(12)}{(9)(10)} = \frac{84}{90} = 0.9333$$

The Adjusted R.E. values are re-computed as:

$$R.E.\,(RCBD, ROW) = (0.86635)(0.9333) = 0.8086 \approx 0.81$$

$$R.E.\,(RCBD, COLUMN) = (3.9333)(0.9333) = 3.67935 \approx 3.68$$

For this trial, a RCBD with Columns as blocks would have been as efficient as a LSD.

11.6.2 A Numerical Example (2) of a Latin Square Design (LSD)

An Agronomist, Department of Crop Science, Uniben, investigated the yields of four varieties of cowpeas on the University Farmland, in a Latin Square Design, and obtained the following data (in kilograms) per plot:

Table 11.6.1: Yields (kg) of four varieties of cowpeas in LSD in a Uniben Farmland

Rows	1	2	3	4	Total
1	C= 10.5	D = 7.7	B = 12.0	A = 13.2	43.4
2	B = 11.1	A = 12.0	C = 10.3	D = 7.5	40.9
3	D = 5.8	C = 12.2	A = 11.2	B = 13.7	42.9
4	A = 11.6	B = 12.3	D = 5.9	C =10.2	40.0
Total	**39.0**	**44.2**	**39.4**	**44.6**	**167.2**

1. The Structural Model:

$X_{ijk} = \mu + \alpha_i + \beta_j + Y_k + \varepsilon_{ijk}$, where:

X_{ijk} = Random experimental unit which is independent and normally distributed with mean, μ, and variance, σ^2. i.e. $\left(X_{ijk} \sim N(\mu, \sigma^2)\right)$

μ = Fixed overall population mean which is independent and normally distributed, and common to all treatments and replications.

α_i = Fixed Row Effect, which is normally distributed with mean zero and variance, σ^2, subject to the restriction that:. $\sum \alpha_i = 0$

β_j = Fixed Column Effect, which is normally distributed with mean zero and variance σ^2 and subject to the restriction that: .$\sum \beta_j = 0$

Y_k = Fixed Treatment Effect which is normally distributed with mean zero and variance, σ^2 and subject to the restriction that: $\sum Y_k = 0$

ε_{ijk} = Random Experimental Error due to uncontrolled sources of variation and normally distributed with mean, zero and variance, σ^2 i.e .$\varepsilon_{ijk} \sim N(o, \sigma^2)$

$\left(\mu + \alpha i + \beta_j + Y_k + \varepsilon_{ijk}\right)$ implies additivity of all model components.

2. The Hypotheses that can be tested are:

i. $H_0 : \alpha_1 = \alpha_2 = \alpha_3 = \alpha_4 = 0$. i.e. Row-blocking effect on cowpea yields is zero..

$H_1 : \alpha_1 \neq \alpha_2 \neq \alpha_3 \neq \alpha_4 \neq 0$. i.e. Row-blocking Effect is not zero, or it has some effect on cowpea yields.

ii. $H_0: \beta_1 = \beta_2 = \beta_3 = \beta_4 = 0$. i.e. Column-blocking has no effect on cowpea yields.

$H_1: \beta_1 \neq \beta_2 \neq \beta_3 \neq \beta_4 \neq 0$. i.e. Column-blocking has some effect on cowpea yields.

iii. $H_0: Y_1 = Y_2 = Y_3 = Y_4 = 0$. i.e. Treatment Effect is zero; has no effect on cowpea yields.

$H_1: Y_1 \neq Y_2 \neq Y_3 \neq Y_4 \neq 0$. i.e. Treatment Effect is not zero; has some effect on cowpea yields.

3. State significance level: $\alpha = 0.05$ or $\alpha = 0.01$.

4. Decision Rule: If F-calculated is greater than F-tabulated, reject H0 and conclude that F is significant; otherwise accept H0 as specified.
5. Computations:

Obtain a Table of Treatment totals:

Treatments

	A	B	C	D
Col 1	11.6	11.1	10.5	5.8
Col 2.	12.0	12.3	12.2	7.7
Col 3.	11.2	12.0	10.3	5.9
Col 4.	13.2	13.7	10.2	7.5
Total	48.0	49.1	43.2	26.9

$$Grand\ Total\ (T\ ...) = \sum_{i=1}^{r} \sum_{j=1}^{c} \sum_{k=1}^{n} X_{ijk} =$$

$$10.5 + 11.1 + 5.8 + 11.6 + \cdots 13.7 + \cdots + 10.2 = 167.2$$

$$Correction\ Factor\ (CF) = \frac{\left(\sum_{i=1}^{r} \sum_{j=1}^{c} \sum_{k=1}^{n} X_{ijk}\right)^2}{r^2} = \frac{(167.2)^2}{16} = \frac{17,955.84}{16}$$

$$= 1,747.24$$

$$SS_{TOTAL} = \sum_{i=1}^{r} \sum_{j=1}^{c} \sum_{k=1}^{n} X_{ijk}^2 - CF$$

$$= 1,837.64 - 7,747.24 = \underline{\textbf{90.40.}}$$

$$SS_{ROWS} = \frac{\sum_{i}^{r} T_{i..}^2}{r} - CF = \frac{43.4^2 + 40.9^2 + 42.9^2 + 40.0^2}{4} - 1,747.24$$

$$= \frac{6,996.78}{4} - 1,747.24 = 1,749.195 - 1747.24 = 1.955$$

$$SS_{COLUMNS} = \frac{\sum_{i}^{r} T_{.j.}^2}{r} - CF = \frac{39.0^2 + 44.2^2 + 39.4^2 + 44.6^2}{4} - 1,747.24$$

$$= \frac{7,016.16}{4} - 1,747.24 = 1,754.04 - 1747.24 = 6.80$$

$$SS_{TREATMENT} = \frac{\sum_{j}^{r} T_{t}^2}{R} - CF = \frac{48.0^2 + 49.1^2 + 43.2^2 + 26.9^2}{4} - 1,747.24$$

$$= \frac{7,304.66}{4} - 1,747.24 = 1826.165 - 1747.24 = 78.925$$

$$SS_{ERROR} = SS_{TOTAL} - SS_{ROWS} - SS_{COLUMNS} - SS_{TRT}$$

228

$$90.40 - 1.955 - 6.80 - 78.925 = \underline{\textbf{2.72}}$$

Table 11.6.1 a: Anova Table for a Latin Square Design on Yield of four Varieties of Cowpeas in Uniben Farm

Source of variation	DF	SS	MS	F_{CAL}	$F_{0.05}$	$F_{0.01}$
Row	$r-1= 3$	1.955	0.651667	$F_1=1.4375^{ns}$	4.76	9.78
Column	$c-1=3$	6.80	2.266667	$F_2 = 5.000^*$	4.76	9.78
Treatment	$t-1 = 3$	78.925	26.308333	$F_3=58.033^{**}$ *	4.76	9.78
Error	$(r-1)(r-2) = 6$	2.72	0.453333			
Total	$r^2-1 = 15$	90.40				

6. **Conclusions:**

Since $F_{ROWS} = 1.44 \ll F_{0.05} = 4.76$, accept H0 and conclude that Row-blocking has no effect on the yields of Cowpeas.

Since $F_{COLUMNS} = 5.00 > F_{0.05} = 4.76$, reject H0, but at F0.01 = 9.78, accept H0.

Since $F_{TRT} = 58.03 \ggg F_{0.05} = 4.76$, and $F_{0.01} = 9.78$, reject H0 and conclude that the treatment is highly significantly different and so significantly affect cowpeas yields.

7. Efficiencies of Row-and Column-Blockings

The efficiencies of both Row-and Column-blockings in a Latin Square Design indicate the gain in precision, relative to either the CRD, or the RCBD.

1. The Relative Efficiency of a LSD as compared to a CRD is given by:

$$R.E.(CRD) = \frac{E_r + E_c + (t-1)E_e}{(t-1)E_e}, where$$

$$E_r = Row\ MS; E_c = Column\ MS;$$

$$E_e = Error\ MS\ in\ the\ Latin\ Square\ Design$$

t = number of Treatments.

From the Anova Table above:

$$R.E.(LSD\ Compared\ to\ CRD): \frac{0.651667 + 2.266667 + (3)(0.453333)}{(5)(0.453333)}$$

$$\frac{2.918334 + 0.1359999}{2.266665} = \frac{3.0543339}{2.266665} = 1.34750$$

This means that the use of LSD increased experimental precision by 135%. This implies that if the CRD had been used, an estimated 1.35 times (or approximately 2 times) more replications would have been required to detect the treatment difference of the same magnitude as that detected with the LSD.

2. The Relative Efficiency of a LSD, as compared to RCDB can be computed in two ways –

i. when Rows are considered as blocks, and (ii) when columns are considered as blocks of the RCBD:

a)
$$R.E.(RCBD, Row) = \frac{E_r + (t-1)E_e}{(t)(E_e)}$$

b) $$R.E.(RCBD, Column) = \frac{E_c + (t-1)E_e}{(t)(E_e)}$$

In our example above:

a)
$$R.E.(RCBD, Row) = \frac{0.651667 + (3)(0.453333)}{(5)(0.453333)}$$

$$\frac{0.651667 + 0.1359999}{1.813332} = \frac{2.011666}{1.813332} = 1.109375448$$

$$R.E.(RCBD, Column) = \frac{2.266667 + (3)(0.453333)}{(4)(0.453333)} = \frac{3.626666}{1.813332} = 2.0000011.$$

Since the Error degree of freedom is only 6, which is less than 20, we must compute the Adjustment Factor, k, to adjust each R.E. as:

$$k = \frac{[(t-1)(t-2)+1][(t-1)^2+3]}{[(t-1)(t-2)+3][(t-1)^2+1]} = \frac{[(3)(2)+1][9+3]}{[(3)(2)+3][9+1]} = \frac{(7)(12)}{(9)(10)} = \frac{84}{90}$$
$$= 0.933333$$

The Adjusted R.E. (RCBD, Row) = (1.109375448)(0.933333) = 1.035416715 ≈ 1.035.

Adjusted R.E. (RCBD, Column) = (2.0000011)(0.933333) = 1.866667027 ≈ ~1.867.

Therefore, using RCBD with Rows as the Blocks would have improved efficiency by only 3.5%, while RCBD with columns as Blocks would have improved efficiency by 86.7%. Therefore, the LSD is still far superior to an RCBD.

Thus, R.E. (LSD compared to CRD) as above, would need to be adjusted as:

Adjusted R.E. = (1.347501241)(0.933333) = 1.257667376 ≈ 1.258.

CRD would have improved efficiency by only 25.8%. Therefore, LSD is still superior to CRD in this experiment.

11.7.0 Nested Design (Hierarchal or Hierarchical Classification)

In Nested Designs, each sample is composed of sub-samples, which in turn, may be composed further of other sub-samples. This repeated sampling and sub-sampling gives rise to Nested, or Hierarchal, or Hierarchical Designs, or Samples-within-Samples.

Nested Classification are slightly more complicated cases of single classification anova. A single classification anova is frequently insufficient to represent the complexity of a given experiment and extract all the relevant information from it.

For example, the Department of Forestry and Wildlife, Uniben, was interested in evaluating the three regional Forestry Training Schools, each with two instructors who teach Field Practical Training (FPT)

Classes. The Department was concerned with the effect of School (Factor A) and Instructors (Factor B) on the learning that the students achieved. To investigate these effects, classes were formed and randomly assigned to one of the two instructors in each school. This was done for two sessions, and at the end of each session, the class was examined on the materials taught and the average class grade recorded.

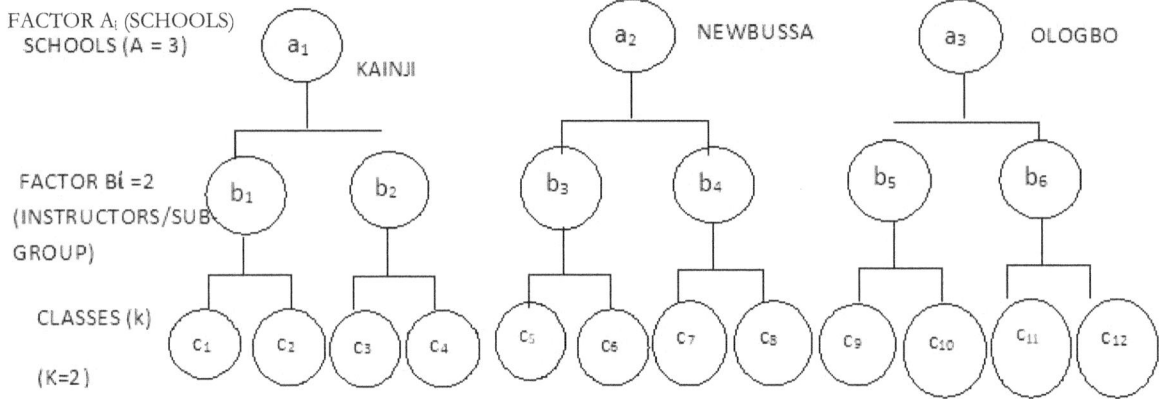

Schematic Representation of the Nested FPT Training

The study is not an ordinary two-factor study because the Instructors in Kainji did not also teach in the other two schools in New Bussa and Ologbo. Each level of Factor B (Instructors) occurs with only one level of factor A (School). Specifically, each Instructor teaches in only one school. Therefore, factor B (Instructors) and the classes are said to be **Nested within Factor A** (Schools).

Nested Design Model:

Let Y_{ijk} denote the kth observation when factor A is at the ith level and Factor B is at the jth level. When Factors A and B have fixed effects, an appropriate nested design model is given by:

$$Y_{ijk} = \mu_{\ldots} + \alpha_i + \beta_{j(i)} + \varepsilon_{k(ij)}, \text{ where: } \quad i = 1, 2, \ldots, a; j = 1, 2, \ldots, b; k = 1, 2, \ldots, n.$$

Y_{ijk} = Random experimental unit which is independent and normally distributed with mean, μ, and variance, σ^2, i.e. $Y_{ijk} \sim N(\mu, \sigma^2)$.

$\mu_{\ldots}$ = Fixed overall population mean, which is independent and normally distributed and common to all treatments and replications.

α_i = Fixed factor A Effect which is normally distributed with mean zero and variance, σ^2, subject to the restriction that: $\sum \alpha_i = 0$

$\beta_{j(i)}$ = Fixed Factor B Effect which is nested within Factor A and subject to the restriction that:

$$\sum \beta_{j(i)} = 0$$

$\varepsilon_{j(ij)}$ = Random Experimental Error due to uncontrolled sources of variation, which are normally distributed with mean zero and variance, σ^2.

Table 11.7.1: Numerical Example of the FPT Training

Factor A$_i$ (Schools)

		Kainji (1)	New Bussa (2)	Ologbo (3)

Factor Bj

(Instructor = Sub-Group)		(1) (2)	(3) (4)	(5) (6)
	(1)	25 14	11 22	17 5
Class (k)	(2)	29 11	6 18	20 2

Sub–Group Total $\sum_{k=1}^{n} Y_{ij}$

54   25   17   40   37   7

$$School = Group\ Total$$

$$= \sum_{i}^{a} \sum_{j}^{b} Y_{ij}$$

79   57   44

$$Grand\ Total = \sum_{i=1}^{a} \sum_{j=i}^{b} \sum_{k=1}^{n} Y_{ijk}$$

180

$$= Y...$$

1. $Y_{ijk} = \mu ... + \alpha_i + \beta_{j(i)} + \varepsilon_{k(ij)}$
2. .The Hypotheses that can be tested in this Nested Design are:

   i. $H_0: \alpha_1 = \alpha_2 = \alpha_3 = 0$ .i.e. There is no School Effect.

   $H_1: \alpha_1 \neq \alpha_2 \neq \alpha_3 \neq 0.$ i.e. School has some effect;

   At least one school effect is not equal to zero.

   ii. $H_0: \beta_1(\alpha_1) = \beta_1(\alpha_2) = \beta_1(\alpha_3) = \beta_2(\alpha_1) = \beta_2(\alpha_2) = \beta_2(\alpha_3) = 0$

$H_1: \beta_1(\alpha_1) \neq \beta_1(\alpha_2) \neq \beta_1(\alpha_3) \neq \beta_2(\alpha_1) \neq \beta_2(\alpha_2) \neq \beta_2(\alpha_3) \neq 0$

3. **State the Significance level**: $\alpha = 0.05\ or\ \alpha = 0.01$
4. **Decision Rule**: If F-calculated is greater than F-tabulated, reject H0 and conclude that F is significant; otherwise accept H$_0$ as stated. $\left(\dfrac{\sum_{i=1}^{a} \sum_{j=i}^{b} \sum_{k=1}^{n} Y_{ijk}}{abn} \right)^2$

5. **Computations:**

$$Grand\ Total = Y... = \sum_{i=1}^{a} \sum_{j=i}^{b} \sum_{k=1}^{n} Y_{ijk} = 25 + 29 + \cdots + 5 + 2 = 180$$

$$\text{Correction Factor } (CF) = \left(\frac{\sum\limits_{i=1}^{a} \sum\limits_{j=i}^{b} \sum\limits_{k=1}^{n} Y_{ijk}}{abn} \right)^2 = \frac{180^2}{3 \times 2 \times 2} = \frac{32,400}{12} = 2,700$$

$$SS_{TOTAL} = \sum_{i=1}^{a} \sum_{j=i}^{b} \sum_{k=1}^{n} Y_{IJK}^2 - CF = 25^2 + 29^2 + 14^2 + \cdots + 5^2 + 2^2 - 2,700$$

$$= 3,466 - 2,700 = 766.$$

$$SS_{School} = \frac{\sum_{i=1}^{a} Y_{i..}^2}{bn} - \frac{Y_{...}^2}{abc} = \frac{79^2 + 57^2 + 44^2}{2 \times 2} - 2,700$$

$$= 2,856.5 - 2,700 = 156.5.$$

$$SS_{Instructors\,Within\,Schools} = \sum_{i}^{a} \sum_{j=1}^{b} \frac{Y_{ij.}^2}{n} - \frac{\sum_{i}^{a} Y_{i..}^2}{bn}$$

$$SS_{B(A)} = \frac{54^2 + 25^2 + 17^2 + 40^2 + 37^2 + 7^2}{2} - 2,856.5.$$

$$= 3,424 - 2,856.5 = 567.5$$

$$SS_{Error} = \sum_{i=1}^{a} \sum_{j=i}^{b} \sum_{k=1}^{n} Y_{ijk}^2 - \frac{\sum_{i}^{a} \sum_{j}^{b} Y_{ij.}^2}{n} = 3,466 - 3,424 = 42$$

$$= SS_{Total} - SS_{B(A)}$$

Note: In Nested Designs, we always test lower levels before upper levels. Thus, we first test: MSSUB-GROUP/MSWITHIN for the significance of $\sigma^2{}_{B(A)}$ and then test MSGROUP/MSSUB-GROUP for the significance of σ_A^2.

Table 11.7.1 a: Anova Table of a Nested Design (Fixed Model)

Source of Variation	DF	SS	MS	F_{CAL}	$F_{0.05}$	$F_{0.01}$	EMS (Fixed Model)
Schools(A)	a-1 = 2	156.5	78.25	$F_1 = 11.2^{**}$	5.14	10.9	$\sigma^2 + bn \dfrac{\sum \alpha_i}{(a-1)}$
Instructors Within Schools (B(A))	A (b-1) = 3	567.5	189.17	$F_2 = 27.0^{**}$	4.76	9.78	$\sigma^2 + n \dfrac{\sum \sum \beta_{j(i)}^2}{a(b-1)}$
Error	ab(n-1) = 6	42.0	7.00				σ^2
Total	abn-1 = 11	766.0					

6. **Conclusions**:
   i. Since F1-calculated = 11.2 >> F0.05 = 5.14 and F0.01 = 10.9, reject H0 and conclude that the FPT-Training Schools are highly significant in the students' learning.

ii. Since F2-calculated = 27.0 >> F0.05 = 4.76 and F0.01 9.78, reject H0 and conclude that the Instructors within the FPT-Training Schools are also highly significant in the learning process of the students.

11.7.2 Numerical Example of a Nested Design (Mixed Model)

A Post-graduate student, Department of Forestry and Wildlife, Uniben, investigated the body weight gains of four snails (Factor B = 4) reared in each of three cages (Factor A = 3) and took two independent measurements of the body weight of each snail. The study was conducted in a Nested Design. The data are presented in Table 11.7.2:

Table 11.7.2: Body Weights of Snails Reared in 3 Cages

Factor A	Factor B (Snails)	No. of Measurements per Snail (n = 2)		Snail Total (Sub-Group)	Cage Total (Group)
		(1)	(2)		
Cage I	S_1	58.5	59.5	118.0	
	S_2	77.8	80.9	158.9	582.7
	S_3	84.0	83.6	167.6	
	S_4	70.1	68.3	138.4	
Cage II	S_1	69.8	69.8	139.6	
	S_2	56.0	54.5	110.5	479.7
	S_3	50.7	49.3	100.0	
	S_4	63.8	65.8	129.6	
Cage III	S_1	56.6	57.5	114.1	
	S_2	77.8	79.2	157.0	536.8
	S_3	69.9	69.2	139.1	
	S_4	62.1	64.5	126.6	

This is an example of a Mixed Model where the Cages (Factor A) are fixed and the Snails (Factor B) are random.

1. The Structural Model: $Y_{ijk} = \mu + \alpha_i + \beta_{j(i)} + \varepsilon_{k(ij)}$. See above Model.
2. Hypotheses to be tested are:
    (i) $H_0: \alpha_1 = \alpha_2 = \alpha_3 = 0$. i.e The Cages have no effect on Snails' Body weight gain.

    $H_1: \alpha_1 \neq \alpha_2 \neq \alpha_3 \neq 0$. i.e The Cages have some effect on Snails' Body weight gain.

    (ii) $H_0: \sigma^2_{B(A_1)} = \sigma^2_{B(A_2)} = \sigma^2_{B(A_3)} = \sigma^2$. i.e The variances of the snails nested within cages are equal to the population variance.

    $H_1: \sigma^2_{B(A_1)} \neq \sigma^2_{B(A_2)} \neq \sigma^2_{B(A_3)} = \sigma^2$ i.e The variances are not all equal.

a) **State the significance level:** $\alpha = 0.05$; $\alpha = 0.01$
b) **Decision Rule:** If F-calculated is greater than F-tabulated, reject H0 and conclude that F-cal is significant; otherwise accept H0 as stated.
c) **Computations:**

Grand Total = Y... = $\sum_{i=1}^{a} \sum_{j=i}^{b} \sum_{k=1}^{n} Y_{ijk}$ = 58.5+59.5+ ... +62.1+64.5 = 1599.2

$$\text{Correction Factor} = \frac{\left(\sum_{i=1}^{\alpha} \sum_{j=1}^{b} \sum_{k=1}^{n} Y_{ijk}\right)^2}{3 \times 4 \times 2} = \frac{1599.2^2}{24} = \frac{2{,}557{,}440.64}{24} \ 106{,}560.0267$$

$$SS_{Total} = \sum_{i=1}^{a}\sum_{j=i}^{b}\sum_{k=1}^{n} Y_{ijk}^2 - CF$$

$$= 58.8^2 + 59.5^2 + \cdots + 62.1^2 + 64.5^2 - 106{,}560.0267$$

$$= 108{,}962.0 - 106560.0267 = \underline{2{,}401.9733.}$$

$$SS_{SNAILS}(or\ SS_{SUB-GROUP}) = \frac{\sum_{i=1}^{a}\sum_{j=1}^{b}\left(\sum_{k=1}^{n} Y_{ijk}\right)^2}{n} - CF$$

$$= \frac{118.0^2 + 158.9^2 + \ldots + 139.1^2 + 126.6^2}{2} - 106{,}560.0267$$

$$= 108{,}946.38 - 106{,}560.0267 = \underline{2{,}386.3533.}$$

$$SS_{CAGES}(or\ SS_{GROUP}) = \frac{\sum_{i=1}^{a}\left(\sum_{j=1}^{b}\sum_{k=1}^{n} Y_{ijk}\right)}{bn} - CF$$

$$= \frac{582.7^2 + 479.7^2 + \ldots + 536.8^2}{4 \times 2} - 106{,}560.0267$$

$$= 107{,}225.7025 - 106{,}560.0267 = \underline{665.6758.}$$

$$SS_{SUB-GROUP\ WITHIN\ GROUP_0}\left(or\ SS_{SNAILS\ WITHIN\ CAGES}\right)$$

$$= \frac{\sum_{i=1}^{a}\sum_{j=1}^{b}(\sum_{k=1}^{n} Y_{ijk})^2}{n} - \frac{\sum_{i=1}^{a}(\sum_{j=1}^{b}\sum_{k=1}^{n} Y_{ijk})^2}{bn}$$

$$= 108{,}946.38 - 107{,}225.7025 = \underline{1{,}720.6775.}$$

$$SS_{WITHIN\ SNAILS_0}\left(or\ SS_{WITHIN\ SUB-GROUPS}\right)or\ SS_{ERROR}$$

$$= \left[\sum_{i=1}^{a}\sum_{j=1}^{b}\sum_{k=1}^{n} Y_{ijk}^2\right] - \frac{\sum_{i}^{a}\sum_{j}^{b}(\sum_{k}^{n} Y_{ijk})^2}{n}$$

$$i.e.\ SS_{TOTAL} - SS_{SNAILS} = 108{,}962.0 - 108{,}946.38$$

$$= \underline{15.62.}$$

Table 11.7.2a: Anova Table for a Mixed Model Nested Design

Source of Variation	DF	SS	MS	F_{CAL}	$F_{0.05}$ (2,9)	$F_{0.01}$ (9,12)
Cages	a-1 = 2	665.6758	332.84	$F_1 = 1.74^{ns}$	4.26	8.02
Snails within cages	a(b-1) = 9	1,720.6775	191.19	$F_2 = 147.07^{***}$	2.80	4.39
Within Snails (Error)	ab(n-1) = 12	15.62	1.30			
Total	**abn-1 = 23**	**2,401.9733**				

d) **Conclusions:**

i. Since F1 = 1.74 < F0.05(2,9) = 4.26, accept H0 and conclude that the cages had no effect on the snails' body weight.

ii. Since F2 = 147.07 >>> F0.05(9,12) = 2.80 and F0.01(9,12), = 4.39 reject H0 and conclude that the variances of the snails within cages are highly significantly different.

Although the snails differed from each other for genetic reasons, or possibly due to environmental influences before being brought into the cages, or during their development, the fact that they were reared in different cages did not add significant variation to their body weights.

11.7.3 A Numerical Example of a Nested Design (Model II)

A Crop Scientist investigated the calcium concentration in pumpkin leaves. He randomly selected four (4) pumpkin plants (A=4) and randomly selected three leaves (B=3) from each plant and made two determinations (n = 2) per leaf and obtained the following data:

Table 11.7.3: Calcium Concentration in Pumpkin Leaves

Plant (A_i)	Leaf $(B)_{ij}$	Determinations $X_{ijk}(n = 2)$		Leaf Total	Plant Total
	1	3.28	3.09	6.37	
1	2	3.52	3.48	7.00	19.05
	3	2.88	2.80	5.68	
	1	2.46	2.44	4.90	
2	2	1.87	1.92	3.79	13.07
	3	2.19	2.19	4.38	
	1	2.77	2.66	5.43	
3	2	3.74	3.44	7.18	17.71
	3	2.55	2.55	5.10	
	1	3.78	3.87	7.65	
4	2	4.07	4.12	8.19	22.46
	3	3.31	3.31	6.62	

1. The Mathematical Model:

$$X_{ijk} = \mu + A_i + \beta_{j(i)} + \varepsilon_{k(ij)}$$ -- since it is a variance components model, where:

$A_i = N(o, \sigma_A^2)$ i.e. Random component with mean, zero, variance σ_A^2.

$\beta_{j(i)} = N(o, \sigma_{B(A)}^2)$ i.e. Random Component with mean zero, variance, $(o, \sigma_{B(A)}^2)$.

$\varepsilon_{x(ij)} = N(o, \sigma_\varepsilon^2)$ i.e. Error component with mean zero and variance, σ_ε^2.

2. Hypotheses to be tested are:

i. $H_0: \sigma_{A_1}^2 = \sigma_{A_2}^2 = \sigma_{A_3}^2 = \sigma_{A_4}^2 = \sigma_{A_\varepsilon}^2$ i.e. Variation between plants is the same.

$H_1: \sigma_{A_1}^2 \neq \sigma_{A_2}^2 \neq \sigma_{A_3}^2 \neq \sigma_{A_4}^2 \neq \sigma_{A_\varepsilon}^2$

$H_0: \sigma_{B_1(A_1)}^2 = \sigma_{B_2(A_1)}^2 = \sigma_{B_3(A_1)}^2 = \sigma_{B_1(A_2)}^2 = \sigma_{B_2(A_2)}^2 = \sigma_{B_3(A_2)}^2 = \cdots =$

ii. $\sigma_{B_1(A_4)}^2 = \sigma_{B_2(A_4)}^2 = \sigma_{B_3(A_4)}^2 = \sigma_\varepsilon^2$

$H_1:$ Not all Equal.

3. **Computations**: Same as our examples above:

$$Grand\ Total = \sum_{i}^{a} \sum_{j}^{b} \sum_{k}^{n} X_{ijk} = 3.28 + 3.52 + 2.88 + \cdots + 3.87 + 4.12 + 3.31$$
$$= 72.29$$

$$CF = \frac{(\sum\sum\sum X_{ijk})^2}{abn} = \frac{(72.9)^2}{4 \times 3 \times 2} = \frac{5225.8441}{24} = 217.743504$$

$$SS_{DETERMINATIONS} = SS_{TOTAL} = \sum_{i}^{a} \sum_{j}^{b} \sum_{k}^{n} X_{ijk}^2 - CF$$

$$3.28^2 + 3.09^2 + \ldots + 3.31^2 + 3.31^2 - 2.17.743504^2 = \underline{\mathbf{10.2704}}$$

$$SS_{LEAVES} = \frac{\sum_{i}^{a} \sum_{j}^{b} \left(\sum_{k}^{n} X_{ijk}\right)^2}{n} - CF$$
$$= \frac{6.37^2 + 7.00^2 + \cdots 8.19^2 + 6.62^2}{2} - 217.743504$$

$$= 10.1905.$$

$$SS_{PLANT} = \frac{\sum_{I}^{a} \left(\sum_{j}^{b} \sum_{k}^{n} X_{ijk}\right)^2}{bn} - CF$$

$$= \frac{19.05^2 + 13.07^2 + 13.71^2 + 22.46^2}{3 \times 2} - 217.743504 = 7.5603$$

$$SS_{LEAVES-WITHIN-PLANTS} = SS_{LEAVES} - SS_{PLANTS} = 10.1905 - 7.5603 = 2.6302$$

$$SS_{DETRMINATION-WITHIN\ LEAVES} = SS_{DETRMINATIONS} - SS_{LEAVES}$$

$$= 10.2704 - 10.1905 = \underline{0.0799}$$

Table 11.7.3 a: Anova Table for a Model II Nested Design

Source of Variation	DF	SS	MS	F_{CAL}	$F_{0.05}$ (3,8)	$F_{0.01}$ (8,12)
Plants	a-1 = 3	7.5603	2.5201	$F_1 = \dfrac{2.5201}{0.3288} = 7.66^{***}$	4.07	7.59
Leaves within Plants	a(b-1) = 8	2.6302	0.3288	$F_2 = \dfrac{0.3288}{0.0067} = 49.0^{***}$	2.85	4.50
Determination within Leaves.	ab(n-1) = 12	0.0799	0.0067			
Total	abn-1 = 23	10.2704				

4. Conclusions:
    i. Highly significant.
    ii. Highly significant.

11.8.0 FACTORIAL EXPERIMENTS

Traditionally, as Winer (1971) observed, Factorials are not regarded as a type of Design but as a **"Method of Experimentation"** and hence they are known as **"Factorial type Experiments"**. Factorial Experiments are known as **"Volume Increasing"** because large numbers of treatments or factors can be handled simultaneously and their interactions investigated. Therefore, the information obtained from factorial experiments is more complete than that obtained from a series of single-factor experiments, in the sense that factorial experiments permit the evaluation of interaction effects. At the end of a factorial experiment, the experimenter has information which permits him to make decisions which have a broad range of applicability.

When the number of treatment combinations is large, it is not possible to measure the basic experimental error and therefore we have to use an interaction term as a measurement of experimental error, and of course, on the assumption that no added interaction effect is present.

Another problem which accompanies a factorial anova, with several main effects, is the large number of possible interactions. For instance, a three-factor factorial has:

- Three first-order or two-factor interactions: AB; AC and BC;
- One second-order or three-factor interaction: ABC.

A four-factor factorial has:

- Six first-order or two-factor interactions: AB; AC; AD; BC; BD and CD;
- Four second-order or three-factor interactions: ABC; ABD; ACD; BCD;
- One third-order or four-factor interaction: ABCD.

Not only is the computation of these interactions tedious but the testing of their significance and, more importantly, their interpretation becomes exceedingly complex.

11.8.1 Terminologies and Notation

Winer (1971) and Steele and Torrie (1981) explained some of the terminologies and notation of factorial experiments:

1. **Factor**

   The term "Factor" is used interchangeably with the terms "treatment" and "experimental variable". More specifically, a factor is a series of related treatments or related classifications. There are two major types of factors- - qualitative and quantitative factors.

2. **Levels of a Factor.**

   The Related treatments making up a factor constitutes the "levels" of that factor. For example, a factor **"Dosage"** may consist of four levels": 1cc, 3cc, 5cc and 7cc. The number of levels within a factor is determined largely by the thoroughness with which an experimenter desires to investigate the factor, or the kind of inferences the experimenter desires to make upon conclusion of the experiment.

3. **Dimensions of a Factorial Experiment**

   The "Dimensions" of a factorial experiment are indicated by the number of factors and the number of levels of each factor. For instance, a factorial experiment in which there are two factors, one having three levels and the other having four levels, is called a "3 by 4 factorial" experiment. In a 2 x 3 x 5 factorial experiment, there are three factors, having respective levels of two, three and five.

A 2 x 3 factorial experiment can be represented schematically as A by B (A x B):

Levels of factor B

Levels of Factor A	b$_1$	b$_2$	b$_3$
a$_1$	ab$_{11}$	ab$_{12}$	ab$_{13}$
a$_2$	ab$_{21}$	ab$_{22}$	ab$_{23}$

Thus, a 2 x 3 factorial experiment, six possible combinations of treatments may be formed.

Generally, in a two-factor experiment, the first factor is denoted A with p-levels and the second factor is denoted B with q-levels. Hence the experiment is designated as a p x q factorial experiment. If n elements are to be observed under each treatment combination in a p x q factorial experiment, a random sample of npq elements from the population is required, assuming no repeated measurements on the same elements. The npq elements are then sub-divided, at random, into pq sub-samples of size n each. These sub-samples are then assigned at random to the treatment combinations.

4. **Fixed and Random Factors**

Fixed Factors: The potential (or population) levels of factor A will be designated by the **symbols** a$_1$, a$_2$ . . ., a$_p$.. When p, the number of levels of factor A included in the experiment is equal to P (the population levels), then factor A is called a "fixed factor". Also, when the selection of the p-levels from the potential P-levels is determined by some systematic, non-random procedure, then factor A is also considered a fixed factor- - because the selection procedure reduces the potential P-levels to p-effective levels.

Random Factors: If the p-levels of factor A included in the experiment represents a random sample from the potential P-levels, then factor A is considered to be a Random Factor. In most practical situations in which random factors are encountered, p is quite small relative to P, and the ratio p/P is quite close to zero.

The ratio of the number of levels of a factor in an experiment to the potential number of levels in the population is called the Sampling Fraction for a factor.

5. **Interaction**

Interaction is a measure of the non-additivity of the main effects. The interaction between level ai and level bj, designated by the symbol $\alpha\beta_{ij}$, is a measure of the extent to which the criterion mean for treatment combination abij cannot be predicted from the sum of the corresponding main effects.

Due to its efficiency and comprehensiveness (Snedecor and Cochran, 1967), factorial experimentation is extensively used in research programs, particularly in industry and agriculture.

Above all, factorial experiments can be used in all other experimental designs.

6. **Main Effects**

Cox (1958) in a two-factor factorial, A and B, each at two levels, the Main Effect of factor A is defined as **the average effect of the factor A over all the levels of B**. Therefore, the main effect

of A depends on the particular levels of B used in the experiment.

In a two-factor factorial, A and B, Snedecor and Cochran (1967) observed that there are four treatment combinations, a1b1, a2b1, a1b2, a2b2, and each replication of this experiment supplies two estimates of the effect of A. The comparison, a2b2 – a1b2, estimates the effect of A, when B is held constant at its higher level, while the comparison, a2b1 – a1b1, estimates the effect of A, when B is held constant at its lower level. The average of these two estimates is called the **Main Effect of A –** that is, an average taken over the levels of the other factor.

239

The main effect of A may be expressed (Snedecor and Cochran, 1967) as:

LA = ½ (a2b2) + ½ (a2b1) – ½ (a1b1)-1/2(a1b1) and

$$\text{Variance of LA} = \frac{\sigma^2}{r}\left[\left(\frac{1}{2}\right)^2 + \left(\frac{1}{2}\right)^2 + \left(\frac{1}{2}\right)^2 + \left(\frac{1}{2}\right)^2\right] = \frac{\sigma^2}{r}.$$

The main Effect of B is the comparison:

$$LB = \frac{1}{2}(a_2 b_2) + \frac{1}{2}(a_1 b_2) - \frac{1}{2}(a_2 b_1) - \frac{1}{2}(a_1 b_1) \; with \; variance = \frac{\sigma^2}{r} \; or \; \frac{\sigma^2}{2}.$$

11.8.2 Structural Model of Two-Factor Factorial: (p x q) Factorial with n Observations per Cell)

This model assumes that the factorial effects, as well as, the experimental error are additive, i.e. that an observation is a linear function of the factorial effects and the experimental error:

$$X_{ijk} = \mu + \alpha_i + \beta_j + \alpha\beta_{ij} + \varepsilon_{ijk}, \text{where:}$$

X_{ijk} = Random experimental unit, which is independent and normally distributed with mean, , and variance, σ^2, i.e. $X_{ijk} \sim N(\mu, \sigma^2)$.

μ = Constant overall population mean, common to all treatments and replications.

α_i = Fixed factor A effect, that is subject to the restriction, $\sum \alpha_i = 0$.

β_j = Fixed factor B effect that is subject to the restriction, $\sum \beta_j = 0$

$\alpha\beta_{ij}$ = Fixed factor AB interaction effect that is subject to the restriction, $\sum \alpha\beta_{ij} = 0$

ε_{ijk} = Random Experimental error, independent, normally and identically distributed, due to uncontrolled sources of variation i.e. $\varepsilon_{ijk} \sim N(o, \sigma^2)$.

$\left(\mu + \alpha_i + \beta_j + \alpha\beta_{ij} + \varepsilon_{ijk}\right)$ implies additivity of model components.

Note: If factor A is fixed, $\overline{\alpha\beta_{ij}} = 0$ and $\sigma^2_{\alpha\beta} = 0$. But if factor A is a random factor, $\sigma^2_{\alpha\beta}$ will be a function of n, p, P.

Hypotheses that can be tested are:

i. $H_0 : \alpha_1 = \alpha_2 = 0$. i.e. The Effect of Factor A is zero i.e. $\sigma^2_\alpha = 0$.

$H_1 : \alpha_1 \neq \alpha_2 \neq 0$. i.e. The Effect of Factor A is not zero. i.e. $\sigma^2_\alpha \neq 0$.

ii. $H_0 : \beta_1 = \beta_2 = 0$. i.e. The Effect of Factor B is zero; $\sigma^2_\beta = 0$.

$H_1 : \beta_1 \neq \beta_2 \neq 0$. i.e. The Effect of Factor B is not zero; $\sigma^2_\beta = 0$.

iii. $H_0 : \alpha\beta_{11} = \alpha\beta_{12} = \alpha\beta_{21} = \alpha\beta_{22} = 0$. i.e. There is no interaction between factors A and B. . $\sigma^2_{\alpha\beta} = 0$

$H_1 : \alpha\beta_{11} \neq \alpha\beta_{12} \neq \alpha\beta_{21} \neq \alpha\beta_{22} \neq 0$. i.e. There are some interaction between factors A and B. i.e. $\sigma^2_{\alpha\beta} \neq 0$

11.8.2.1 A Numerical Example of a 2 X 3 Factorial Experiment with 3 Observations per Cell

A Mensurationist, Department of Forestry and Wildlife, Uniben, was interested in evaluating the relative effect of three spacings (0.5m, 1.0m, 1.5m) and the age (1wk and 2 wks) of seedling transplantation after

240

germination, on the survival rates of the seedlings, after transplanting them on Nursery beds. A random sample of 9 1-wk old seedlings were selected and transplanted three per cell, under the three spacings; similarly, 9 2-wk old seedlings were randomly selected and transplanted, three seedlings per cell, under the three spacings as shown in Table 11.8.2.

The Hypotheses to be tested are:

    i.   The Spacing of the Seedlings has no effect on their survival rate.
    ii.   The Age of the Seedlings has no effect on their survival rate.

Table 11.8.2: Effect of Spacing and Seedling Age on Survival Rates

Spacing of the Seedlings

Age of Seedlings	0.5m (b_1)			1.0m (b_2)			1.5m (b_3)			
										n= no. of obs per cell = 3.
1-wk old (a_1)	8	4	0	10	8	6	8	6	4	p = levels of A = 2.
2-weeks old (a_2)	14	10	6	4	2	0	15	12	9	q = levels of B = 3.

Procedure:

The same procedure for all Anovas will still apply:

Structural Model: $X_{ijk} = \mu + \alpha_i + \beta_j + \alpha\beta_{ij} + \varepsilon_{ijk}$.

1. State the various hypotheses that can be tested.
2. State the significance level: $\alpha = 0.05 \ or \ 0.01$
3. Decision Rule: If F-calculated is greater than F-tabulated, reject H0 and conclude that Fcalculated is significant; otherwise accept H0 as stated.
4. Computations:

The first computation is to obtain AB Summary Table as:

Table 11.8.2 a: AB Summary Table.

	0.5m (b_1)	1.0m (b_2)	1.5m (b_3)	Total A_i
1-week old (a_1)	12	24	18	54
2-week old (a_2)	30	6	36	72
Total B_j	42	30	54	126

$$Gand\ Total = T\ldots = \sum_{i=1}^{p}\sum_{j=1}^{q}\sum_{k=1}^{n} X_{ijk} = 8 + 4 + 0 + \cdots + 15 + 12 + 9 = 126$$

$$Correction\ Factor\ (CF) = \frac{\left(\sum_{i=1}^{p}\sum_{j=1}^{q}\sum_{k=1}^{n} X_{ijk}\right)^2}{pqn} = \frac{(126)^2}{2\times3\times3} = \frac{15,876}{18} = 882.0$$

$$SS_{TOTAL} = \sum_{i=1}^{P}\sum_{j=1}^{q}\sum_{k=1}^{n} X_{ijk}^2 - CF = 8^2 + 4^2 + 0^2 + \cdots + 12^2 + 9^2 - 882.0$$

$$= 1198 - 882.0 = \underline{316.0}.$$

$$SS_A = \frac{\sum_i^{p} T_{i.}^2}{qn} = \frac{\sum_{i=1}^{p}\left(\sum_{j=1}^{q}\sum_{k=1}^{n} X_{ijk}\right)^2}{qn} = \frac{54^2 + 72^2}{3 \times 3} = \frac{8100}{9} - CF$$

241

$$= 900 - 882 = \underline{18.0}.$$

$$SS_B = \frac{\sum_i^p T_{.j.}^2}{pn} - CF = \frac{\sum_{j=1}^q \left(\sum_{i=1}^p \sum_{k=1}^n X_{ijk} \right)^2}{pn} - CF = \frac{42^2 + 30^2 + 52^2}{2 \times 3} - 882$$

$$= \frac{5{,}580}{6} - 882 = 930 - 882 = 48.0.$$

$$SS_{AB} = \frac{\sum_k^n T_{ij.}^2}{n} - CF - SS_A - SS_B = \frac{\sum_{i=1}^p \sum_{j=1}^q \left(\sum_{k=1}^n X_{ijk}^2 \right)^2}{n} - CF - SS_A - SS_B.$$

$$= \frac{12^2 + 30^2 + 24^2 + 6^2 + 18^2 + 36^2}{3} - 882 - 18 - 48$$

$$= \frac{3{,}276}{3} - 882 - 18 - 48 = 1092 - 882 - 18 - 48$$

$$= \underline{144.0}.$$

$$SS_{ERROR} = SS_{TOTAL} - SS_A - SS_B - SS_{AB}$$

$$= 316 - 18.0 - 48.0 - 144.0$$

$$= \underline{106.0}.$$

Table 11.8.2 b: Anova Table for a 2x3 Factorial

Source of variation	DF	SS	MS	F_{cal}	$F_{0.05}$	$F_{0.01}$
A (Seedling Age)	p-1 = 1	18	18	$F_1 = 2.038^{ns}$	4.75	9.33
B (Spacing)	q-1 = 2	48	24	$F_2 = 2.717^{ns}$	3.89	6.98
AB Interaction	(p-1)(q-1) = 2.	144	72	$F_3 = 8.151^{***}$	3.89	6.93
Error	pq(n-1) = 12	106	8.883			
Total	npq-1 = 17	316				

Conclusions:

i. Since F1cal = 2.04 < F0.05 = 4.75, accept H0 that the Age has no effect on the seedlings survival rate.

ii. Since F2cal = 2.72 < F0.05 = 3.89, accept H0 that spacing has no effect on the seedlings' survival rate.

iii. Since F3-cal = 8.15 >> F0.05 = 3.89 and F0.01 6.93, reject H0 and conclude that the interaction between Age and Spacing is highly significant (p<<0.005).

The Nature of Interaction Effects

To investigate the nature of interaction effects, we need to compute cell means, or cell totals, and plot these cell means, or cell totals, in a Geometric Representation. Geometric representation is a valuable tool in the interpretation of interactions. Where there is no interaction, the lines are parallel. When interaction is significant, main effects must be interpreted with caution.

Spacings of the Seedlings

Table 11.8.2 c: Cell Means:

Source of variation	DF	SS	MS	F_{cal}	$F_{0.05}$	$F_{0.01}$
A (Seedling Age)	p-1 = 1	18	18	$F_1 = 2.038^{ns}$	4.75	9.33
B (Spacing)	q-1 = 2	48	24	$F_2 = 2.717^{ns}$	3.89	6.98
AB Interaction	(p-1)(q-1) = 2.	144	72	$F_3 = 8.151^{***}$	3.89	6.93
Error	pq(n-1) = 12	106	8.883			
Total	npq-1 = 17	316				

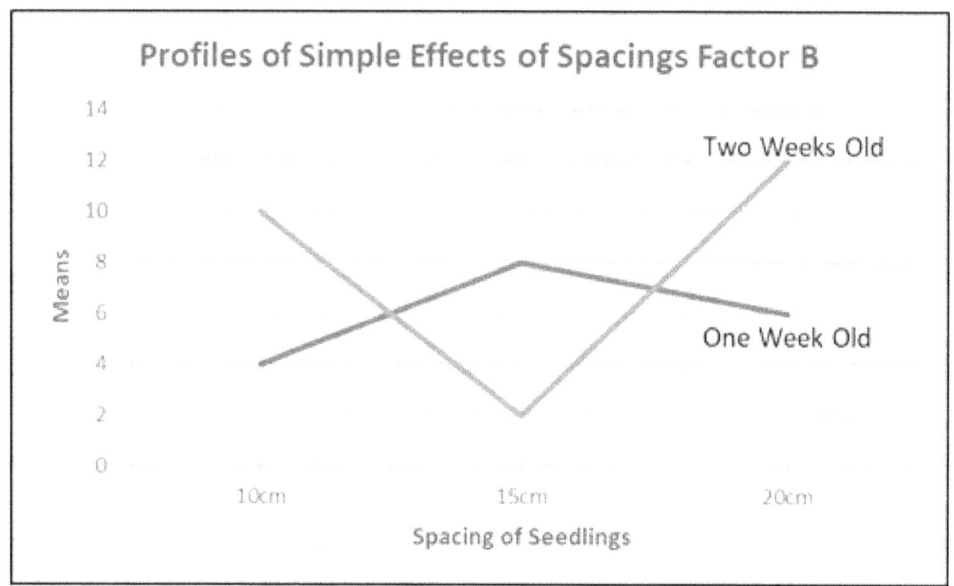

Fig. 11.8.2.2a: A Geometric Representation of Interaction

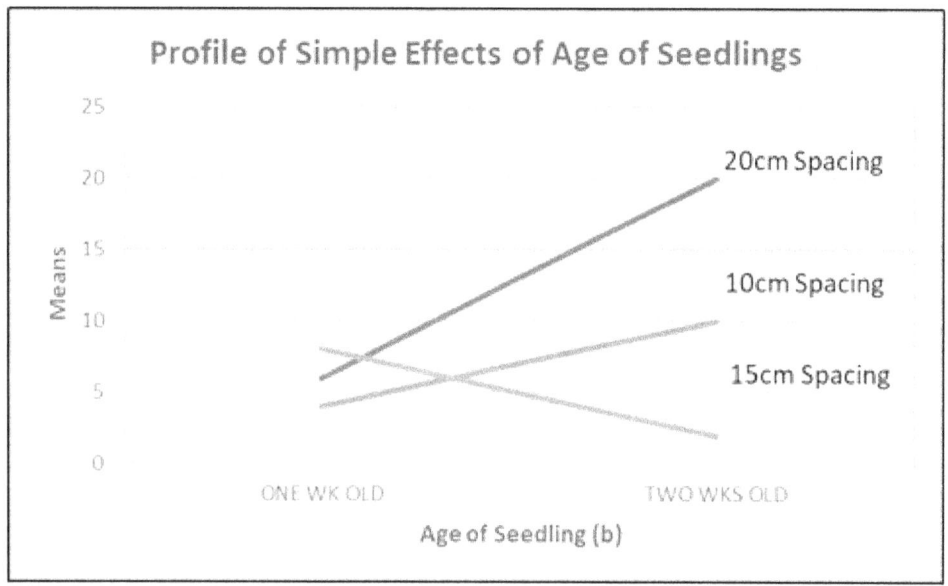

Figure 11.8.26: A Geometric Representation of Interaction.

11.8.3 A Three-Factor Factorial (p x q x r) Experiment with n Observations per Cell

In a three-factor experiment, factor A has p-levels; factor B has q-levels and factor C has r-levels and n observations per cell. There are three two-factor, or first-order interactions: AB, AC, BC; and one three-factor (or second-order) interaction, ABC. Therefore, each of these interactions will have a summary table for computing the interactions, as well as, the main effects.

For a special case of 2 x 3 x 2 factorial experiment, there are pqr = (2)(3)(2) = 12 treatment combinations as illustrated in the following table where a typical treatment combination is designated by the notation abcijk- - where i indicates the level of factor A; j, the level of factor B; k, the level of factor C.

Table 11.8.3: ABC Summary Table:

Instructors

Training Method		C_1			C_2			
		b_1	b_2	b_3	b_1	b_2	b_3	Total
Educational level	a_1	ABC_{111}	ABC_{121}	ABC_{131}	ABC_{112}	ABC_{122}	ABC_{132}	A_1
	a_2	ABC_{211}	ABC_{221}	ABC_{231}	ABC_{212}	ABC_{222}	ABC_{232}	A_2
Total		BC_{11}	BC_{21}	BC_{31}	BC_{12}	BC_{22}	BC_{32}	G

The column totals in this ABC summary Table have the general form:

$$BC_{jk} = \sum_{i}^{p} ABC_{ijk} = \sum_{i}^{p} \sum_{m}^{n} X_{ijkm}$$

i.e. a column total represents the sum of all observations under treatment combination , bc_{jk} the levels of factor A being disregarded.

The BC summary table has the following form:

Table 11.8.3a: BC Summary Table

	b_1	b_2	b_3	Total
C_1	BC_{11}	BC_{21}	BC_{31}	C_1
C_2	BC_{12}	BC_{22}	BC_{32}	C_2
Total	B_1	B_2	B_3	G

Treatment combination ab_{ij} is defined to be the collection of treatment combinations: $abc_{ij1}, abc_{ij2}, .$ $abc_{ij3} ... abc_{ijr}$ The sum of all observations at level ab_{ij} is thus:

$$AB_{ij} = \sum_{k}^{r} ABC_{ijk} = \sum_{k}^{r} \sum_{m}^{n} X_{ijkm}$$

i.e. AB11 = ABC111 + ABC112.

AB23 = ABC231 + ABC232.

Table 11.8.3b: AB Summary Table

	b_1	b_2	b_3	Total
a_1	AB_{11}	AB_{12}	AB_{13}	A_1
a_2	AB_{21}	AB_{22}	AB_{23}	A_2
Total	B_1	B_2	B_3	G

The symbol AC_{ik} is used to designate the sum of all observations at level ac_{ik} Thus,

$$AC_{ik} = \sum_{j}^{q} ABC_{IJK} = \sum_{j}^{q} \sum_{m}^{n} X_{ijkm}$$

$i.e.\ AC_{12} = ABC_{112} + ABC_{122} + ABC_{132}.$

Table 11.8.3c: AC Summary Table

	C_1	C_2	Total
a_1	AC_{11}	AC_{12}	A_1
a_2	AC_{21}	AC_{22}	A_2
Total	C_1	C_2	G

The sum of all observations at level a1 may be obtained as follows:

$$A_1 = \sum_{j}^{q} AB_{ij} = \sum_{k}^{r} AC_{ik} = \sum_{j}^{q}\sum_{k}^{r} ABC_{ijk} = \sum_{j}^{q}\sum_{k}^{r}\sum_{m}^{n} X_{ijkm}$$

The sum of all observations at level bj is given by:

$$B_j = \sum_{i}^{p} AB_{ij} = \sum_{k}^{r} BC_{jk} = \sum_{i}^{p}\sum_{k}^{r} ABC_{ijk} = \sum_{i}^{p}\sum_{k}^{r}\sum_{m}^{n} X_{ijkm}$$

$$Similarly, C_k = \sum_{i}^{p} AC_{ik} = \sum_{j}^{q} BC_{JK} = \sum_{I}^{P}\sum_{J}^{Q} abc_{ijk} = \sum_{i}^{p}\sum_{j}^{q}\sum_{m}^{n} X_{ijkm}.$$

11.8.3.1 Definition of Computation Symbols

1. $G^2/npqr$ = Correction Factor (CF).
2. $\sum X_{ijkm}^2$ = Total SS
3. $(\sum A_i^2)/nqr$ = Sum of Squares A.
4. $(\sum B_j^2)/npr$ = Sum of Squares B.
5. $(\sum C_k^2)/npq$ = Sum of Squares C.
6. $\left[\left(\sum AB_{ij}\right)^2\right]/nr$ = Sum of Squares AB.
7. $\left[\sum(AC_{ik})^2\right]/nq$ = Sum of Squares AC.
8. $\left[\sum(BC_{jk})^2\right]/np$ = Sum of Squares BC.
9. $\left[\sum(ABC_{ijk})^2\right]/n$ = Sum of Squares ABC.

Note: In $\sum A_i$, the subscripts j, k and m are missing. The number of levels corresponding to these missing subscripts are respectively q, r, n. The number of observations summed to obtain Ai is the product of these missing subscripts, qrn. In Bj, the subscripts, i, k, m are missing. Therefore, the number of observations summed is prn and so on.

11.8.3.2 A Numerical Example of 2 x 3 x 2 Factorial Experiment

A Mensurationist in the Department of Forestry and Wildlife, Uniben, evaluated the relative effect of three Factors on the height growth of Garcinia cola seedlings. He employed two levels of watering the seedlings (Factor A) once a week and twice a week; three spacings (Factor B) at 10cm, 15cm and 20cm, and two age levels (Factor C) at planting (1 week old and 2-weeks old). The Mensurationist randomly selected 10 seedlings in each grouping and obtained the following data in a 2 x 3 x 2 factorial experiment:

Table 11.8.3.2: ABC Summary Table:

Watering levels	C₁ (1-week old)			C₂ (2-weeks old)			
	Spacing of seedlings						
	b_1	b_2	b_3	b_1	b_2	b_3	Total
a_1	20	30	12	16	33	8	119
a_2	36	38	40	40	44	42	240
Total	56	68	52	56	77	50	359

Test the hypotheses that the treatments and their interactions have no effect on the height growth of the seedlings at $\alpha = 0.05$ and 0.01

1. The Mathematical Model of a 2 x 3 x 2 Factorial Experiment is:

$X_{ijkm} = \mu + \alpha_i + \beta_j + \gamma_k + \alpha\beta_{ij} + \alpha\gamma_{ik} + \beta\gamma_{jk} + \alpha\beta\gamma_{ijk} + \varepsilon_{ijkm}$, where:

X_{ijkm} = Random experimental unit which is independent and normally distributed with mean

μ, and variance, σ^2 i.e. $X_{ijkm} \sim N(\mu, \sigma^2)$.

μ = Constant overall population mean, common to all treatments and replications.

α_i = Fixed Factor A effect, subject to the restriction, $\sum \alpha_i = 0$.

β_j = Fixed Factor B effect, subject to the restriction, $\sum \beta_j = 0$.

γ_k = Fixed Factor C effect, subject to the restriction, $\sum \gamma_k = 0$

$\alpha\beta_{ij}$ = Fixed factor AB interaction Effect, subject to the restriction, $\sum \alpha\beta_{ij} = 0$.

$\alpha\gamma_{ik}$ = Fixed factor AC interaction Effect, subject to the restriction, $\sum \alpha\gamma_{ik} = 0$.

$\beta\gamma_{jk}$ = Fixed factor BC interaction Effect, subject to the restriction, $.\sum \beta\gamma_{jk} = 0$

$\alpha\beta\gamma_{ijk}$ = Fixed factor ABC interaction Effect, subject to the restriction,

$\sum \alpha\beta\gamma_{ijk} = 0$
.

ε_{ijkm} = Random experimental error, independent, normally and identically distributed with mean

zero and variance, σ^2 i.e. $\varepsilon_{ijkm} \sim N(0, \sigma^2)$.

$\left(\mu + \alpha_i + \beta_j + \gamma_k + \alpha\beta_{ij} + \alpha\gamma_{ik} + \beta\gamma_{jk} + \alpha\beta\gamma_{ijk} + \varepsilon_{ijkm}\right)$ implies additivity of model components.

2. Hypotheses that can be tested here:

   i. $H_0: \alpha_1 = \alpha_2 = 0$.. i.e The Effect of Factor A is zero on seedling height growth.

   $H_1: \alpha_1 \neq \alpha_2 \neq 0$. i.e The Effect of Factor A is not zero on seedling height growth.

   ii. $H_0: \beta_1 = \beta_2 = \beta_3 = 0$. .i.e. The Effect of Factor B is zero on seedling height growth.

   $H_1: \beta_1 \neq \beta_2 \neq \beta_3 \neq 0$. i.e. The Effect of Factor B is not zero on seedling height growth.

   iii. $H_0: \gamma_1 = \gamma_2 = 0$. .i.e. The Effect of Factor C is zero on seedling height growth.

$H_1: Y_1 \neq Y_2 \neq 0$. i.e. The Effect of Factor C is not zero on seedling height growth.

iv. $H_0: \propto \beta_{11} = \propto \beta_{12} = \propto \beta_{13} = \propto \beta_{21} = \propto \beta_{22} = \propto \beta_{23} = 0$..i.e. There is no interaction between Factors A and B.

$H_1: \propto \beta_{11} \neq \propto \beta_{12} \neq \propto \beta_{13} \neq \propto \beta_{21} \neq \propto \beta_{22} \neq \propto \beta_{23} \neq 0$. i.e. There is some interaction between Factors A and B.

v. $H_0: \propto Y_{11} = \propto Y_{12} = \propto Y_{21} = \propto Y_{22} = 0$. .i.e. There is no interaction between Factors A and C.

$H_1: \propto Y_{11} \neq \propto Y_{12} \neq \propto Y_{21} \neq \propto Y_{22} \neq 0$. i.e. There is some interaction between the Factors A and C.

vi. $.H_0: \beta\gamma_{11} = \beta\gamma_{12} \dots = \beta\gamma_{32} = 0$ i.e. There is no interaction between Factors B and C.

$\beta\gamma \neq \beta\gamma_{22} \neq \beta\gamma_{32} \neq 0$ i.e. There is some interaction between Factors B and C.

vii. $.H_0: \alpha\beta\gamma_{111} = \alpha\beta\gamma_{112} = \dots = \alpha\beta\gamma_{232} = 0$ i.e. There is no interacion among the three Factors.

$.H_0: \alpha\beta\gamma_{111} \neq \alpha\beta\gamma_{112} \neq \alpha\beta\gamma_{121} \neq \dots \neq \alpha\beta\gamma_{232} \neq 0$ i.e. There is interaction among the three Factors.

3.   **State the Significance level:** $\propto = 0.05$ and 0.01.
4.   **Decision Rule:** If any of the F-calculated is greater than F-tabulated reject H0 and conclude that the F is significant, otherwise accept H0 as stated.

Computations:
1.   Obtain the AB, AC and BC summary Tables as:

Table 11.8.3.2a   AB Summary Table:

	b_1	b_2	b_3	Total (A_i)
a_1	36	63	20	119
a_2	76	82	82	240
Total (B_j)	112	145	102	359

Table 11.8.3.2b   AC Summary Table:

	c_1	c_2	Total (A_i)
a_1	62	57	119
a_2	114	126	240
Total (C_k)	176	183	359

Table 11.8.3.2c   BC Summary Table:

	b_1	b_2	b_3	Total (C_k)
c_1	56	68	52	176
c_2	56	77	50	183
Total (B_j)	112	145	102	359

$$Grand\ Total = G = \sum_{i=1}^{p}\sum_{j=1}^{q}\sum_{k=1}^{r}\sum_{m=1}^{n} X_{ijkm} = 20: 36 + \dots + 44 + 42 = 359$$

$$Correction\ Factor\ (CF) = \frac{\left(\sum_{i=1}^{p}\sum_{j=1}^{q}\sum_{k=1}^{r}\sum_{m=1}^{n}X_{ijkm}\right)^2}{p\quad q\quad r\quad n} = \frac{(359)^2}{2\ x\ 3\ x\ 2\ x\ 10}$$

$$= \frac{128,881}{120}$$

$$= \underline{1,074.008333}.$$

$$SS_{TOTAL} = \sum_{i=1}^{p}\sum_{j=1}^{q}\sum_{k=1}^{r}\sum_{m=1}^{n}X_{ijkm}^2\ (Not\ available\ in\ the\ data\ above) - CF$$

$$= 1,360 - 1,074.008333 = 285.991667$$

$$SS_A = \frac{\sum_i^p\left(\sum_j^q\sum_k^r\sum_m^n X_{ijkm}\right)^2}{qrn} - CF = \frac{(119)^2 + (240)^2}{3\ x\ 2\ x\ 10} - 1,074.008333$$

$$= \frac{71,761}{60} = 1,074.008333 = 1,196.016667 - 1,074.008333$$

$$= \underline{122.0088337}.$$

$$SS_B = \frac{\sum_{j=1}^{q}\left(\sum_{i=1}^{p}\sum_{k=1}^{r}\sum_{m=1}^{n}X_{ijkm}\right)^2}{prn} - CF$$

$$= \frac{(112)^2 + (245)^2 + (102)^2}{2\ x\ 2\ x\ 10} - 1,074.008333$$

$$\frac{43,973}{40} - 1,074.008333 = 1,099.325 - 1,074.008333$$

$$= \underline{25.316667}.$$

$$SS_C = \frac{\sum_{k=1}^{r}\left(\sum_{i=1}^{p}\sum_{k=1}^{q}\sum_{m=1}^{n}X_{ijkm}\right)^2}{pqn} - CF = \frac{(176)^2 + (183)^2}{2\ x\ 3\ x\ 10} - 1,074.008333$$

$$\frac{64,465}{60} - 1,074.008333 = 1,074.416667 - 1,074.008333.$$

$$= \underline{0.408333666}.$$

$$SS_{AB} = \frac{\sum_{i=1}^{p}\sum_{j=1}^{q}\left(\sum_{k=1}^{r}\sum_{m=1}^{n}X_{ijkm}\right)^2}{rn} - CF - SS_A - SS_B =.$$

$$= 1,244.45 - 1,074.008333 - 122.008334 - 25.316667$$

$$= \underline{23.116666}.$$

$$SS_{AC} = \frac{\sum_{i=1}^{r}\sum_{k=1}^{p}\left(\sum_{j=1}^{q}\sum_{m=1}^{n}X_{ijkm}\right)^2}{q\ n} - CF - SS_A - SS_C.$$

$$= \frac{62^2 + 57^2 + 114^2 + 126^2}{3 \times 10} - CF - SS_A - SS_C.$$

$$\frac{35,965}{30} - 1,074.008333 - 122.008334 - 0.4083337.$$

$$= 1,198.833333 - 1,074.008333 - 122.008334 - 0.4083337.$$

$$= \underline{2.4083326}.$$

$$SS_{BC} = \frac{\sum_{j=1}^{q} \sum_{k=1}^{r} \left(\sum_{i=1}^{p} \sum_{m=1}^{n} X_{ijkm}\right)^2}{p\, n} - CF - SS_B - SS_C.$$

$$\frac{56^2 + 68^2 + 52^2 + 56^2 + 77^2 + 50^2}{2 \times 10} - 1,074.008333 - 25.316667 - 0.4083337.$$

$$\frac{22,029}{20} - 1,074.008333 - 125.31666 - 0.4083337.$$

$$= 1,101.45.833333 - 1,074.008333 - 25.316667 - 0.4083337.$$

$$= \underline{1.7166663}.$$

$$SS_{ABC} = \frac{\sum_{i=1}^{p} \sum_{j=1}^{q} \sum_{k=1}^{r} \left(\sum_{m=1}^{n} X_{ijkm}\right)^2}{n} - CF - SS_A - SS_B - SS_C - SS_{AB}$$
$$- SS_{AC} - SS_{BC}$$

$$\frac{20^2 + 30^2 + \cdots + 44^2 + 42^2}{10} - 1,074.008333 - 122.0083337 - 25.316667$$
$$- 0.4083337 - 23.11666 - 2.4083326 - 1.7166663$$

$$= \frac{12,493}{10} - 1,074.008333 - 122.008337 - 25.316667 - 0.4083326 - 1.7166663$$

$$= 1,249.3 - 1,074.008333 - 122.0083337 - 25.316667 - 0.4083326 - 23.116666.$$

$$= 2.4083326 - 1.7166663 = \underline{0.3166677}.$$

$$SS_{ERROR} = SS_{TOTAL} - SS_A - SS_B - SS_C - SS_{AB} - SS_{AC} - SS_{BC} - SS_{ABC}.$$

$$285.991667 - 122.0083337 - 25.316667 - 0.4083337 - 23.116666 - 2.4083326 - 1.7166663$$

$$- 0.3166677 =$$

$$= \underline{108.9833355}.$$

Anova Table for a Three-Factor Factorial

Source of Variation	DF	SS	MS	F_{CAL}	$F_{0.05}$	$F_{0.01}$
A (Rate of Watering)	p-1 = 1	122.0083337	122.0083337	$F = 120.9075^{***}$	3.96	6.97
B(Spacing of Seedlings)	q-1 = 2	25.316667	12.658334	$F_2 = 12.544^{***}$	3.11	4.89
C(Age of Seedlings)	r-1 = 1	0.4083337	0.4083337	$F_3 = 0.4046^{ns}$		
AB Interaction	(p-1)(q-1) = 2	23.116666	11.558333	$F_4 = 11.454^{***}$		
AC Interaction	(p-1)(r-1) = 1	2.4083326	2.4083326	$F_5 = 2.387^{ns}$		
BC Interaction	(q-1)(r-1)=2	1.7166663	0.858333	$F_6 = 0.851^{ns}$		
ABC Interaction	(p-1) (q-1) (r-1)=2	0.3166677	0.158334	$F_7 = 0.157^{ns}$		
Error	Pqr(n-1) = 108	108.9833355	1.009105			
Total	Pqrn − 1 = 119	285.991667				

Conclusions:

i. Since F1 = 120.908 >>> F05;1,108 = 3.96 and F01,2,108 = 6.97, reject H0 and conclude that the rate of watering was highly beneficial to the height-growth of the seedlings.

ii. Since F2 = 12.544 >>> F05,2,108 = 3.11 and F012,108 = 4.89, reject H0 and conclude that the spacing was also highly beneficial to the height-growth of the seedlings.

iii. Since F3 = 0.405 < F05 1,108=3.96, accept H0 and conclude that the Age of the seedlings had no effect on their growth in height.

iv. Since F4 = 11.454 >>> F05, 2,108 = 3.11 and F01, 2,108 = 4.89, reject H0 and conclude that the AB interaction was highly significant and hence we need to be cautious in interpreting the main Effects.

v. Since F5, F6 and F7 are not significant, we accept H0 and conclude that all other interactions have no effect on the height-growth of the seedlings.

11.8.3.3 Computations of Standard Errors for testing Main Effects and Interactions

In order to compare main effects, as well as, interaction effects in factorial experiments, we need to compute standards errors $(S_{\bar{x}})$ that can be used in the least significant difference (LSD) and the Least Significant Studentized Ranges (LSR).

Using the three-factor factorials of the 2 x 3 x 2 numerical example above and Alika (2003), we outline the procedure for computing the standard errors:

1. Obtain the structural Model:

$$X_{ijkm} = \mu + \alpha_i + \beta_j + Y_k + \alpha\beta_{ij} + \alpha Y_{ik} + \beta Y_{jk} + \alpha\beta Y_{ijk} + \varepsilon_{ijkm}$$

2. Use the "Definition of Computation Symbols" in section 11.8.3.1.and the **"missing subscripts"** principle to obtain the divisors for the various standard errors.

3. Obtain the value of MSE from the Anova Table and use the "missing subscripts" principle as the divisors for the MSE.

   a) Standard Errors (s.e.) for the Main Effects are computed as:

i. $\text{S.E.} (\alpha_i) = \sqrt{\dfrac{MSE}{qrn}} = \sqrt{\dfrac{1.009105}{3 \times 2 \times 10}}$

$$= \sqrt{\dfrac{1.009105}{60}} = \sqrt{0.016818416} = 0.129685838 \approx 0.130$$

Since j, k, m, are the missing subscripts for dividing MSE.

ii. $\text{S.E.} (\beta_j) = \sqrt{\dfrac{MSE}{prn}}$, since i, k, m, are the missing subscripts for dividing MSE.

$$\text{S.E.} = = \sqrt{\dfrac{1.009105}{2 \times 2 \times 10}} = \sqrt{0.025227635} = 0.158832065 \approx 0.159$$

iii. $\text{S.E.} (\gamma_k) = \sqrt{\dfrac{MSE}{pqn}} = \sqrt{\dfrac{1.009105}{2 \times 3 \times 10}} = \sqrt{0.016818416} = 0.129685838$

iv. Standard Errors for the two-way Interaction Effects are computed as:

b) $\text{S.E.} (\alpha\beta_{ij}) = \sqrt{\dfrac{MSE}{rn}} - -$ since k and m are the missing subscripts in $\alpha\beta_{ij}$.

$$= \sqrt{\dfrac{1.009105}{2 \times 10}} = \sqrt{0.05045525} = 0.224622461 \approx \mathbf{0.225}.$$

c) $\text{S.E} (\alpha\gamma_{ik}) = \sqrt{\dfrac{MSE}{qn}} - -$ since j and m are the missing subscripts in $\alpha\gamma_{ik}$.

$$= \sqrt{\dfrac{1.009105}{3 \times 10}} = \sqrt{0.033633833} = 0.183395292 \approx \mathbf{0.183}.$$

d) $\text{S.E} (\beta\gamma_{jk}) = \sqrt{\dfrac{MSE}{pn}} - -$ since i and m are the missing subscripts in $\beta\gamma_{jk}$.

$$= \sqrt{\dfrac{1.009105}{2 \times 10}} = \sqrt{0.05045525} = 0.224622461 \approx \mathbf{0.225}.$$

v. Standard Error for three-way Interaction Effect is computed as:

$\text{S.E} (\alpha\beta\gamma_{ijk}) = \sqrt{\dfrac{MSE}{n}} - -$ since m and is the only subscript missing subscripts in $\alpha\beta_{ijk}$.

$$= \sqrt{\dfrac{1.009105}{10}} = \sqrt{0.1009105} = 0.31766413 \approx \mathbf{0.318}.$$

11.8.3.4 A Numerical Example of a 3x3 Factorial in RCBD

A Mensurationist, Department of Forestry and Wildlife, UNIBEN, Benin, investigated the effect of three spacings (factor B) on the height growth of the seedlings of three indigenous tree species (Factor A) of

Terminalia ivorensis, Terminatia superba and Milicia excelsa, for a period of one year. The experiment was conducted in a 3 x 3 factorial arranged in RCBD as presented in Table 11.8.3.4.

Table11.8.2.2: One-year Height-growth (cm) of three tree species in A 3 x 3 Factorial in RCBD

Species (A_i)	Spacings (B_j)	BLOCKS				Total
		1	2	3	4	
	10cm x 10cm	56	45	43	46	190
M. excelsa	15cm x 15cm	60	50	45	48	203
	20cm x 20cm	66	57	50	50	223
	10cm x 10cm	65	61	60	63	249
T. superba	15cm x 15cm	60	58	56	60	234
	20cm x 20cm	53	53	48	55	209
	10cm x 10cm	60	61	50	53	224
T. ivorensis	15cm x 15cm	62	68	67	60	257
	20cm x 20cm	73	77	77	65	292
Total	-	555	530	496	500	2,081

1. Mathematical Model of a 3 x 3 Factorial Arranged in RCBD:

$X_{ijk} = \mu + r_k + \alpha_i + \beta_j + \alpha\beta_{ij} + \varepsilon_{ijk}$, where:

X_{ijk} = Random Experimental unit, which is independent and randomly distributed with mean, μ, and variance, σ^2, i.e. $X_{ijk} \sim N(\mu, \sigma^2)$

μ = Fixed overall population mean, common to all treatments and replications.

r_k = Fixed Block Effect, subject to the restriction, $\sum r_k = 0$

α_i = Fixed Species Effect, subject to the restriction, $\sum \alpha_i = 0$

β_j = Fixed Spacing Effect, subject to the restriction, $\sum \beta_j = 0$

$\alpha\beta_{ij}$ = Fixed Species–Spacing Interaction, subject to the restriction, $\sum \alpha\beta_{ij} = 0$

ε_{ijk} = Random Experimental Error, independent, normally and identically distributed, and due to uncontrolled sources of variation, i.e. $\varepsilon_{ijk} \sim N(0, \sigma^2)$

$\left(\mu + r_k + \alpha_i + \beta_j + \alpha\beta_{ij} + \varepsilon_{ijk}\right)$ implies additivity of all model components.

Hypotheses that can be tested in a 3 x 3 factorial arranged in RCBD:

$H_o: r_1 = r_2 = r_3 = r_4 = 0.$ i.e. The Block Effect is equal to zero: or Blocking has no effect.

$H_1: r_1 \neq r_2 \neq r_3 \neq r_4 \neq 0$ i.e. The Block Effect is not equal to zero or Blocking has some effect.

$H_o: \alpha_1 = \alpha_2 = \alpha_3 = 0$ i.e. There is no difference among the Species' height growth.

$H_1: \alpha_1 \neq \alpha_2 \neq \alpha_3 \neq 0$ i.e. There are differences among species' height growth.

$H_o: \beta_1 = \beta_2 = \beta_3 = 0$ i.e. Spacing has no effect on the species' height growth.

$H_1: \beta_1 \neq \beta_2 \neq \beta_3 \neq 0$ i.e. Spacing has some effect on the species' height growth.

252

$H_o: \alpha\beta_{11} = \alpha\beta_{12} = \alpha\beta_{13} = \alpha\beta_{21} = \alpha\beta_{22} = \alpha\beta_{23} = \alpha\beta_{31} = \alpha\beta_{32} = \alpha\beta_{33} = 0.$ i.e. There is no interaction.

H_1: At least one interaction is not equal to zero: There are significant interactions.

2. **State the significance level**: $\alpha = 0.05 \; or \; \alpha = 0.01$
3. **Decision Rule:** If any of the F-calculated is greater than F-tabulated, reject H0 and conclude that the F-calculated is significant, otherwise accept H0 as stated.
4. **Computations** for Testing these Hypotheses:

First, obtain the AB Summary Table as:

SPACINGS

Table 11.3.4a: AB (pq) Summary Table:

Species	10cm x 10cm	15cm x 15cm	20cm x20cm	
M. excelsa	190	203	223	616
T. superba	249	234	209	692
T. worensis	224	257	292	773
Total	**663**	**694**	**724**	**2,081**

$$Grand\ Total = G = \sum_{k-1}^{r}\sum_{i=1}^{p}\sum_{j=1}^{q} X_{ijk} = 56 + 60 + 66 + \cdots + 53 + 60 + 65 = 2,081$$

$$Correction\ Factor\ (CF) = \frac{\left(\sum_{k-1}^{r}\sum_{i=1}^{p}\sum_{j=1}^{q} X_{ijk}\right)^2}{pqr} = \frac{(2,081)^2}{3\times3\times4} = \frac{4,330,561}{36}$$

$\underline{= 120,293.361111}$

$$SS_{TOTAL} = \sum_{k=1}^{r}(\sum_{i=1}^{p}\sum_{j=1}^{q} X_{ijk})^2 \frac{56^2+60^2+66^2+\cdots++53^2+60^2+65^2}{36} \; CF = -CF$$

$= 122,921 - 120,293.361111 = 2,627.64$

$$SS_{BLOCKS} = \frac{\sum_{k=1}^{r}(\sum_{i=1}^{p}\sum_{j=1}^{q} X_{ijk})^2}{pq}$$
$$= \frac{555^2 + 530^2 + 496^2 + 500^2}{3\times3} - 120,293.36$$

$= \frac{1,081,941}{9} - 120,293.36 = 120,549 - 120,293.36 = \underline{255.64}$

$$SS_A\left(or\ SS_{SPECIES}\right) = \frac{\sum_{k=1}^{r}(\sum_{i=1}^{p}\sum_{j=1}^{q} X_{ijk})^2}{pr} - CF = \frac{616^2+692^2+773^2}{3\times3}$$

$= \frac{1,455,849}{12} - 120,293.36$

$$= 121,320.75 - 120,293.36 = 1,027.39.$$

$$SS_B(\text{or } SS_{SPACING}) = \frac{\sum_{k=1}^{q}(\sum_{i=1}^{p}\sum_{j=1}^{r}X_{ijk})^2}{pr} - CF = \frac{636^2 + 694^2 + 724^2}{3 \times 4} - 120,293.36.$$

$$= \frac{1,455,381}{12} - 120,293.36 = 120,448.416666 - 120,293.36$$

$$= 155.0567.$$

$$SS_{AB}(SS_{SPECIES \times SPACING}) = \frac{\sum^{P}\sum^{q}(\sum^{r}X_{ijk})^2}{r} - CF - SS_A - SS_B$$

$$= \frac{190^2 + 203^2 + 223^2 + \cdots + 257^2 + 292^2}{4} - 120,293.36 - 1,027.39 - 155.0567$$

$$= 122,241.25 - 120,293.36 - 1,027.39 - 155.057 = \underline{765.443}.$$

$$SS_{ERROR} = SS_{TOTAL} - SS_{BLOCKS} - SS_{SPECIES} - SS_{SPACING} - SS_{SPECIES \times SPACING}$$

$$= 22,627.64 - 255.64 - 1,027.39 - 155.057 - 765.443 = \underline{424.11}.$$

Table 11.8.3.4 b: Anova Table for a 3x3 Factorial Arranged in RCBD

Source of Variation	DF	SS	MS	F_{cal}	$F_{0.05}$	$F_{0.01}$
Blocks	r-1 = 3	255.64	85.2133	$F_1 = 4.8221$**	3.01	4.72
Species	p-1 = 2	1,027.39	513.695	$F_2 = 29.0695$***	3.40	5.61
Spacings	q-1 = 2	155.057	77.5285	$F_3 = 4.3873$*	3.40	5.61
Species x spacings	(p-1)(q-1) = 4	765.443	191.36075	$F_4 = 10.8289$***	2.78	4.22
Error	rp(q-1)=24	424.11	17.67125			
Total	**rpq-1 = 35**	**2,627.64**				

* Significant at 5% level; ** Significant at 1% level; *** Significant at 0.01% level

From the Anova Table above, we can draw the following conclusions:

Conclusions:

i. Since $F_1 = 4.82 \gg F_{0.05} = 3.01$ and $F_{0.01} = 4.72$, reject H0 and conclude that the Blocks are highly significant (P<0.01).

ii. Since $F_2 = 29.07 \gg F_{0.05} = 3.40$ and $F_{0.01} = 5.61$, reject H0 and conclude that the Tree Species are very highly significant in their height growth.

iii. Since $F_3 = 4.39 > F_{0.05} = 3.40$ and $F_{0.01} = 5.61$, reject H0 at $\alpha = 0.05$ and accept H0 at $\alpha = 0.01$ – spacing is significant at only, $\alpha = 0.05$ but not at $\alpha = 0.01$

iv. Since $F_4 = 10.83 \gg F_{0.05} = 2.78$ and $F_{0.01} = 4.22$, reject H0 and conclude that Species x Spacing Interactions are highly significant and therefore, the Main Effects (Species and Spacing) should be interpreted with caution.

v. Since Interactions are significant, let us calculate the cell means to investigate them through geometric diagrams.

Table 11.8.3.4 c: Cell Means for Investigating Interactions

	10cm x 10cm	15cm x 15cm	20cm x 20cm	Total
M. excels	47.5	50.8	55.8	51.3
T. superba	62.3	58.5	52.3	57.7
T. ivorensis	56.0	64.3	73.0	64.4
Total	**55.3**	**57.8**	**60.3**	**57.8**

Comparison of Treatment Means:

$$Compute \ LSD_{0.05} = t_{0.025} \ (df) \sqrt{\frac{2MSE}{n}}$$

$$= 2.064 \sqrt{\frac{2(17.67125)}{3}} = 2.064 \sqrt{11.780833}$$

$$= (2.064)(3.4323218) = 7.0843121 \approx 7.08$$

a) **Spacings means: 10cm x 10cm   15cm x 15cm   20cm x 20cm**

   Ordered Means: 55.3   57.8   60.3

   Comparisons:

$60.3 - 55.3 = 5.0 < LSD = 7.08$ - - ns

$60.3 - 57.8 = 2.5 < 7.08$ - - ns

$57.8 - 55.3 = 2.5 < 7.08$ - - ns

All the spacings are not statistically significant. But the spacing of 20cm x 20cm has the largest mean and therefore is the best spacing.

b) **Species Means: M. excelsa T. superba T. ivorensis**

   Ordered Means: 51.3   57.7   64.4

   Comparisons:

   $64.4 - 51.3 = 13.1 > LSD = 7.08$ - - significant

   $64.4 - 57.7 = 6.7 < LSD = 7.08$ - - ns

   $57.7 - 51.3 = 6.4 < LSD = 7.08$ - - ns

   Hence: 51.3   57.7   64.4

          b   b   a

T. ivorensis has the best performance in height growth in this spacing trial because it has the largest mean.

The interplay between species and spacing is very important in this study because the Anova Table shows that the F-calculated is 10.83 which is highly significant at $\alpha = 0.05 \ and \ 0.01$

c) Geometric interpretations of Interactions between Species and Spacings

Geometric interpretations of Spacings x Species interactions

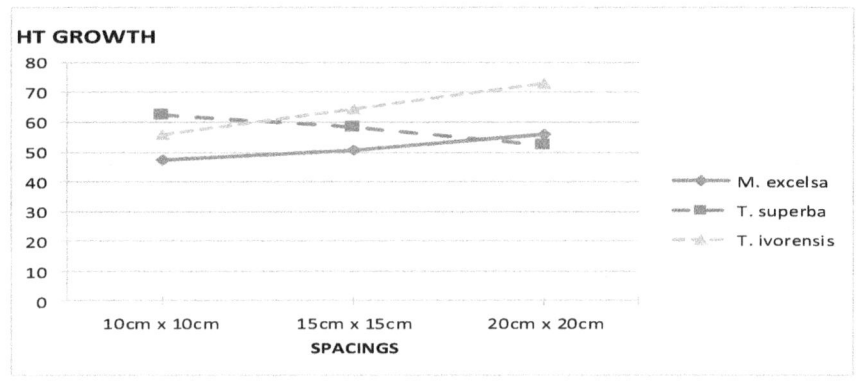

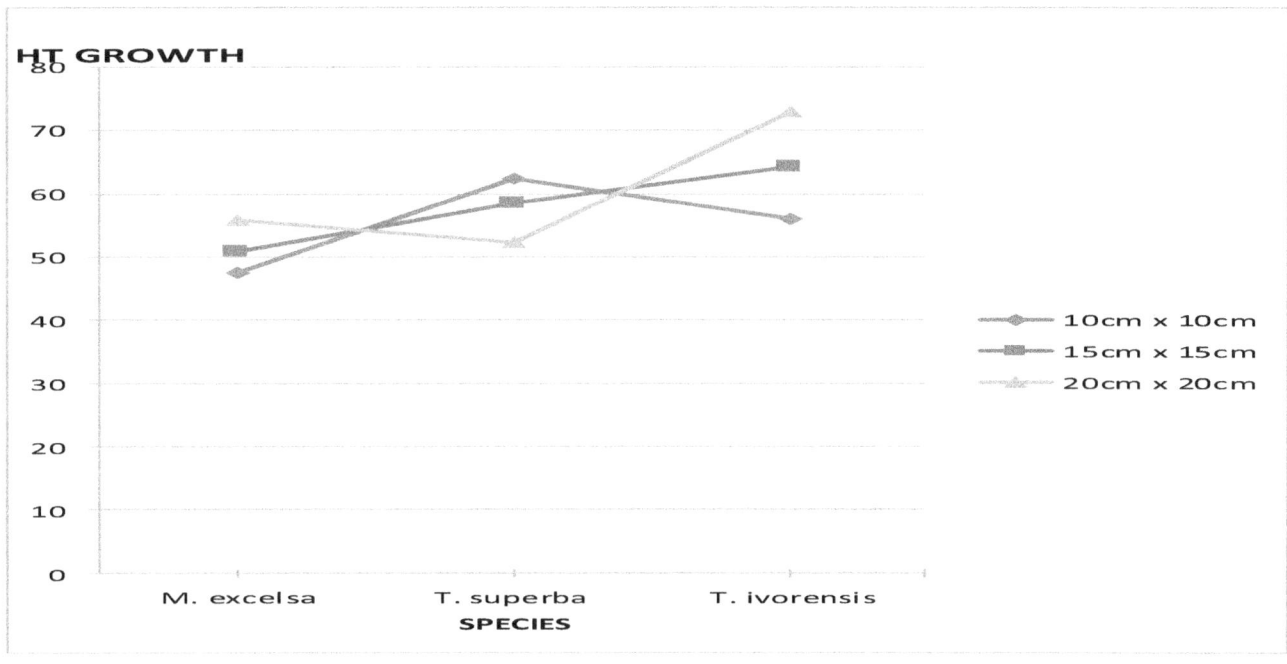

11.8.4 THE SPLIT-PLOT DESIGN

The Split-Plot Design (Gomez and Gomez, 1984) is specifically suited for a two-factor experiment that has more treatments than can be accommodated by a Complete Block Design. The term "Split-Plot", according to Winer (1971), comes from agricultural experimentation in which a single level of one treatment is applied to a relatively large plot of ground called the **Whole Plot, or Main Plot,** but all levels of a second treatment are applied to sub-plots within the Whole plot. Thus, each main Plot becomes a block for the sub-plot treatment.

In a Split-plot design, the precision for the measurements of the effects of the main-plot factor is sacrificed to improve that of the sub-plot factor. It often improves the precision for comparing the average effects of treatments assigned to sub-plots and, when interactions exist, for comparing the effects of subplot treatments for a given main plot treatment. This arises from the fact that experimental error for the main plots is usually larger than the experimental error used to compare subplot treatments. Frequently, the error term for subplot treatments is smaller than would be obtained if all treatment combinations were arranged in a randomized complete block design.

The basic Split-plot design involves assigning the treatments of one factor to main plots arranged in a Completely Random, Randomized Complete Block, or a Latin Square Design (Little and Hills, 1972). These authors also asserted that Split-plot designs are frequently used for factorial experiments where the nature of the experimental material, or the operations involved make it difficult to handle all factor combinations in the same manner.

Each variation of the Split-plot design imposes certain restrictions as to the error term that may be used to test treatment effects. It is important therefore, to assign factors in a manner to give the greatest precision for comparing the interactions and average treatment effects in which we are most interested.

Randomization

The randomization of the treatments assigned to Main Plots is carried out separately for each block. Subplot treatments are then randomized within each main plot, a separate randomization being made for each main plot.

11.8.5.1 Field Layout of a Split-Plot in a RCBD

An Agronomist laid out an experiment to test the effect of pre-treatment of organic fertilizers on the subsequent production of cucumber at two levels of NPK fertilization in a Split-plot design:

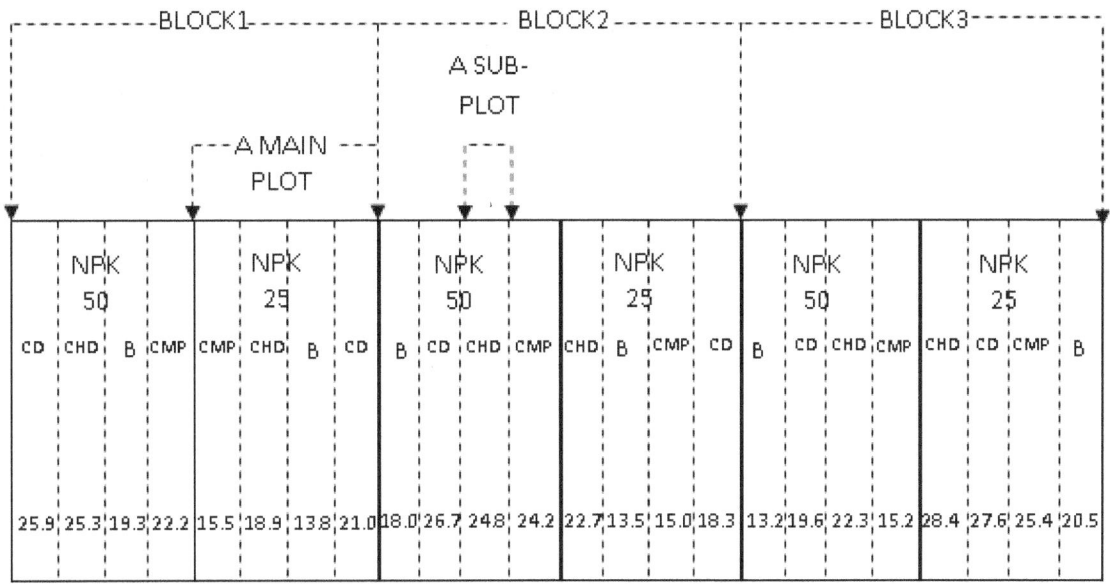

Figure 11.8.5: Split-plot Design. Main Plots (NPK 50, NPK 25 – levels): Sub-plots CD, CHD, B, CMP are organic fertilizer treatments. All plots are laid out in strips on the field. Plot yields of cucumber in kg.

Organic Fertilizers:

CD = Cow Dung

CHD = Chicken Droppings

B = Bare (Control)

CMP = Compost

Before we organize the Split-plot layout for analysis, let us provide the statistical model and the form of Analysis of variance to establish the sources of variation.

Mathematical Model of Split-plot Design in RCBD:

$X_{ijk} = \mu + r_k + \alpha_i + \delta_{ij} + \beta_j + \alpha\beta_{ij} + \varepsilon_{ijk}$, where:

X_{ijk} = Random independent experimental sub-plot unit, normally distributed with mean μ and variance, σ^2

μ = Fixed overall population mean.

r_k = Fixed Block, or replication effect, subject to the restriction that, $\sum rk = 0$.

α_i = Constant main-plot (Factor A) effect, subject to the restriction, $\sum \alpha i = 0$.

δ_{ik} = Random Main-plot Error of Factor A— $\delta_{ik} \sim N(0, \sigma_R^2)$.

β_j = Constant Sub-plot (Factor B Effect) subject to the restriction, $\sum \beta jb = 0$.

$\alpha\beta_{ij}$ = Constant interaction between the main-plot and sub-plot treatments subject to the restriction, $\sum \alpha\beta ij$

258

$= 0.$

$\varepsilon_{ijk} =$ Random independent Sub-plot Error of factor B — $\varepsilon_{ijk} \sim N(0,\sigma^2)$

Table 11.8.5 a: Anova Table for a Split-Plot Design.

Source of Variation	DF	SS	MS	F$_{CAL}$	F$_{05}$	F$_{01}$
Blocks (Reps)	r-1	SS$_{REP}$	MS$_{REP}$			
Main Plot (Factor A)	a-1	SS$_A$	MS$_A$	MS$_A$/MSE(A)		
Main Plot Error (A)	(r-1)(a-1)	SS$_{ERROR}$(A)	MSE(A)			
Sub-Plot (Factor B)	b-1	SS$_B$	MS$_B$	MS$_B$/MSE(B)		
AB Interaction	(a-1)(b-1)	SS$_{AB}$	MS$_{AB}$	MS$_{AB}$/MSE(B)		
Sub-Plot Error (B)	a(r-1)(b-1)	SS$_{ERROR}$ (B)	MSE(B)			
Total	rab-1	SS$_{TOTAL}$				

Computations for Testing Hypotheses:

1. $GRAND\ TOTAL = T_{..}(G) = \sum_{k=1}^{r} \sum_{i=1}^{p} \sum_{j=1}^{q} X_{ijk} = X_{111} + X_{112} + \ldots + X_{rpq}$

2. Correction Factor (CF) $= \dfrac{\left(\sum\limits_{k=1}^{r} \sum\limits_{i=1}^{p} \sum\limits_{j=1}^{q} X_{ijk}\right)^2}{rpq}.$

3. $SS_{TOTAL} = \sum\limits_{k=1}^{r} \sum\limits_{i=1}^{p} \sum\limits_{j=1}^{q} X_{ijk}^2 - C$

4. $SS_{MAIN-PLOT} = \dfrac{\sum_{k=1}^{r} \sum_{i=1}^{p} \left(\sum_{j=1}^{q} X_{ijk}\right)^2}{q} - C$

5. $SS_{BLOCKS}\ (SS_{REPS}) \quad \dfrac{\sum_{k=1}^{r} \left(\sum_{i=1}^{p} \sum_{j=1}^{q} X_{ijk}\right)^2}{pq} - C$

6. $SS_A = \dfrac{\sum_{i=1}^{p} \left(\sum_{k=1}^{r} \sum_{j=1}^{q} X_{ijk}\right)^2}{rq} - C$

7. $SS_{ERROR}(A) = SS_{MAIN-PLOT} - SS_{BLOCKS} - SS_A$

8. $SS_{SUB-PLOT}\ (SS_B) = \dfrac{\sum_{j=1}^{q} \left(\sum_{i=1}^{p} \sum_{k=1}^{r} X_{ijk}\right)^2}{rp} - C$

9. $SS_{AB} = \dfrac{\sum_{i=1}^{p} \sum_{j=1}^{q} \left(\sum_{k=1}^{r} X_{ijk}\right)^2}{r} - C - SS_{MAIN-PLOT} - SS_{SUB-PLOT}$

10. $SS_{ERROR}(B) = SS_{TOTAL} - SS_{MAIN-PLOT} - SS_B - SS_{AB}$

11.8.5 . Numerical Example Using the Field Layout of Cucumber Above.

Table 11.8.5 b: Yield (tons/ha) Organized by Blocks, Main plots and Treatments.

Main-plot (A) (NPK Fertilizers)	Organic Fertilizers	Blocks			Treatment Totals (Sub-plot Totals)
		I	II	III	
NPK_{25}	Cow dung	21.0	18.3	19.6	58.9
	Chicken Droppings	18.9	22.7	22.3	63.9
	Bare Plot	13.8	13.5	13.2	40.5
	Compost	15.5	15.0	15.2	45.7
	Main-Plot Total	69.2	69.5	70.3	
NPK_{50}	Cow Dung	25.9	26.7	27.6	80.2
	Chicken Droppings	25.3	24.8	28.4	78.5
	Bare Plot	19.3	18.0	20.5	57.8
	Compost	22.2	24.2	25.4	71.8
	Main-Plot Total	92.7	93.7	101.9	
	Block Totals	161.9	163.2	172.2	

Table 11.8.5 c: Table of Treatment Totals for Calculating A and B effects.

NPK FERTILIZERS	ORGANIC FERTILIZERS				NPK Totals
	Cow Dung	Chicken Droppings	Bare Plot	Compost	
25	58.9	63.9	40.5	45.7	209.0
50	80.2	78.5	57.8	71.8	288.3
Organic-Fertilizer Totals	139.1	142.4	98.3	117.5	497.3

$$Correction\ Factor\ (C) = \frac{(497.3)^2}{3 \times 2 \times 4} = \frac{247,307.29}{24} = 10,304.47042.$$

$$SS_{TOTAL} = 21.0^2 + 18.9^2 + \ldots + 20.5^2 + 25.4^2 - C$$

$$= 10,820.59 - 10,304.47042$$

$$= \underline{\mathbf{274.92208}}$$

$$SS_{MAIN-PLOT}\ (or\ SS_{NPK}) = \frac{69.2^2 + 69.5^2 + 70.3^2 + 92.7^2 + 93.7^2 + 101.9^2}{4} - C$$

$$= \frac{42,317.57}{4} - 10,304.47042$$

$$= 10,579.3925 - 10,304.47042$$

$$= \underline{\mathbf{274.922.08}}$$

$$SS_{BLOCKS} = \frac{161.9^2 + 163.2^2 + 172.2^2}{2 \times 4} - 10,304.47042$$

$$= \frac{82,498.69}{8} - 10,304.47042$$

260

$$= 10{,}312.33625 - 10{,}304.47042 = \mathbf{\underline{7.86583.}}$$

$$SS_A = SS_{NPK} = \frac{209.0^2 + 288.3^2}{3 \times 4} - 10{,}304.47042$$

$$= \frac{126{,}797.89}{12} - 10{,}304.47042$$

$$= 10{,}566.49083 - 10{,}304.47042 = \mathbf{\underline{262.0204133.}}$$

$$SS_{ERROR}(A) = SS_{MAIN-PLOT} - SS_{BLOCKS} - SS_{NPK}$$

$$= 274.92208 - 7.86583 - 262.0204133$$

$$= \mathbf{\underline{5.0358367.}}$$

SS Sub-Plot:

$$SS_B = SS_{ORG.\ FERT} = \frac{139.1^2 + 142.4^2 + 98.3^2 + 117.5^2}{2 \times 4} - 10{,}304.47042$$

$$\frac{63{,}095.71}{6} - 10{,}304.47042 = 10{,}515.95 - 10{,}304.47042$$

$$= \mathbf{\underline{211.4812467.}}$$

$$SS_{NPK \times ORG.FERT} = \frac{58.9^2 + 80.2^2 + \ldots + 45.7^2 + 71.8^2}{3} - C - SS_{NPK} - SS_{ORG.\ FERT}$$

$$= \frac{32{,}371.53}{3} - 10{,}304.47042 - 262.0204133 - 211.4812467$$

$$10{,}790.51 - 10{,}304.47043 - 262.0204133 - 211.4812467$$

$$= \mathbf{\underline{12.53792.}}$$

$$SS_{ERROR\ (SUB-PLOT)(B)} = SS_{TOTAL} - SS_{MAIN\ PLOT} - SS_{SUB-PLOT} - SS_{INTERACTION}$$

$$= 516.11958 - 274.92208 - 211.4812467 - 12.53782$$

$$= \mathbf{\underline{17.1783333.}}$$

Note: $SS_{TOTAL} = SS_{BLOCK} + SS_A + SS_{ERROR\ (A)} + SS_B + SS_{AB} + SS_{ERROR\ (B)}$

Table 11.8.5 d: Anova Table for a Split-Plot Arranged in RCBD.

Source of Variation	DF	SS	MS	F_{CAL}	F_{05}	F_{01}
Blocks	$r-1 = 2$	7.86583	3.932915			
Main Plot (A)	$a-1 = 1$	262.020413	262.020413	$F_1 = 104.0623^{***}$	18.50	98.50
Main-Plot Error (A)	$(r-1)(a-1) = 2$	5.0358367	2.51791835			
Sub-Plot (B)	$b-1 = 3$	211.4812467	70.4937489	$F_2 = 49.24372^{***}$	3.49	5.95
AB Interaction	$(a-1)(b-1) = 3$	12.53792	4.17930667	$F_3 = 2.919473^{ns}$	3.49	5.95
Sub-Plot Error (B)	$a(r-1)(b-1) = 12$	17.1783333	1.43152778			
Total	$rab-1 = 23$	516.11958				

Hypotheses to be Tested:

1. $H_0: \alpha_1 = \alpha_2 = 0$. i.e. There is no difference between the Main-plots or NPK Treatments.

   $H_1: \alpha_1 \neq \alpha_2 \neq 0$. i.e. There is some difference between the Main-plot or NPK Treatments.

Conclusion:

Since the Main-plot (NPK) = F1 = 104.06 >> F0.05(1, 2) = 18.50 and F0.01,(1, 2) = 98.50, reject H0 and conclude that the main-plot treatments are highly significant at $\alpha = 0.05$ and $\alpha = 0.01$

2. $H_0: \beta_1 = \beta_2 = \beta_3 = \beta_4 = 0$..i.e. The Sub-Plot Treatments (Organic Fertilizers) have no effect on cucumber production. (Yields)

   $H_1: \beta_1 \neq \beta_2 \neq \beta_3 \neq \beta_4 \neq 0$. i.e. The Organic Fertilizers have some effect on cucumber production. (Yields)

Conclusion:

Since F2 = 49.24 >> F0.05,3,12 = 3.49 and F0.01,3,12 = 5.95, reject H0 and conclude that the organic fertilizers have a highly significant effect on the cucumber production.

3. $H_0: \alpha\beta_{11} = \alpha\beta_{12} = \alpha\beta_{13} = \alpha\beta_{14} = \cdots = \alpha\beta_{23} = \alpha\beta_{24} = 0$. i.e. There is no interaction between the Main-plot and Sub-plot factors.

   $H_1: \alpha\beta_{11} \neq \alpha\beta_{12} \neq \alpha\beta_{13} \neq \alpha\beta_{14} \neq \cdots \neq \alpha\beta_{23} \neq \alpha\beta_{24} \neq 0$. i.e. There is some interaction between the Main-plot and Sub-plot factors.

Conclusion:

Since F3 = 2.92 < F.05, 3,12 = 3.49 and F.01 3,12 = 5.95, accept H0 and conclude that there is no interaction between the main-plot factor (NPK) and the sub-plot factor (Organic Fertilizers).

11.8.5.3. Comparisons of Treatments in a Split-Plot Design

Like any other experimental design, comparisons of treatments of the main-plot and the sub-plots can also be carried out provided we use the appropriate standard errors. The appropriate standard errors for mean comparisons are presented below:

To Compare Treatments Between	Standard Error to Use
1. Two Main-plot (or A) Means.	$\sqrt{\dfrac{2\,Error(A)}{rq}}$
2. Two Sub-plot (or B) Means.	$\sqrt{\dfrac{2\,Error(B)}{rq}}$
3. Two Sub-plot Means (B) at the same level of Main-plot (A).	$\sqrt{\dfrac{2\,Error(B)}{r}}$
4. Two Main-plot Means (A) at the same level or different levels of Sub-plot (B).	$\sqrt{\dfrac{2(b-1)\,Error(B) + Error(A)}{rq}}$

From our cucumber example above, the LSD values to compare the

   i.   Main-plot means of NPK is calculated as:

$$LSD0.05 = t_{0.025}\,(df)\sqrt{\frac{2\,ERROR\,(A)}{rq}}$$

$$= t_{0.025}(2)\sqrt{\frac{2\,(2.51791835)}{(3)(4)}}$$

$$= 4.303\sqrt{\frac{5.0358367}{12}} = 4.303\sqrt{0.419653058}$$

$$= (4.303)(0.647806343) = 2.7875\ = 2.79.$$

   ii.   To compare two Sub-Plot (B) Means, the LSD is:

$$LSD0.05 = t_{0.025}\,(df)\sqrt{\frac{2\,ERROR\,(B)}{rp}}$$

$$= t_{0.025}(12)\sqrt{\frac{2\,(1.43152778)}{(3)(4)}} = t_{0.025}\sqrt{\frac{2.86305556}{6}}$$

$$= (2.179)\left(\sqrt{0.477175926}\right) = (2.179)(0.690779217)$$

$$= 1.505207914\ = 1.51.$$

11.8.5.4.A Numerical Example of the Vegetative Growth of Thaumatococcus danielli (Benn) in a Split-Plot Design Arranged in RCBD

An Ecologist in the Department of Forestry and Wildlife, UNIBEN, conducted a study to evaluate the vegetative growth of Thaumatococcus danielli in a split-plot design with two levels of mulching (Factor A) and five levels of shading (Factor B). The growth parameters of plant height, collar girth, crown diameter, leaf area, and leaf number were measured bi-weekly, as shown in the plant height data in Table 11.8.5.4:

Table 11.8.5.4: Effect of Shade on the Growth of Plant Height (cm) of Mulched/Bare Plots of Thaumatococcus danielli.

Treatments	Replications - - Weeks After Planting (WAP)									
	3	5	7	9	11	13	15	17	19	21
T_1 Mulch	5.87	9.77	11.22	13.78	15.84	17.41	17.49	17.52	17.60	20.30
Bare	7.96	8.15	11.23	11.67	15.10	15.39	15.86	16.69	17.34	18.64
T_2 Mulch	3.37	6.30	8.20	8.90	13.50	13.78	16.63	19.50	20.00	20.60
Bare	5.70	7.00	7.48	9.00	10.41	11.77	13.00	13.79	15.68	19.35
T_3 Mulch	4.40	8.03	13.00	18.32	16.39	19.29	18.72	19.20	20.43	21.01
Bare	3.28	5.35	13.49	16.84	16.11	16.85	16.70	16.62	19.17	18.99
T_4 Mulch	6.10	9.96	11.39	12.63	12.86	14.14	12.95	14.47	17.69	19.36
Bare	2.99	4.99	8.90	8.12	10.84	12.63	12.12	14.15	15.36	23.55
T_5 Mulch	4.22	6.04	5.03	7.33	7.06	7.50	7.91	8.33	8.50	10.23
Bare	1.98	5.40	5.30	7.10	5.30	5.40	5.95	5.55	5.60	5.40

Main-plots (Mulch, Bare) - - A

Sub-plots: T_1 = Dense Shade - - B

T_2 = Open-Gap

T_3 = Medium Shade

T_4 = Low Shade

T_5 = Bare (control)

Table 11.8.5.4 a: Thaumatococcus danielli Height Growth (cm) Organized by Blocks, Main-Plots and Treatments.

Mulch (A)	Shade (B)	Replications (WAP)										Treatment Totals
		3	5	7	9	11	13	15	17	19	21	
	T_1	5.87	9.77	11.22	13.78	15.84	17.41	17.49	17.52	17.60	20.30	146.80
	T_2	3.37	6.30	8.20	8.90	13.50	13.78	16.63	19.50	20.00	20.60	130.78
Mulch	T_3	4.40	8.03	13.00	18.32	16.39	19.29	18.72	19.20	20.43	21.01	158.79
	T_4	6.10	9.96	11.39	12.63	12.86	14.14	12.95	14.47	17.69	19.36	132.05
	T_5	4.22	6.04	5.03	7.33	7.06	7.50	7.91	8.33	8.50	10.23	72.15
Mulch Sub-Total		23.96	40.10	48.84	60.96	65.65	72.12	73.70	79.02	84.22	92.00	
	T_1	7.96	8.15	11.23	11.67	15.10	15.39	15.86	16.69	17.34	18.64	138.03
	T_2	5.70	7.00	7.48	9.00	10.41	11.77	13.00	13.79	15.68	19.35	113.18
Bare	T_3	3.28	5.35	13.49	16.84	16.11	16.85	16.70	16.62	19.17	18.99	143.40
	T_4	2.99	4.99	8.90	8.12	10.84	12.63	12.12	14.15	15.36	23.55	113.65
	T_5	1.98	5.40	5.30	7.10	5.30	5.40	5.95	5.55	5.60	5.40	52.98
Bare Sub-Total		21.91	30.89	46.40	52.73	57.76	62.04	63.63	66.80	73.15	85.93	
Replication Total		45.87	70.99	95.24	113.69	123.41	134.16	137.33	145.82	157.37	177.93	1,201.81

Table 11.8.5.4 b: Table of Treatment Totals for Calculating A and B Effects

Mulch Levels (A)	Levels of Shading (B)					
	T_1	T_2	T_3	T_4	T_5	Main-Plot Totals
Mulch (A_1)	146.80	130.78	158.79	132.05	72.15	640.57
Bare (A_2)	138.03	113.18	143.40	113.65	52.98	561.24
Sub-plot Totals	284.83	243.96	302.19	245.70	125.13	1,201.81

1. **Structural Model of Split-plot Design Arranged in RCBD:**

$$X_{ijk} = \mu + r_k + \alpha_i + \delta_{ik} + \beta_j + \alpha\beta_{ij} + \varepsilon_{ijk}$$, where

X_{ijk} = Random independent experimental sub-plot unit, normally distributed with mean, μ and variance, σ^2 μ =. Constant overall population mean.

r_k = Constant Block Effect, subject to the restriction, $\sum$rk= 0

α_i = Constant Main-plot, or Factor A Effect, subject to the restriction,

$\sum$αi= 0.

δ_{ik} = Random independent Main-Plot Error (Factor A Error) - -

$\delta_{ik} \sim N(0, \sigma_A^2)$.

β_j = Constant Sub-plot, or Factor B Effect, subject to the restriction,

$\sum \beta j = 0$.

$\alpha\beta_{ij}$ = Constant Interaction between the Main-Plot and Sub-Plot

Treatments, subject to the restriction $\sum \alpha\beta ij = 0$.

ε_{ijk} = Random, independent Sub-Plot Error, (or Factor B Error) - -

$\varepsilon_{ijk} \sim N(0, \sigma^2)$

2. Hypotheses To Be Tested:

i. $H_0: \alpha_1 = \alpha_2 = 0$. i.e. There is no difference between the Main-Plot (Mulch) Treatments.

$H_1: \alpha_1 \neq \alpha_2 \neq 0$. i.e. There is some difference between the Main-Plot (Mulch) Treatments.

ii. $H_0: \beta_1 = \beta_2 = \beta_3 = \beta_4 = \beta_5 = 0$. i.e. The Sub-Plot Treatments of Shading have no effect on the height growth of Thaumatococcus danielli plants.

$H_1: \beta_1 \neq \beta_2 \neq \beta_3 \neq \beta_4 \neq \beta_5 \neq 0$. i.e. The Sub-Plot Treatments of Shading have some effect on the height growth of Thaumatococcus danielli plants.

iii. $H_0: \alpha\beta_{11} = \alpha\beta_{12} = \alpha\beta_{21} = \alpha\beta_{22} = \cdots = \alpha\beta_{15} = \alpha\beta_{25} = 0$. i.e There is no interaction between the Main-Plot and Sub-Plot Factors.

$H_0: \alpha\beta_{11} \neq \alpha\beta_{12} \neq \alpha\beta_{21} \neq \alpha\beta_{22} \neq \cdots \neq \alpha\beta_{15} \neq \alpha\beta_{25} \neq 0$ i.e There is some interaction between the Main-Plot and Sub-Plot Factors.

3. Computations For Testing The Hypotheses:

Grand Total = $T_{...}(G) = \sum_{k=1}^{r} \sum_{i=1}^{p} \sum_{j=1}^{q} X_{ijk} += 5.87 + 9.77 + 11.22$

$13.78 + \cdots + 5.55 + 5.60 + 5.40 = 1,201.81$

4. Correction Factor (CF) = $\dfrac{\left(\sum_{k=1}^{r} \sum_{i=1}^{p} \sum_{j=1}^{q} X_{ijk}\right)^2}{rpq} = \dfrac{(1,201.81)^2}{10 \times 2 \times 5}$

$\dfrac{1,444,347.276}{100} = 14,443.47276$.

$SS_{TOTAL} = \sum_{k=1}^{r} \sum_{i=1}^{p} \sum_{j=1}^{q} X_{ijk}^2 - CF = 5.87^2 + 9.77^2 + 11.22^2 + \ldots + 5.55^2 + 5.60^2 + 5.40^2$

$= 17,316.7727 - 14,443.47276 = 2,873.29994$

5. Complete Main-Plot Analysis:

i. $SS_{MAIN-PLOT} = \dfrac{\sum_{k=1}^{r} \sum_{i=1}^{p} \left(\sum_{j=1}^{q} X_{ijk}\right)^2}{q} - CF \equiv \dfrac{\sum^r \sum^p T_{ij.}^2}{q} - CF$

$$= \frac{23.96^2 + 40.10^2 + 48.84^2 + \ldots + 66.80^2 + 73.15^2 + 85.93^2}{5} - CF$$

$$\frac{79,826.3711}{5} - 14,443.47276 = 15,965.27422 - 14,443.47276$$

$$= 1,521.80146.$$

ii. $SS_{BLOCKS} = \frac{\sum_{k=1}^{r}\left(\sum_{i=1}^{p}\sum_{j=1}^{q} X_{ijk}\right)^2}{pq} - CF \equiv \frac{\sum^p \sum^q T_{..}^2}{pq} - CF$

$$= \frac{45.87^2 + 70.99^2 + 95.24^2 + 113.69^2 + 123.41^2 + 134.16^2 + 137.33^2 + 145.82^2 + 157.37^2 + 177.93^2}{2 \times 5}$$

$$\frac{158,916.0475}{10} - 14,443.47276 = 15,916.60475 - 14,443.47276$$

$$= 1,448.13199.$$

iii. $SS_A(SS_{MULCHING}) = \frac{\sum_{i=1}^{p}\left(\sum_{k=1}^{r}\sum_{j=1}^{q} X_{ijk}\right)^2}{rq} - CF \equiv \frac{\sum^r \sum^q T_{j.}^2}{rq} - CF$

$$= \frac{640.57^2 + 561.24^2}{10 \times 5} - 14,443.47276$$

$$= \frac{725,320.2625}{50} - 14,443.47276 = 14,506.40525 - 14,443.47276$$

$$= 62.93249.$$

iv. $SS_{ERROR}(A) = SS_{MAIN-PLOT} - SS_{BLOCKS} - SS_A$

$$= 1,521.8014 - 1,448.13199 - 62.93249$$

$$= 10.73698.$$

6. **Complete Sub-Plot Analysis:**

v. $SS_{SUB-PLOT}(B) = \frac{\sum^q\left(\sum^r \sum^p X_{ijk}\right)^2}{rp} - CF \equiv \frac{\sum^r \sum^p T_{..k}^2}{rp} - CF$

$$= \frac{284.83^2 + 243.96^2 + 302.19^2 + 245.70^2 + 25.13^2}{10 \times 2} - 14,443.47276$$

$$= \frac{307,989.4135}{20} - 14,443.47276 = 15,399.47068 - 14,443.47276$$

$$= 955.997915$$

vi. $SS_{AB} = \frac{\sum^p \sum^q\left(\sum^r X_{ijk}\right)^2}{r} - CF - SS_A - SS_B \equiv \frac{\sum^p \sum^q T_{.jk}^2}{r} - CF$

$$-\frac{148.80^2 + 130.78^2 + \cdots + 113.65^2 + 52.98^2}{10 \times 2} - CF - SS_a - SS_B$$

$$= \frac{154,659.4937}{10} - 14,443.47276 - 62.93249 - 955.997915$$

$$= 15,465.94937 - 14,443.47276 - 6293249 - 955.997915$$

$$= 3.546205$$

vii. $$SS_{ERROR}(B) = SS_{TOTAL} - SS_{MAIN-PLOT} - SS_{SUB-PLOT} - SS_{INTERACTION}$$

$$= 2,873.29994 - 1,521.80146 - 955.997915 - 3.546205$$

$$= 391.95436.$$

Table 11.8.5.4 c: Anova Table for a Split-Plot Arranged in a RCBD

Source of Variation	DF	SS	MS	Fcal	F_{05}	F_{01}
Blocks	r-1 = 9	1,448.13199	160.903554	$F_1 = 134.8733$	3.18	5.35
Main-Plot (A)	a-1 = 1	62.93249	62.93249	$F_2 = 52.7516^{**}$	5.12	10.56
Main-Plot Error (A)	(r-1)(a-1) = 9	10.73698	1.19299778			
Sub-Plot (B)	b-1 = 4	955.997915	238.999479	$F_3 = 43.9030^{***}$	2.49	3.57
AB Interaction	(a-1)(b-1) = 4	3.546205	0.8865513	$F_4 = 0.62855^{ns}$	2.49	3.57
Sub-Plot Error (B)	a(r-1)(b-1) = 72	391.95436	5.44381056			
Total	rab-1 = 99	2,873.29994				

Conclusion (1):

Since Fcal1 = 134.87>>> F.05 = 3.18 and F0.01 =5.35, reject H0 and concluded that the main (Mulch) Treatments are highly significant on the height growth of the Thaumatococcus plants.

Conclusion (2):

Since Fcal2 = 52.75 < F.05, 1, 9 = 5.12 and F0.01, 1,9 = 10.56, reject H0 that there is a highly significant difference in the height growth of mulched and bare plots.

Conclusion (3):

Since Fcal3 = 43.90 >> F0.05, 4, 72 = 2.49 and F0.01, 4,72 = 3.57, reject H0 and conclude that the Sub-Plot Treatment of Shading has a highly significant effect on the height growth Thaumatococcus danielli plants.

Conclusion (4):

Since Fcal 4= 0.629<< F0.05, 4,72 = 2.49, accept H0 and conclude that there is no interaction between the Main-Plot and Sub-Plot Factors.

Comparisons of Treatment Means in a Split-Plot Design:

Table 11.8.5.4 d Treatment Means in a Split – Plot Design

MULCHING	SHADING					
	T_1	T_2	T_3	T_4	T_5	Main-Plot Means
Mulched Plots (A_1)	14.68	13.08	15.88	13.21	7.22	12.81–A_1
Bare Plots (A_2)	13.80	11.32	14.34	11.37	5.30	11.22-A_2
Sub-Plot Means (B)	14.24	12.20	15.11	12.29	6.26	

To use both LSD and Duncan's Multiple Range Tests, the standard errors based on the variability within experimental units to which treatments are applied are calculated as follows:

	Means to Compare	Standard Error
1	Main-Plot Treatments: A_1-A_2	$\sqrt{\dfrac{ERROR(A)}{rq}}$
2	Sub-Plot Treatments: B_1-B_2	$\sqrt{\dfrac{ERROR(B)}{rp}}$
3	Sub-Plot Treatments for the same Main-Plot A_1B_1-A_1B_2	$\sqrt{\dfrac{ERROR(B)}{r}}$
4	Sub-Plot Treatments for different Main-Plots A_1B_1-A_2B_1 or A_1B_1-A_2B_2	$\sqrt{\dfrac{(b-1)\,ERROR(B)+ERROR(A)}{rq}}$

i.   .LSD for Differences Between Main-Plot Treatments:

$$LSD0.05 = t0.025\,(df)\,\sqrt{\frac{2\,ERROR(A)}{rq}}\,, \text{ where } df = 9$$

$$= 2.262\,\sqrt{\frac{2(1.19299778)}{10\times5}} = 2.262\,\sqrt{0.047719112}$$

$$= 2.62 \times 0.218447046 = 0.494127218$$

$$= 0.494.$$

ii.   .LSD for Differences Between Sub-Plot Treatments:

$$LSD0.05 = t0.025\,(df)\,\sqrt{\frac{2\,ERROR(B)}{rp}}\,, \text{ where } df = 72$$

$$= 1.960\,\sqrt{\frac{25.44381056}{10\times2}} = 1.960\,\sqrt{0.544381056}$$

$$= 1.960 \times 0.737821832 = 1.446130791$$

$$= 1.446.$$

iii.   LSD for Differences Between Sub-plot Treatments for the same main-plot Treatments

$$:LSD0.05 = t0.025\,(df)\,\sqrt{\frac{2\,ERROR(B)}{r}}$$

$$1.960 \sqrt{\frac{2(5.44381056)}{10}} = (1.960) \left(\sqrt{1.088762112}\right)$$

$$= (1.960)(1.043437642) = 2.045137778$$

$$= 2.045.$$

iv. **LSD for Differences Between Sub-Plot Treatments for Different Main-Plot Treatments.**

$$\text{LSD}0.05 = t0.025 \text{ (df)} \sqrt{\frac{2\left[(b-1)Error\ (B) + Error\ (A)\right]}{rq}}$$

$$= 1.960 \sqrt{\frac{2[(4)(5.44381056) + 1.19299778]}{10 \times 2}}$$

$$= 1.960 \sqrt{\frac{45.9368004}{50}} = (1.960)\left(\sqrt{0.9187296}\right)$$

$$= (1.960)(0.958503834) = 1.878667516$$

$$= 1.879.$$

11.8.6 SPLIT-SPLIT-PLOT DESIGN

In a Split-plot design, there are generally two factors. But in a split-split-plot design, a third factor is added by splitting sub-plots of a split-plot design. The design is quite useful for a three-factor experiment to facilitate field operations and keep treatment combinations together (Little and Hills, 1972). The additional restriction on randomization makes it necessary to compute a third error term that is used to test for the main effects of the factor applied to the second split and for all interactions involving this factor.

The randomization procedure is the same for the split-plot design with the sub-plots being split into sub-sub-plots, equal in number to the levels of factor three, to which the third factor is randomly assigned separately and independently for each of the subplots using a new randomization for each of the sub-subplots.

12 PROBLEM DATA

Problem data are any set of data that does not satisfy the implied, or the stated conditions for a valid analysis of variance. The two major groups of problem data are:

1. Missing Data.
2. Data that violate some assumptions of the analysis of variance.

12.1 MISSING DATA

A missing data situation occurs whenever a valid observation is not available for any one of the experimental units. The occurrence of missing data results in two major difficulties:

i.   Loss of information and
ii. Non-applicability of the standard analysis of variance.

12.2 Common Causes of Missing Data

Missing data may arise from several causes (Snedecor and Cochran, 1967; Gomez and Gomez, 1984):

a) **Improper Treatment:**

   Improper treatment is declared when an experiment has one or more experimental plots that do not receive the intended treatment.

b) **Destruction of Experimental Plants:**

   Most field experiments aim for a perfect stand in all experimental plots, but that is not always achieved. Poor germination, physical damage during crop culture and pest damage are common causes of the destruction of experimental plants. The destruction of the experimental plants must not be the result of the treatment effect. If a plot has no surviving plants because it has been grazed by stray cattle, or vandalized by thieves, each of which is clearly not treatment related, missing data should be appropriately declared.

c) **Loss of Harvested Samples:**

   Harvested samples may require additional processing before the required data can be measured. For example, grain yield of rice can be measured only after drying, threshing and cleaning are completed.

   Some characters may involve long sampling and measurement processes, or may require specialized and elaborate measuring devices. Leaf area, grain weight and protein content are generally measured in a laboratory, instead of in the field.

   It is not uncommon for some portion of the samples to be lost between the time of harvesting and the actual data recording. Because no measurement of such characters is possible, missing data should be declared.

d) **Illogical Data:**

   Data may be considered illogical if their values are too extreme to be considered within the logical range of the normal behavior of the experimental materials. For example, if the height of a human being is recorded as 3.5 meters among other heights that range from 1.1 meters to 1.7 meters, the height that is recorded as 3.5 meters will be considered illogical.

   Therefore, illogical data are misread observations, incorrect transcription and improper application of sampling techniques, or the improper reading of the measuring instrument.

   Data that a researcher suspects to be illogical should not be treated as missing simply because they do not conform to the researcher's preconceived ideas or hypothesis.

12.3 MISSING DATA FORMULA TECHNIQUE

It must be pointed out that an estimate of a missing value does not supply additional information to the experimenter but only facilitates the analysis of the remaining data.

However, Steel and Torrie (1981) reported a method, developed by Yates, which is commonly used for estimating such missing data for various experimental designs.

When an experiment has one or more observations missing, the standard computational procedures of the analysis of variance for the various experimental designs, except the CRD, no longer apply.

An estimate of a single missing observation is provided through an appropriate formula according to the experimental design used. This estimate is used to replace the missing value and the augmented set is then subjected, with some slight modifications, to the standard analysis of variance. It is important to note that once the data is lost, no amount of statistical manipulations can retrieve it.

12.3.1. Completely Randomized Design (CRD) with a single missing value.

A missing value in CRD is simply treated as "treatments with unequal sample sizes".

12.3.2 (a). Randomized Complete Block Design (RCBD) With a Single Missing Value.

Where a single value is missing in a Randomized Complete Block experiment, we calculate an estimate of the missing value by the formula:

$$X = \frac{rB_0 + tT_0 - G_0}{(r-1)(t-1)}$$

where: X = estimate of the missing value.

t = number of treatments.

r = number of replications.

B_0 = total of observed values of the replication that contains the missing data.

T_0 = total of observed values of the treatment that contains the missing data.

G0 = Grand total of all observed values.

Procedure:

1. Replace the missing data by its estimated value computed by the formula above, and do the analysis of variance of the augmented data set.
2. Make the following modifications to the analysis of variance obtained in (1) above and subtract one (1) from both the Total and Error degrees of freedom.
3. **Compute the Correction Factor for BIAS (B) as:**

Bias (B) = $\frac{[B_0 - (r-1)X]^2}{t(t-1)}$, because the treatment Sum of Squares is inflated or biased upward by an amount

B. **Subtract the computed B value from the Treatment SS and the Total SS.**

4. For pair comparison of treatment means, where one of the treatments has the missing data, compute the standard error of the mean difference, $S_{\bar{d}}$ as:

$$S_{\bar{d}} = \sqrt{S^2\left(\frac{2}{r} + \frac{t}{r(r-1)(t-1)}\right)},$$

where:

S^2 = MSE from the analysis of variance.

r = number of replications

t = number of treatments.

The computed $S_{\bar{d}}$ is appropriate for use either in the computation of LSd values, or DMRT values.

12.3.2 (b) A Numerical Example of RCBD with One Missing Value.

Table 12.1.3.2: Yield of Four Varieties of Cowpeas in Five Randomized Blocks with one missing value.

Varieties	Blocks 1	2	3	4	5	Varieties Total
A	32.3	34.0	34.3	35.0	36.5	172.1
B	33.3	33.0	36.3	36.8	34.5	173.9
C	30.8	34.3	35.3	32.3	35.8	168.5
D	- - - -	26.0	29.8	28.8	28.8	112.6
Block Total	96.4	127.3	135.7	132.1	135.6	627.1

i.   A single missing value in RCBD is estimated by the formula:

$$X = \frac{aT + bB - S}{(a-1)(b-1)}, \text{ (Snedecor and Cochran, 1967):}$$

where:

a = number of treatments.

b = number of blocks.

B = total of items in same block as missing item.

S = Grand total of all observed items.

In the example above:

T = 112.6; B = 96.4; S = 627.1; α = 4; b = 5.

$$X = \frac{4(112.6) + 5(96.4) - 627.1}{(3)(4)} = \frac{932.4 - 627.1}{12}$$

$$\frac{305.3}{12} = 25.4416667 \approx 25.4$$

ii.   Substitute the 25.4 for the missing value and carry out the Anova as usual.
iii. Reduce the degrees of freedom in the Total SS and Error SS by 1.
iv.   The Treatment MS in the Anova is slightly inflated, or biased and must be corrected and subtracted from the mean square. The Bias is estimated as:

$$\text{Bias (B)} = \frac{[B - (a-1)X]^2}{a(a-1)} \text{ (Snedecor and Cochran, 1967)} = \frac{[B96.4 - (3)(25.4)]^2}{(4)(3)(3)} = \frac{(20.2)^2}{36} = \frac{408.04}{36}$$

$$= 11.334444 \approx 11.33$$

Table 12.1.3.2 a: Anova Table of RCBD with One Missing Observation.

Source of Variation	DF	SS	MS	F_{cal}	$F_{0.05}$	$F_{0.01}$
Blocks	(5-1) = 4	35.39	8.8475	5.626*	3.36	5.67
Varieties	(4-1) = 3	171.36	57.12-11.33 = 45.79	29.0647**	3.20	5.32
Error	(4)(3)-1 = 11	17.33	1.57545			
Total	(ab-1)-1=19-1 =18	224.08				

For Pair Comparisons the standard error (s.e.) is:

$$s.e = \sqrt{S^2\left(\frac{2}{b} + \frac{a}{b(b-1)(a-1)}\right)} = \sqrt{(1.575)\left[\frac{2}{5} + \frac{4}{(5)(4)(3)}\right]}$$

$$= \sqrt{(1.575)(0.4 + 0.066666} = \sqrt{0.735} = 0.8573$$

12.1.3.3. LATIN SQUARE DESIGN (LSD) WITH A SINGLE MISSING VALUE.

The missing data in a Latin Square Design is estimated as:

$$X = \frac{t(R_0 + C_0 + T_0) - 2G_0}{(t-1)(t-2)} = \frac{r(R_0 + C_0 + T_0) - 2G_0}{(r-1)(r-2)}, \text{ since in an LSD, t = r = c. where:}$$

t = number of treatments (r = number of rows; c = number of columns).

R_0 = total of observed values of the row that contains the missing data.

C_0 = total of observed values of the column that contains the missing data.

T_0 = total of observed values of the treatment that contains the missing data.

G_0 = Grand total of all observed values.

- Compute the estimate of the missing data.
- Subtract one from both the Total and Error d.f.

Procedure For Handling LSD Missing Data:

1. Compute the estimate of the missing data using the formula above.
2. Replace the missing data with its estimated value computed from the formula in (1) and perform the usual Analysis of variance on the augmented data set with the following modifications:
   - Subtract one from both the Total and Error degrees of freedom.
   - Compute the correction factor for Bias (B) as:

a) $B = \dfrac{[G_0 - R_0 - C_0 - (t-1)T_0]^2}{[(t-1)(t-2)]^2}$ and subtract this B-value from the Treatment SS and Total SS (Gomez and Gomez, 1984).

b) $B = \dfrac{[S - R - C(c-1)T]^2}{(a-1)^3(a-2)^2}$ and deduct this from Treatments Means Squares for Bias (Snedecor and Cochran, 1967).

3. For pair Comparisons of Treatment Means where one of the treatments has missing data, compute the standard error of the mean difference as:

$$S_{\bar{d}} = \sqrt{S^2\left(\frac{2}{t} + \frac{1}{(t-1)(t-2)}\right)}$$

12.3.3. A Numerical Example of A Latin Square Design with One Missing Value.

Suppose we have a 3 x 3 LSD with a missing value as shown in Table 12.1.3:

Table 12.1.3.3: LSD with one missing value

Rows	Columns 1	2	3	Row Totals	Treatment Totals
I	A -- (512)	B = 885	C = 940	1,825	A = 1,477
II	B = 715	C = 1,087	A 766	2,568	B = 2,432
III	C = 844	A = 711	B = 832	2,387	C = 2871
Column Totals	1,559	2,683	2,538	6,780	G = 6,780

Source: Snedecor and Cochran, 1967.

i. Compute the estimated missing data as:

$$X = \frac{t(R_0 + C_0 + T_0) - 2G_0}{(t-1)(t-2)} = \frac{3(1825 + 1559 + 1477) - 2(6780)}{(3-1)(3-2)}$$

$$\frac{3(4,861) - 13,560}{(2)(1)} = \frac{14,583 - 13560}{2} = \frac{1,023}{2} = 511.5 \approx 512$$

Estimate Bias (B) as:

$$\textbf{\textit{Estimate Bias (B) as: B}}\ \frac{[S - R - C\,(a-1)T]^2}{(a-1)^3(a-2)^2} = \frac{[6780 - 1825 - 1559 - 2(6780)]^2}{(2^3)(1^2)}$$

$$= \underline{\textbf{24,420.}}$$

ii. Replace the missing value in the data with the estimated value and carry out the Anova; and subtract the B-value (24,420) from the Treatment MS in the Anova table. Then subtract 1 from Error df and 1 from

275

Source of Variation	DF	SS	MS	F_{cal}	$F_{0.05}$	$F_{0.01}$
Rows	$a-1 = 2$	9,847	4,923.5	$F_1 = 1.776^{ns}$	200	4,999
Columns	$a-1 = 2$	68,185	34,092.5	$F_2 = 12.294^{ns}$	200	4,999
Treatments	$a-1 = 2$	129,655	$64,828 - 24,420 = 40,408$	$F_3 = 40,408/2773 = 14.572^{ns}$	200	4,999
Error	$(a-1)(a-2)-1=1$	2,773	2,773			
Total	$(a^2 - 1)-1 = 7$	210,460				

Total df, as shown in the Anova Table above.

iii. **Standard Error for Pair Comparisons:**

$$S_{\bar{d}} = \sqrt{S^2\left(\frac{2}{t} + \frac{1}{(t-1)(t-2)}\right)} = S_{\bar{d}} = \sqrt{2,773\left(\frac{2}{3} + \frac{1}{(2)(1)}\right)}$$

$$= \sqrt{2,773(1.1666667)} = \sqrt{3,235.166667} = \mathbf{56.8785}$$

12.3.4. Iterative Procedure for Two or More Missing Values

When two or more values are missing, we require more complicated methods like an iterative scheme for estimation (Snedecor and Cochran, 1967). Using an RCBD with two missing values in the data (Table 12.1.3.4):

Table 12.1.3.4: RCBD with Two Missing Values

Treatments	Blocks 1	2	3	Total
A	6	5	4	15
B	15	X_{22}	8	23
C	X_{31}	15	12	27
Total	21	20	24	$G = 65$

$$\bar{X}.. = \frac{65}{7} = 9.2857 \approx 9.3$$

Procedure of iteration

i. Guess a reasonable value for one of the missing data - - since the grand mean, $\bar{X}... = 9.3$,both the block and treatment means are above average, hence we guess $X_{22} = 10.5$.

ii. Estimate the missing value, X_{31} as: $X_{31} = \frac{aT + bB - S}{(a-1)(b-1)}$

276

where:

a = number of Treatments.

b = number of Blocks.

T = sum of items with same treatment as missing item.

B = sum of items in same block as missing item.

S = sum of all observed items.

$$X_{31} = \frac{3(27) + 3(21) - 75.5}{(3-1)(3-1)} = \frac{144 - 75.5}{4} = \frac{68.5}{4} = 17.125 \approx 17.1$$

iii. Substitute X_{22} = 17.1 in the table and now get a better estimate of X_{22} by using the formula for X_{22} missing:

$$X_{22} = \frac{3(23) + 3(20) - 82.1}{(2)(2)} = \frac{129 - 82.1}{4} = \frac{46.9}{4} = 11.725 \approx 11.7$$

iv. With this revised X_{22}, re-estimate X_{31}:

$$X_{31} = \frac{3(27) + 3(21) - 76.7}{(2)(2)} = \frac{144 - 76.7}{4} = \frac{67.3}{4} = 16.825 \approx 16.8$$

v. With the re-estimated X_{31} in the table, re-estimate X_{22}:

$$X_{22} = \frac{3(23) + 3(20) - 81.8}{(2)(2)} = \frac{129 - 81.8}{4} = \frac{47.2}{4} = 11.8 \;\text{— no change}$$

vi. Using the new X_{22} = 11.8, re-estimate X_{31}

$$X_{31} = \frac{3(27) + 3(21) - 76.8}{(2)(2)} = \frac{144 - 76.8}{4} = \frac{67.2}{4} = 16.8 \;\text{— no change}$$

vii. Using the values $X_{22} = 11.8 \;and\; X_{31} = 16.8$, carry out the Anova and subtract 2 degrees of freedom from the Total and Error Sum of Squares. The Treatment SS and MS are biased or inflated upwards.

viii. To obtain the correct Treatment SS, re-analyze the data in Table 12.1.3.4, ignoring the Treatments and missing values, treat the data as a CRD with unequal sample sizes and the blocks used as the classes.

ix. Subtract from the CRD Error SS, the Error you obtained in the RCBD analysis of the completed data and subtract their degrees of freedom.

The same method applies to a Latin Square Design with two missing values.

12.3.5 : Split-Plot Design with a Single Missing Value

The missing data in a Split-plot Design is estimated as:

$$X = \frac{rM_0 + bT_0 - P_0}{(b-1)(r-1)}$$

where:

b = level of subplot factor.

r = number of replications.

M_0 = total of observed values of the specific main plot that contains the missing data.

T_0 = total of observed values of the treatment combination that contains the missing data.

P_0 = total of observed values of the main-plot treatment that contains the missing data.

Gomez and Gomez (1984) observed that the missing data formula for a Split-plot design is the same as that of the RCBD with the Main-plot replacing Replication.

The procedure for handling missing data is the same for split-plot, RCBD and LSD.

For pair comparisons of treatment means where one of the treatments has the missing data, the standard errors $(S_{\bar{d}})$ are computed as follows:

Table 12.1.3.5: Standard Errors of the Mean Difference $(S_{\bar{d}})$ in a Split-plot Design with Missing Data.

	Comparison	Measured as	Standard Error of Difference
1	Difference between two A means	$a_i - a_j$	$\sqrt{\dfrac{2(E_a + f E_b)^{\bullet}}{rb}}$
2	Difference between two B means	$b_i - b_j$	$\sqrt{\dfrac{2E_b\left(1 + \frac{fb}{a}\right)}{ra}}$
3	Difference between two B means at the same level of A	$a_i b_j - a_i b_k$	$\sqrt{\dfrac{2E_a\left(1 + \frac{fb}{a}\right)}{r}}$
4	Difference between two A means		
(a)	At the same level of B	$a_i b_j - a_k b_j$	$\sqrt{\dfrac{2E_a + 2E_b[(b-1) + fb^2]}{rb}}$
(b)	At different levels of B	$a_i b_j - a_k b_i$	

For one missing value, $f = \dfrac{1}{\{2(r-1)(b-1)\}}$

For more than one missing value: $f = \dfrac{k}{\{2(r-d)(b-k+c-1)\}}$

where:

278

r = number of replications.

a = number of main-plot treatments.

b = number of sub-plot treatments

c = number of blocks containing missing values.

d = number of values in the sub-unit treatment ajbk that is affected most.

k = number of missing values.

12.4. EXPECTED MEAN SQUARES.

The linear additive mathematical model in Anova gives an insight into the contributions of the various components to the total variation. The mean square values obtained in Anova table are unbiased estimates $(S_A^2 \ and \ S^2)$ of the variance components among treatments (σ_A^2) and within treatments, σ^2, respectively.

For example, in the Completely Randomized Design (a Single Factor Experiment), the linear model is: $X_{ij} = \mu + \alpha_j + \varepsilon_{ij}$. X_{ij} is an observation of the independent variable, while the terms on the right-hand side are the parameters underlying the independent variables. These terms cannot be observed directly, but the data from an experiment will give unbiased estimators of these parameters.

The Expected values of the Mean Squares: (1) show how to obtain unbiased estimates of Error for the comparisons of interest (2) in the studies of variability, they provide estimates of the contributions made by the different sources to the variance of a measurement (Snedecor and Cochran, 1967).

The parameters of the model, α_j, in the completely randomized design above, may be fixed or random, depending upon how the experimental units and the treatments were selected - - either from finite or infinite populations; or whether the inference will be on the treatments in the experiment only, or also extended to other treatments not included in the experiment.

The decision must be made prior to the conduct of the experiment whether the levels of the factor considered are to be set at fixed values, or are to be chosen at random from many possible levels (Hicks,1973). When all levels are fixed, the mathematical model of the experiment is called a **fixed model**; when all levels are chosen at random, the model is called **a random model**; when several factors are involved with some at fixed levels and others at random levels, the model is called a **mixed model.**

12.4.1 :SINGLE-FACTOR MODELS.

For a Completely Randomized Design, the factor is referred to as a treatment effect whose model is:

$$X_{ij} = \mu + \alpha_j + \varepsilon_{ij}$$

where:

$\mu =$ is assumed to be a fixed constant.

$\varepsilon_{ij} =$ are normal and independently distributed (NID) with mean zero and variance, .

$\sigma^2 \left(\varepsilon_{ij} \sim N(0, \sigma^2) \right)$

$\alpha_j =$ assumption depends on whether α_j is fixed or random.

S/N	Fixed Model	Random Model
1	Assumptions: α_j's are fixed constants $\sum_{j=1}^{k} \alpha_j = \sum_{j=1}^{k}(\mu_{.j} - u) = 0$.	1. Assumption: α_j's are random variables and are NID $(0, \sigma_\alpha^2)$ are the variances among α_j's.
2	Analysis: Procedures of Anova for computing SS.	2. Analysis: Same as fixed model.
3	Hypothesis tested: $H_0: \alpha_j = 0$	3. Hypothesis tested: $H_0: \sigma_\alpha^2 = 0$
4	EMS:	4. EMS:

Anova Table (Fixed Model)

Source	DF	EMS
α_j	(k-1)	$\sigma_\varepsilon^2 + n\varnothing_\alpha^*$
ε_{ij}	k(n-1)	σ_ε^2

Anova Table (Random Model)

Source	DF	EMS
α_j	(k-1)	$\sigma_\varepsilon^2 + n\varnothing_\alpha^2$
ε_{ij}	k(n-1)	σ_ε^2

$$* \ \varnothing_\alpha = \frac{\sum A_i}{\alpha - 1}$$

Note:

For the fixed model, Hicks (1973) observed that if the hypothesis is true that $\sigma_j = 0$ for all j and all the k fixed treatments are equal, $\sum \sigma_j^2 = 0$, the EMS for σ_j and ε_{ij} are both σ_ε^2. Hence the observed mean squares for treatments and error mean squares are both estimates of the error variance which can be compared by an F-test. If the F-test is a significantly high value, it means that $\frac{n \sum \sigma_j^2}{k-1} = \varnothing_\sigma$ is not zero and therefore the hypothesis is rejected.

12.4.2: TWO-FACTOR MODELS.

The general case for a two-factor linear model is:

$X_{ijk} = \mu + A_i + B_j + AB_{ij} + \varepsilon_{k(ij)}$ with i = 1,2, ..., p; j = 1,2, ..., q and k = 1, 2, ..., n.

In this model,

i.   Both A and B are at fixed levels - - a fixed model.
ii. Both A and B are at random levels - - a random model.
iii. One is at fixed levels and the other at random levels - - a mixed model.

If the number of elements observed in an experiment is small relative to the number of potential elements in the population, that is, the ratio $\frac{n}{N}$ is practically equal to zero, then the coefficient $(1 - \frac{n}{N})$ is assumed to be equal to unity.

i.   If factor A and B are fixed:

A - - the ratio $\frac{p}{P}$ will be equal to unity and the coefficient $(1 - \frac{p}{P}) = 0$

B - - the ratio $\frac{q}{Q}$ will be equal to unity and the coefficient $(1 - \frac{q}{Q}) = 0$

ii. If both factors A and B are random:

A - - the ratio $\frac{p}{P}$ will be equal to zero and the coefficient $(1-\frac{p}{P}) = 0$

B - - the ratio $\frac{q}{Q}$ will be equal to zero and the coefficient $(1 - \frac{q}{Q}) = 0$

iii.   If one factor is fixed and the other random, the ratios in (i) and (ii) apply.

S/N	Fixed Model	Random Model	Mixed Model
1	Assumptions: A_i's are fixed constants and $\sum A_i = 0$. B_j's are fixed constants and $\sum B_j = 0$. AB_{ij}'s are fixed constants and $\sum_i^p \sum_j^q AB_{ij} = 0$.	1. Assumptions: A_i's are NID $(0, \sigma_A^2)$ B_j's are NID $(0, \sigma_B^2)$. AB_{ij}'s are $(0, \sigma_{AB}^2)$.	1. Assumptions: A_i's are fixed and $\sum A_i = 0$. B_j's NID $(0, \sigma_B^2)$. AB_{ij}'s are NID $(0, \sigma_{AB}^2)$ but $\sum_i^p AB_{ij} = 0$; $\sum_i^p AB_{ij} \neq 0$; for A fixed and B random.
2	Analysis: Procedures for SS	2. Analysis: Same procedures	Analysis: Same procedures
3	Hypotheses Tested: $H_i : A_i = 0$ for all i. $H_2 : B_j = 0$ for all j. $H_3 : AB_{ij} = 0$ for all i and j.	3. Hypotheses Tested: $H_1 : \sigma_A^2 = 0$. $H_2 : \sigma_B^2 = 0$. $H_3 : \sigma_{AB}^2 = 0$.	3. Hypotheses Tested: $H_i : A_i = 0$. for all i. $H_2 : \sigma_B^2 = 0$. $H_3 : \sigma_{AB}^2 = 0$.
4	Ems (Fixed)	Ems (Random)	Ems (Mixed)

Anova Source			
Source df			
A_i (p-1) B_j (q-1) $AB_{ij} = $ (p-1)(q-1) $\varepsilon_{k(ij)} = $ pq(n-1)	$\sigma_\varepsilon^2 + nq\emptyset A$ $\sigma_\varepsilon^2 + nq\emptyset B$ $\sigma_\varepsilon^2 + nq\emptyset AB$ σ_ε^2	$\sigma_\varepsilon^2 + n\sigma_{AB}^2 + nq\sigma_A^2$ $\sigma_\varepsilon^2 + n\sigma_{AB}^2 + np\sigma_B^2$ $\sigma_\varepsilon^2 + n\sigma_{AB}^2$ σ_ε^2	$\sigma_\varepsilon^2 + n\sigma_{AB}^2 + nq\emptyset A$ $\sigma_\varepsilon^2 + np\sigma_B^2$ $\sigma_\varepsilon^2 + n\sigma_{AB}^2$ σ_ε^2

For the fixed model, the mean squares for A, B and AB are each compared to the error mean square to test the respective hypotheses (Hicks, 1973).

For the random model, the hypothesis of no interaction is tested by comparing the mean square for interaction to the mean square for error. But the mean square for the main effects Ai and Bj are compared with the mean square interaction to test the hypotheses.

But for the mixed model, the interaction hypothesis is tested by comparing the interaction mean square with the error mean square; similarly, the random effect Bj is also tested by comparing its means square with the error mean square. But the fixed effect Ai is tested by comparing its mean square with the interaction mean square.

12.4.3:RULES FOR DERIVING EXPECTED MEAN SQUARES (EMS)

Due to the importance of EMS column in determining what tests of significance are to be run, it is quite useful to have a simple method of deriving the values from the model for a given experiment. According to Hicks (1973) and Winer (1971), the following are a set of Rules to enable rapid determination of EMS without recourse to their derivation:

RULE 1:

Write the appropriate model for the design, making explicit in the notation those effects which are nested.

Model: $X_{ijk} = \mu + A_i + B_j + AB_{ij} + \varepsilon_{k(ij)}$

RULE 2:

Construct a two-way table in which the terms in the model (except the grand mean) are the row headings and the subscripts appearing in the model are the column headings; over each subscript write F, if the factor levels are fixed; R, if the factor levels are random. Also write the number of observations each subscript is to cover:

	p	q	n
	F	**R**	**R**
	i	j	k
A_i		q	n
B_j	p		n
AB_{ij}			n
$\varepsilon_{k(ij)}$			

RULE 3:

For each row (each term in the model) copy the number of observations under each subscript, provided that the subscript does not appear in the row heading.

For example, for row Ai the column j number of observations 'q' is copied under subscript j and 'n' is copied under subscript k'; similarly, for row Bj, the number of observations 'p' is copied under subscript i and 'n' under subscript k'; also for row ABij, copy the number of observations 'n' under the subscript k, as shown above.

RULE 4:

For any bracketed subscripts in the model, place a 1 under those subscripts which are inside the brackets; fill the remaining cells with a 0 or 1, depending on whether the subscript represents a fixed (F) or a random (R) factor as:

	p	q	n
Source	F	R	R
	i	j	k
Ai	0	q	n
Bj	p	1	n
ABij	0	1	n
$\varepsilon_{k(ij)}$	1	1	1

RULE 5:

The expected mean square for any term in the model, for example, the main effect of factor A is a weighted sum of the variances due to all effects which contain the subscript i.

a) For A_i, cover column i and other columns containing non-bracketed subscripts in letter i; for , $\varepsilon_{k(ij)}$ cover column k.

b) Multiply the remaining numbers in each row. The sum of these coefficients multiplied by the variance of their corresponding terms $(\theta_A \, or \, \sigma_A^2 (\emptyset_A \, or \, \sigma_A^2)$ is the EMS. For A_i, when column i is covered, the products of the remaining coefficients are q_n, n, n, and 1, but the first n is not used since there is no i in B_j. Thus, the EMS of A_i is:

$$nq\emptyset A + n\sigma_{AB}^2 + n\sigma_\varepsilon^2$$

c) Multiply the remaining numbers in each row

d) We must start at the last row with coefficients 1.1. σ_ε^2 and add this to row AB_{ij} and add row AB_{ij} to row A_i:

	p	q	n	
Source	F	R	R	
	i	j	k	EMS
Ai	0	q	n	$\sigma_\varepsilon^2 + n\sigma_{AB}^2 + qn\emptyset A$
Bj	p	1	n	$\sigma_\varepsilon^2 + pn\sigma_B^2$
ABij	0	1	n	$\sigma_\varepsilon^2 + n\sigma_{AB}^2$
$\varepsilon_{k(ij)}$	1	1	1	σ_ε^2

EMS for a Four – Factor Factorial (6 x 4 x 5 x 2) factorial with 6 Replications) in CRD.

The four factors:

O_i has 6 levels (i =1, 2, . . . 6), where O_i is random.

A_j has 4 levels (j = 1,2, 3, 4), where A_j is fixed.

C_k has 5 levels (k = 1,2, . . ., 5), where C_k is fixed.

L_m has 2 levels (m = 1,2), where L_m is fixed.

n = 6 and n is random.

The mathematical Model is:

$$X_{ijkmq} = \mu + O_i + A_j + C_k + L_m + OA_{ij} + OC_{ik} + AC_{jk} + OL_{im} + AL_{jm} + CL_{km}$$

$$+ OAC_{ijx} + OAL_{ijm} + OCL_{ikm} + ACL_{ikm} + OACL_{ijkm} + \varepsilon q_{(ijkm)}$$

285

	p	q	r	i	n	
	6	4	5	2	6	
Source	R	F	F	F	R	
	i	j	k	m	q	EMS
O_i	1	4	5	2	6	$\sigma_\varepsilon^2 + 240\sigma_0^2$
A_j	6	0	5	2	6	$\sigma_\varepsilon^2 + 60\sigma_{0A}^2 + 360\emptyset A$
C_k	6	4	0	2	6	$\sigma_\varepsilon^2 + 480\sigma_{0C}^2 + 288\emptyset c$
L_m	6	4	5	0	6	$\sigma_\varepsilon^2 + 120\sigma_{0L}^2 + 720\emptyset L$
$0A_{ij}$	1	0	5	2	6	$\sigma_\varepsilon^2 + 60\sigma_{0A}^2$
$0C_{ik}$	1	4	0	2	6	$\sigma_\varepsilon^2 + 48\sigma_{0C}^2$
$0L_{im}$	1	4	5	0	6	$\sigma_\varepsilon^2 + 120\sigma_{0L}^2.$
AC_{jk}	6	0	0	2	6	$\sigma_\varepsilon^2 + 12\sigma_{0AC}^2 + 72\emptyset AC.$
AL_{jm}	6	0	5	0	6	$\sigma_\varepsilon^2 + 30\sigma_{0AL}^2 + 180\emptyset AL$
CL_{km}	6	4	0	0	6	$\sigma_\varepsilon^2 + 24\sigma_{0CL}^2 + 144\emptyset CL$
$0AC_{ijk}$	1	0	0	2	6	$\sigma_\varepsilon^2 + 12\sigma_{0AC}^2.$
$0AL_{ijm}$	1	0	5	0	6	$\sigma_\varepsilon^2 + 30\sigma_{0AL}^2$
$0CL_{ikm}$	1	4	0	0	6	$\sigma_\varepsilon^2 + 24\sigma_{0CL}^2.$
ACL_{jkm}	6	0	0	0	6	$\sigma_\varepsilon^2 + 6\sigma_{0ACL}^2 + 36\emptyset ACL$
$0ACL_{ijkm}$	1	0	0	0	6	$\sigma_\varepsilon^2 + 6\sigma_{0ACL}^2.$
$\varepsilon q_{(ijkm)}$	1	1	1	1	1	σ_ε^2

13 REGRESSION ANALYSIS.

Regression Analysis is a statistical tool that utilizes the relation between two or more quantitative variables so that one variable can be predicted from the other or others. In regression analysis, developed by Galton in 1888, attention is centered on the dependence of one variable, Y, on another variable, X. In mathematics, Y is called a function of X, but in

statistics, the term regression (Snedecor and Cochran, 1967) is generally used to describe the relationship.

The variable whose value is determined first (Ross, 2005) is called the input, or independent, or concomitant variable and the other variable is called the response, or dependent variable.

Mathematical models may be classified as **Deterministic or Probabilistic**:

A. Functional Relation between two Variables - -Deterministic Models.

A functional relation between two variables is expressed by a mathematical formula. If X is the independent variable and Y the dependent variable, a functional relation is of the form:

$Y = f(X)$.

For example, the relation between the Naira sales (Y) of timber sold at a fixed price and number of units sold (X), if the selling price is two naira (N2) per unit, the relation is expressed by the equation: $Y = 2X$.

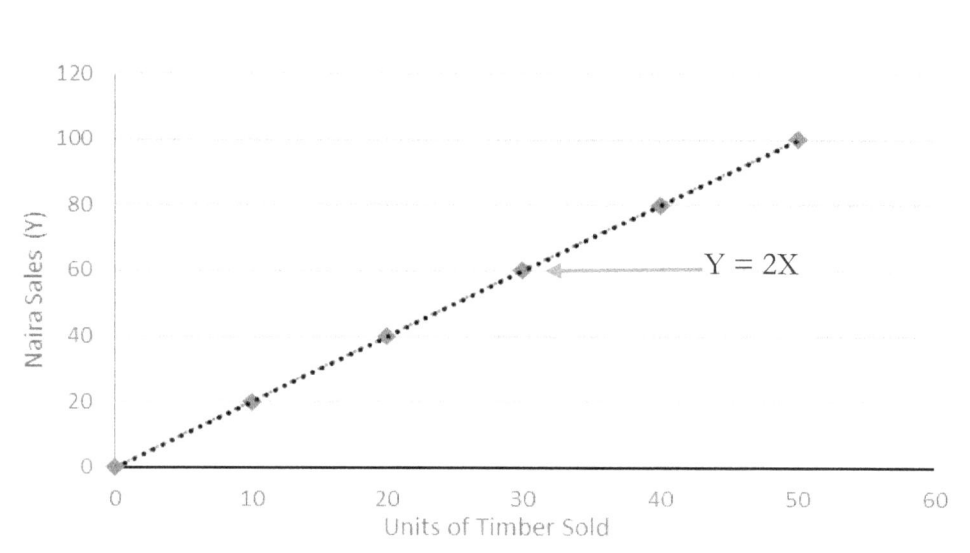

Figure 13:1.1: Relationship of the Price and quantity sold.

mathematical model is considered accurate, that is, a good model of nature, if it is able to predict some variable and do so with an error that will be negligible for practical purposes. For instance, Newton's law relating the force of a moving body to its mass and acceleration, $F = Ma$, is a deterministic model that predicts with little error of prediction for all practical purposes.

B. Statistical Relation between two Variables - - Probabilistic or Stochastic Models.

A statistical relation, unlike a functional relation, is not a perfect one. In general, the observations for a statistical relation do not fall directly on the curve of relationship.

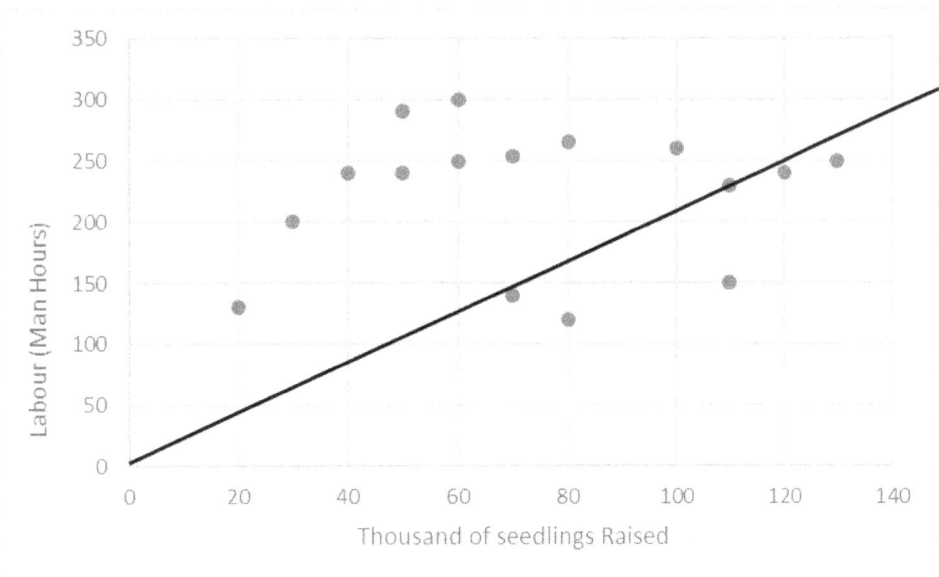

Figure 13.1.2: Statistical Relation between number of Tree Seedlings (in 1000s) raised versus man-hours Required

A Model of the form:

$y = \beta_0 + \beta_1 X$, is the equation of a straight line.

The model $= \beta_0 + \beta_{1x}$, provides a deterministic mathematical model for the relationship between y and x $(x = X_i - \bar{X})$

Probabilistic models contain one or more random components that are intended to explain the apparent random variability of y for a given value of x. For example, we might use the probabilistic model:

$y = \beta_0 + \beta_{1x} + \varepsilon$, where ε is a random variable representing the variability of y about the basic

relationship: $y = \beta_0 + \beta_{1x}$

13.1. Simple Linear Regression

According to Neter et al. (1983), a regression model is a formal means of expressing the two essential ingredients of a statistical relation:

i. A tendency of the dependent variable, Y, to vary with the independent variable or variables in a systematic fashion. This systematic relationship is called the regression function of Y on X.

ii. A scattering of observations around the curve of statistical relationship - -the means of these probability distributions of Y, vary in some systematic fashion with X.

Units of Timber Sold (X).

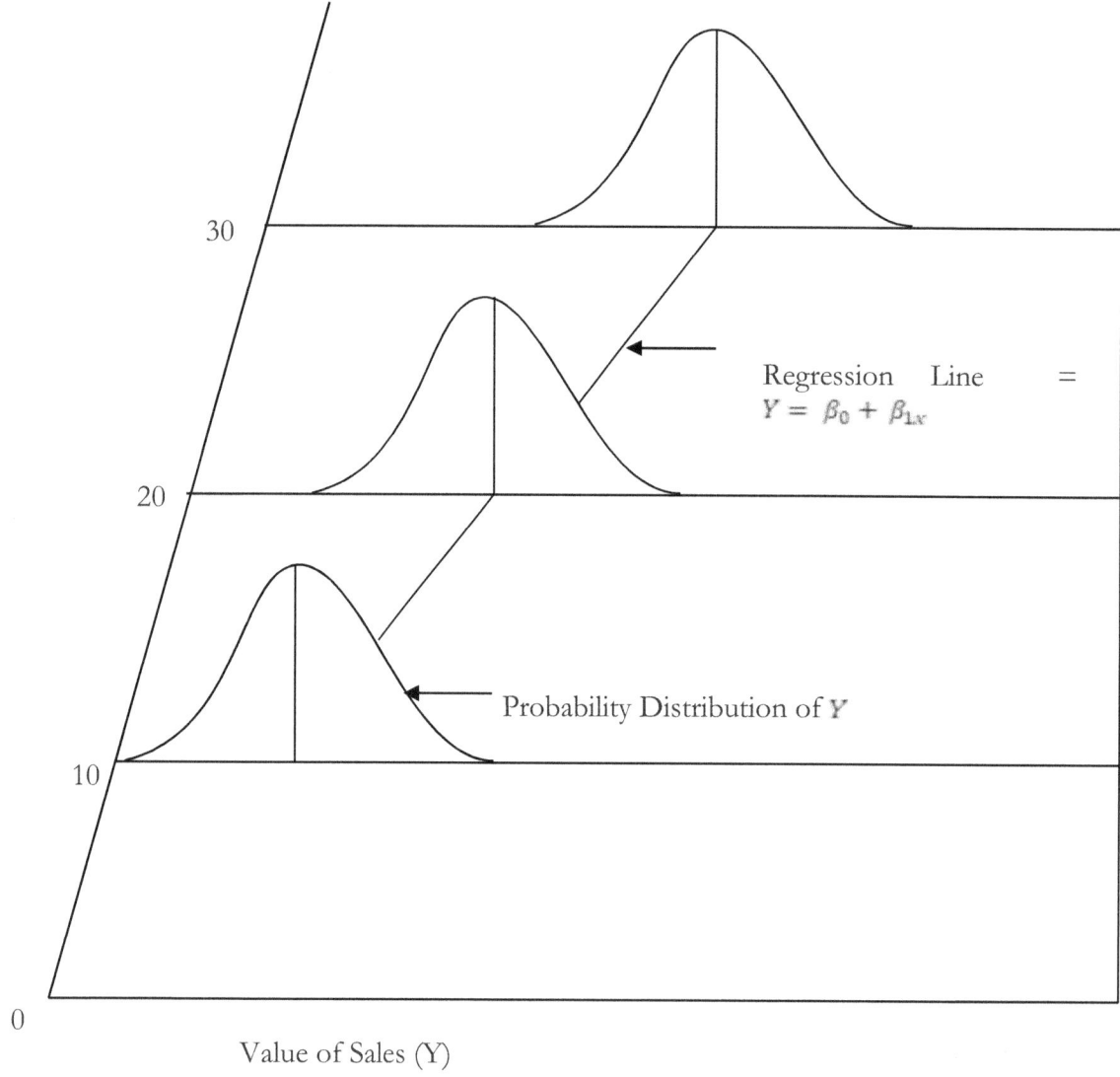

Figure 13.1.3: Pictorial Representation of Linear Regression Model.

A curve of regression of Y on X is said (Walpole, 1974) to be a Linear Regression, if and only if,

$\mu_{y/x} = \alpha + \beta_x$, for α and β real numbers $\alpha, \beta \neq 0$.

The model: $Y_i = \beta_0 + \beta_1 X_i + \varepsilon_i$, or $Y_i = \alpha + \beta X_i + \varepsilon_i$

where:

Y_i the value of the response variable in the ith trial.

β_0 and β_1 are unknown parameters ($\alpha = \beta_0$) to be estimated from sample data.

β_0 (or α) = the Y-intercept of the regression line, i.e. a constant since it is a parameter.

β_1 (or β) = the slope (or regression coefficient) which is also a constant parameter.

X_i = a known constant, which is the value of the independent variable in the ith trial.

ε_i = a random error term with mean zero and variance, $\sigma_{\varepsilon i}^2 = \sigma^2$

ε_i and ε_j are uncorrelated and hence the covariance, $\sigma\left(\varepsilon_i, \epsilon_j\right) = 0$.

The model is said to be **simple, linear in the parameters, and linear in the independent variable** (Neter et al., 1983). It is **"simple"** because there is only one independent variable; it is **"linear in the parameters"** because no parameter appears as an exponent, or is multiplied, or divided by another parameter, and

"linear in the independent variable" because this variable appears only in the first power. This type of model is also called a **first-order model** which takes the form of a straight line.

According to Gomez and Gomez (1984), the simple linear regression analysis deals with the estimation and tests of significance concerning the two parameters:

$$Y_i = \alpha + \beta X_i + \varepsilon_i \ (1)$$

$$Y_i = \beta_0 + \beta_i X_i + \varepsilon_i \ (2)$$

It should be noted that the simple linear regression analysis is performed under the assumption that there is a linear relationship between X and Y.

The simplest type of a linear regression equation in forestry is Y = X, used to show the relation between the number of growth-rings on a coniferous tree as a function of the age in years. It is obvious that whatever the value of X (age in years), the value of Y (number of growth rings), will be of corresponding magnitude. Thus, if the tree is ten years old, it will have ten growth rings; a fifteen-year old coniferous tree will have fifteen growth rings, as shown in the graph below:

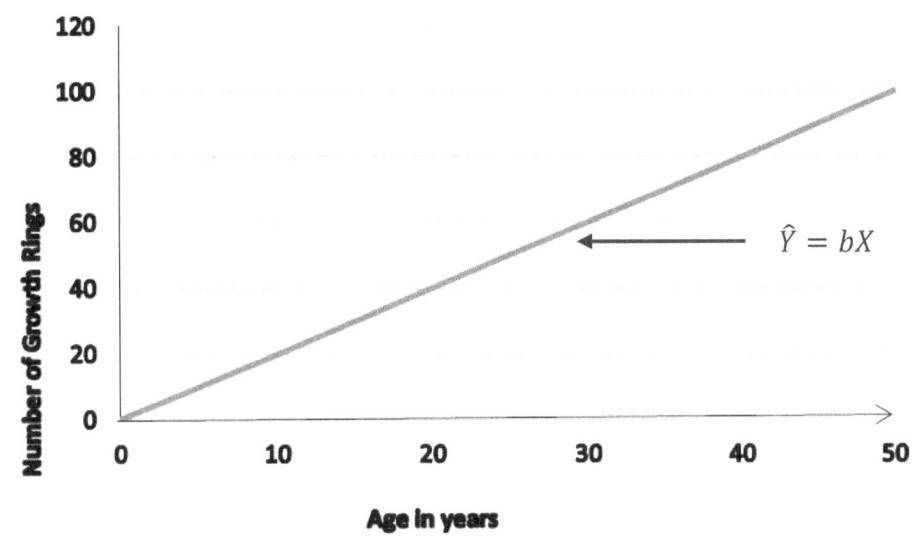

Figure 13.1.4: Number of Growth Rings of a Tree as a function of Age in years.

A general functional relationship above is of the form: Yi = bXi represents a regression line which passes through the origin and therefore has no intercept. The coefficient, b, of the independent variable, X, is called the slope of the regression equation.

Note:

If the data used for the functional relationship consist of all the elements of the population, then, Cangelosi et al. (1976) suggested that the computed values should be treated as **parameters**, unless these values will be used for prediction or projection in future. But if the data used for the functional relationship is considered as a sample, then the computed values should be treated as estimators.

Thus, using sample data, we will determine the point estimates of α and β denoted as 'a' and 'b', and form the regression equation as: $\hat{Y} = a + bX$.

13.2. MEANING OF REGRESSION PARAMETERS

For the model: $Y_i = \beta_0 + \beta_i X_i + \varepsilon_i$ or $Y_i = \alpha + \beta X_i + \varepsilon_i$, the parameters β_0 (α) and β_1, are called **Regression Coefficients** (Neter et al., 1983). Is β_1 the slope of the regression line which indicates the amount of change in Y for each unit change in X. In other words, β_1 indicates the change in the mean of the probability distribution of Y per unit increase in X. β_0 (or α) is the Y-intercept of the regression line. If the scope of the model includes X = 0, β_0 gives the mean of the probability distribution of Y at X = 0. When the slope of the model does not cover X = 0, Neter et al. (1983) say that β_0 does not have any particular meaning as a separate term in the regression model. ε_i is the random deviation from the true regression line.

13.3. ESTIMATING THE REGRESSION PARAMETERS

Before estimating the regression parameters, we need to plot the raw data of their pairs of fixed (X_i, Y_i) observations in a scatter diagram **(scattergram)** to provide a picture of any relationship between the independent and dependent variables (Cangelosi et al., 1976):

   i.   The scatter diagram (scatter gram) will generally demonstrate whether or not there is an apparent relationship between the two variables.
  ii.   If there is a relationship, it may suggest whether it is linear or non-linear.
 iii.   If the relationship is linear, the scattergram will show whether the relationship is positive or negative - - if both Y and X increase, the relationship is positive, but if Y increases while X decreases, the relationship is negative.

13.3.1. Method of Least Squares

According to Daniel and Wood (1971), the least-squares (LS) method says: **"Find the values of the constants in the chosen equation that minimize the sum of the squared deviations of the observed values from those predicted by the equation"**. This statement is justified by the Gauss-Markov theorem which states that the estimates obtained are **the most precise unbiased estimates that are linear functions of the observations** - - assuming that the correct form of the equation has been chosen.

Suppose we have available 'n' sets of observations (X1, Y1,), (X2, Y2), …., (Xn, Yn), then we can write (Draper and Smith, 1966):

$$Y_i = \beta_0 + \beta_1 X_i + \varepsilon_i, \text{ - - - 13.3.1.}$$

so that the sum of squares of deviations from the true line is:

$$S = \sum_{i=1}^{n} \varepsilon_i^2 \sum_{i=1}^{n} (Y_i - \beta_0 - \beta_i X_i)^2 \dots 13.3.2$$

$$S = \sum_{i=1}^{n} \varepsilon_1^2 \sum_{i=1}^{n} (Y_i - \beta_o - \beta_1 X_1)^2 \dots 13.3.2$$

We choose our estimates b_0 and b_1 to be the values which when substituted for β_0 and β_1 in equation 13.3.2, produce the least possible value of S. We can determine b_0 and b_1 by differentiating equation 13.3.2 with respect to β_0 and β_1 respectively, and setting the values equal to zero. Hence the estimates b_0 and b_1 are given by:

$$\sum_{i=1}^{n} (Y_i - b_0 - b_1 X_i) = 0$$

$$\sum_{i=1}^{n} X_i (Y_i - b_0 - b_1 X_i) = 0 \quad \ldots \ldots 13.3.3.$$

From these, we obtain:

$$\sum_{i=1}^{n} Y_i = nb_0 - b_1 \sum_{i=1}^{n} X_i \ldots \ldots \ldots 13.3.5$$

$$\sum_{i=1}^{n} X_i Y_i - b_0 \sum_{i=1}^{n} X_i - b_1 \sum_{i=1}^{n} X_i^2 = 0 \quad \ldots \ldots 13.3.4.$$

Or

$$\sum_{i=1}^{n} Y_i = nb_0 + b_1 \sum_{i=1}^{n} X_i \ldots \ldots \ldots 13.3.5$$

$$\sum_{i=1}^{n} X_i Y_i = b_0 \sum_{i=1}^{n} X_i + b_1 \sum_{i=1}^{n} X_i^2$$

These are called **the Normal Equations** which can be solved simultaneously to obtain the values of b_1 and b_0 as:

$$b_1 = \sum_{i=0}^{n} X_i Y_i - \frac{(\sum_{i=1}^{n} X_i)(\sum_{i=1}^{n} Y_i)}{n} = \frac{\sum(X_i - \bar{X})(Y_i - \bar{Y})}{\sum(X_i - \bar{X})^2}$$

$$b_0 = \bar{Y} - b_1 \bar{X}.$$

13.3.2: Short-Cut Methods of Computations in Regression

Let us examine the short-cut methods of computations in Regression:

a) **Uncorrected Sums And Cross-Products:**

The following quantities are computed from the raw sample data:

$$\sum X_i, \sum Y_i, \sum X_i Y_i = \text{sums of X and Y and sum of cross-products, } X_i Y_i;$$

$$\bar{X} = \frac{\sum X_i}{n}; \text{Y} = \sum Y_i/n, \text{ are the means of X and Y respectively.}$$

b) **Calculate Uncorrected sums of Squares and Cross-Products as:**

i. $\sum_{i=1}^{n} Y_i^2$.

ii. $\sum_{i=1}^{n} X_i^2$

iii. $\sum_{i=1}^{n} X_i Y_i$

c) Calculate Corrected Sums of Squares and Cross-Products as:

The Corrected sums of squares, simply means that the variables X_i, Y_i and $X_i Y_i$ have been corrected for their means as:

i. $y_i = (Y_i - \bar{Y}): \sum_{i=1}^{n} y_i^2 = \sum_{i=1}^{n}(Y_i - \bar{Y})^2$

$\sum_{i=1}^{n} y_i^2 =$ Total sum of squares, i.e. the total variation in Yi.

$$\sum_{i=1}^{n} y_i^2 = \sum_{i=1}^{n}(Y_i - \bar{Y})^2 = \sum_{i=1}^{n} Y_i^2 - \frac{(\sum_{i=1}^{n} Y_i)^2}{n}$$

ii. $x_i = (X_i - \bar{X})$

$$\sum_{i=1}^{n} x_i^2 = \sum_{i=1}^{n}(X_i - \bar{X})^2 = \sum_{i=1}^{n} X_i^2 - \frac{(\sum_{i=1}^{n} X_i)^2}{n}$$

iii. $x_i y_i = (X_i - \bar{X})(Y_i - \bar{Y})$

$$\sum_{i=1}^{n} x_i y_i = \sum_{i=1}^{n}(X_i - \bar{X})(Y_i - \bar{Y}) = \sum_{i=1}^{n} X_i Y_i - \frac{(\sum_{i=1}^{n} X_i)(\sum_{i=1}^{n} Y_i)}{n}$$

d) Calculate Regression Coefficient (b1) and the Intercept (b0) as:

$$b_1 = \frac{\sum_{i=1}^{n} x_i y_i}{\sum x^2}$$

$$b_0 = \bar{Y} - b_1 \bar{X}$$

Note: b0 and b1 may be positive or negative. If Y increases as X increases, b1 will be positive; if Y decreases as X increases, b1 will be negative.

e) The Estimated Regression Equation is ($\hat{Y}_i$):

$$\bar{Y}_i = b_0 + b_1 X_i$$

If we substitute $\bar{Y} - b_1 \bar{X}$ for b_0 , the estimated regression equation becomes:

$$\hat{Y}_i = \bar{Y} - b_1 \bar{X} + b_1 X_i = \bar{Y} + b_1 X_i - b_1 \bar{X} = \bar{Y} + b_1(X_i - \bar{X}). .$$

f) Calculate the Sum of Squares Regression (or the Explained SS) $(\sum_{i=1}^{n} \hat{y}_i^2)$ as:

The calculated sum of squares regression (SSREG) is also called the "Explained sum of squares" denoted by $\sum_{i=1}^{n} \hat{y}_i^2$ and calculated as:

$$\sum_{i=1}^{n} \hat{y}_i^2 = \frac{(\sum_{i=1}^{n} x_i y_i)^2}{\sum_{i=1}^{n} x_i^2} = SS_{REGRESSION}$$

g) **Calculate the Sum of Squares Residuals (or Unexplained SS)** $\left(\sum d_{y.x}^2\right)$ **as:**

The sum of squares deviations $\left(\sum d_{y.x}^2\right)$ are the unexplained sum of squares which is the difference between the total sums of squares and the regression sums of squares:

$$\sum_{i=1}^{n} d_{y.x}^2 = \sum_{i=1}^{n} y_i^2 - \frac{(\sum x_i y_i)^2}{\sum x_i^2}$$

$$\sum_{i=1}^{n} d_{y.x}^2 = \sum_{i=1}^{n} y_i^2 - \sum_{i=1}^{n} \hat{y}_i^2$$

From the sum of squares deviations, $\left(\sum_{i=1}^{n} d_{y.x}^2\right)$, the error due to regression is estimated as:

$$SSE = S_{y.x}^2 = \frac{\sum_{i=1}^{n} d_{y.x}^2}{n-2}$$, where, (n-2) is the error degree of freedom.

The residual $\left(e_i = Y_i - \hat{Y}_i\right)$ is the observed vertical deviation of Yi from the fitted regression line and residuals are very useful for studying whether a given regression model is appropriate for the data under investigation.

Thus, $e_i = Y_i - \hat{Y}_i = Y_i - b_0 - b_1 X_i$

The magnitudes of the residuals are shown by the vertical lines between an observation and the fitted value on the estimated regression line (Neter et al., 1983).

Before we plot the estimated regression line and the residuals, let us take a look at the properties of the estimated regression line as suggested by Neter et al. (1983, 1990):

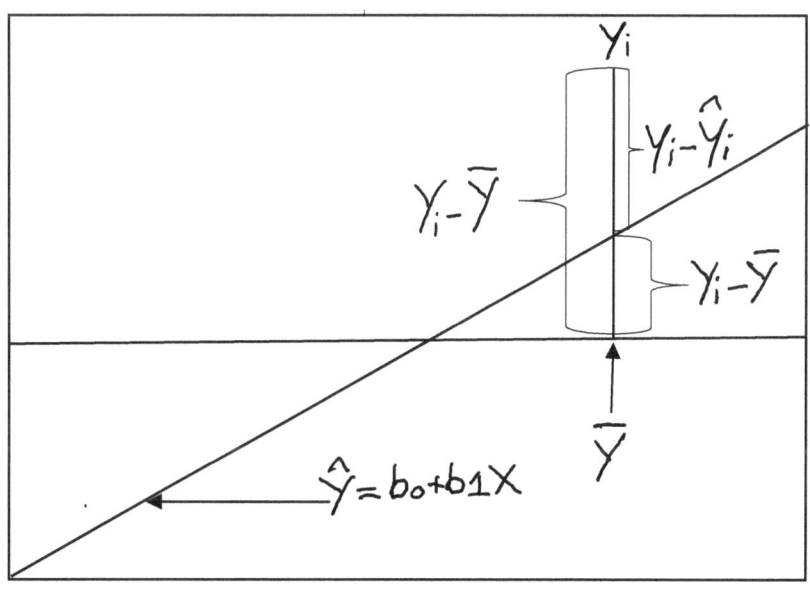

Figure 13.3.2: Partitioning of Deviations $(Y_i - \bar{Y})$

1. The sum of the residuals is zero: $\sum_{i=1}^{n} e_i = 0$ - - of course, rounding errors may be present in any particular case.

2. The sum of the squared residuals, $\sum_{i=1}^{n} e_i^2$, is a minimum, according to the Gauss-Markov theorem on least squares estimators.

3. The sum of the observed values, Y_i, equals the sum of the fitted values, $\hat{Y}_i$:

$$\sum_{i=1}^{n} Y_i = \sum_{i=1}^{n} \hat{Y}_i \text{ , as can easily be seen from the first normal equation:}$$

$$\sum_{i=1}^{n} Y_i = nb_0 + b_1 \sum_{i=1}^{n} X_i$$

$$= \sum b_0 + b_1 \sum X_i = \sum (b_0 + b_1 X_1) = \sum_{i=1}^{n} \hat{Y}_i = \sum_{i=1}^{n} Y_i$$

Therefore, the mean of the Y_i is the same as that of the $\hat{Y}_i$ i.e. $\bar{Y}$.

4. The regression line always goes through the point $(\bar{X}, \bar{Y})$. This can be seen from (3) above. If , we $X_i = \bar{X}$ have:

$$\hat{Y} = \bar{Y} + b_1(X_i - \bar{X}) = \bar{Y} + b_1(\bar{X} - \bar{X})) = \bar{Y}$$

5. The sum of the weighted residuals is zero, when weighted by the level of the independent variable and when weighted by the fitted value of the response variable:

i. $\sum_{i=1}^{n} X_i e_i = 0$

$$i.e \sum_{i=1}^{n} X_i e_i = \sum_{i=1}^{n} X_i (Y_i - b_0 - b_1 X_i)$$

$$\sum X_i Y_i - b_0 \sum X_i - b_1 \sum X_i^2 = 0$$

ii. $\sum_{i=1}^{n} \hat{Y}_i e_i = 0$

$$SSE = S_{y.x}^2 = SS_{TOTAL} - SS_{REGRESSION}$$

Draper and Smith (1966) observed that the SS due to regression can be computed from a single function of Y1, Y2 . . . , Yn, namely b1 since:

$$\sum (\hat{Y}_i - \bar{Y})^2 = b_1^2 \sum (X_i - \bar{X})^2, \text{ which has one degree of freedom.}$$

Consider the deviation $Y_i - \bar{Y}$, the basic quantity measuring the variation of the observations Y_i - - can be decomposed as:

$$Y_i - \bar{Y} = \hat{Y}_i - \bar{Y} + Y_i - \hat{Y}_i$$

Total deviation = Deviation of fitted Regression value plus Deviation around regression line. That is:

i. The deviation of the fitted value $\hat{Y}_i$ around the mean, $\bar{Y}, plus$
ii. The deviation of Y_i around the regression line.

Thus:

$$\sum(Y_i - \bar{Y})^2 = \sum(\hat{Y}_i - \bar{Y})^2 + \sum(Y_i - \hat{Y})^2$$

$$SS_{TOTAL} = SS_{REGRESSION} + SS_{ERROR}$$

13.3.2 a. ANALYSIS OF VARIANCE APPROACH TO REGRESSION ANALYSIS.

The breakdowns of the total sum of squares and associated degrees of freedom are displayed in the form of an analysis of variance table (Anova Table):

Anova Table for a Simple Linear Regression Analysis.

Source of Variation	DF	SS	MS	F
Regression (SS_R) (or Explained Variation) $(Y_i - \hat{Y}_i)$ (SSE)	1	$\sum_{i=1}^{n}(\hat{Y}_i - \bar{Y})^2 = \sum_{i=1}^{n} y_i^2$ $= \frac{(\sum xy)^2}{\sum x_i^2} . = \sum \hat{y}^2$	$\frac{S_{\hat{y}}^2}{1}$	$\frac{M_{SR}}{MSE}$
Error or Unexplained Variation $(Y_i - \hat{Y}_i)$	n-2	$\sum_{i=1}^{n}(Y_i - \hat{Y}_i)^2 = \sum_{i=1}^{n} y_i^2 - \sum_{i=1}^{n} \hat{y}_i^2$ $= d_{y.x}^2 = SSE.$	$\frac{S_{\hat{y}}^2}{n-2}$ MS.	
Total $(Y_i - \bar{Y})$ (Total Variation)	n-1	$\sum_{i=1}^{n} y_i^2 = \sum_{i=1}^{n} Y_i^2 - \frac{(\sum Y_i)^2}{n}$		

13.3.3 INFERENCES IN REGRESSION ANALYSIS

The Regression Model is: $Y_i = \beta_0 + \beta_1 X_i + \varepsilon_i$.

Before we can make inferences concerning the parameters of the Regression Model, β_0 and β_1 , we need to estimate the standard errors of their estimated values, b_0 and b_1 to test the hypotheses.

   i.   **Inferences Concerning β_1:**

The hypothesis concerning β_1:

   H_0 : $\beta_1 = 0$. i.e. no linear relationship between Y and X variables.

   H_1 : $\beta_1 = 0$. i.e. there is a linear relationship between Y and X variables.

   ii.   **Standard Error of β_1**

Since $b_0 \ and \ b_1$ re point estimators of β_0 and β_1 we must estimate the variances of the sampling distributions of these estimators. The variance of b_1 is:

$$V(b_1) = \frac{\sigma^2}{\sum(X_i - \bar{X})^2}$$

Substituting the unbiased estimator of σ^2 which is MSE:

$$S^2(b_1) = \frac{MSE}{\sum(X_i - \bar{X})} = \frac{S_{y.x}^2}{\sum x_i^2}$$

$$S.E.(b_1) = S_{b1}\sqrt{\frac{S_{y.x}^2}{\sum x_i^2}}$$

b_1 is normally distributed and hence the standardized statistic:

$\frac{b_1 - \beta_1}{\sigma(b_1)}$ is a standard normal.

By substituting $S(b_1)$ for $\sigma(b_1)$, the standardized statistic $\frac{b_1 - \beta_1}{S(b_1)}$ is distributed as t(n-2).

Thus, $\left|t_{b_1}^*\right| > \frac{b_1 - \beta_1}{S_{b_1}} = \frac{b_1}{S_{b_1}}$, *since* $\beta_1 = 0$. The t_b^* value is compared with $t_{\frac{\alpha}{2}, n-2}$ in the t-Table.

Decision Rule: Reject H$_0$, if the $\left|t_{b_1}^*\right| > t_{0.025, n-2}$

Conclusion: Since, $|t_b^*| > t_{0.025, n-2}$ we conclude that there is a strong linear relationship between Y and X, and can therefore be used for estimation and prediction. But if $|t_b^*| \leq t_{0.025, n-2}$, conclude H$_0$

iii. **Confidence Interval for β_1**

Since $\frac{b_1 - \beta_1}{S_{b_1}}$ has a t-distribution,

$$P\left(t_{\alpha/2}, n-2 \leq \frac{b_1 - \beta_1}{S_{b_1}} \leq t_{\alpha/2}, n-2\right) = 1 - \alpha$$

Thus, $P\left(b_1 - S_{b1}\left(t_{\alpha/2}, n-2\right)\right) \leq \beta_1 \leq \left(b_1 + S_{b1}\left(t_{\alpha/2}, n-2\right)\right) = 1 - \alpha$

$(1 - \alpha)$ 100% confidence limits for β_1 are:

$$b_1 \pm S_{b1}\left(t_{\alpha/2}, n-2\right)$$

iv. **Standard Error of the Intercept, β_0 (or α):**

To test the hypothesis that:

$H_0 : \beta_0 = 0$ i.e. the intercept passes through the origin.

$H_1 : \beta_0 = 0$ i.e. the intercept does not pass through the origin,

we need to estimate the standard error of the point estimator, b_0.

$b_0 = \bar{Y} - b_1\bar{X}$

$Var(b_0) = Var(\bar{Y}) + \bar{X}^2 Var(b_1)$

i.e. $\sigma^2(b_0) = \sigma^2(\bar{Y}) + \bar{X}^2\sigma^2(b_1)$

From 13.3.3(i) above:

$$\sigma^2(b_0) = \frac{\sigma^2}{n} + \bar{X}^2\sigma^2(b_1)$$

$$\frac{\sigma^2}{n} + \frac{\bar{X}^2\sigma^2}{\Sigma(X_i - \bar{X})^2}.$$

Substituting the unbiased estimator of $\sigma^2 = MSE$, we have:

$$S^2_{(b_0)} = \frac{MSE}{n} + \frac{\bar{X}^2 MSE}{\Sigma(X_i - \bar{X})^2} = MSE\left(\frac{1}{n} + \frac{\bar{X}^2}{\Sigma(X_i - \bar{X})^2}\right)$$

Note that MSE $= S^2_{y \cdot x}$. Hence,

$$S^2_{(b_0)} = S^2_{y \cdot x}\left(\frac{1}{n} + \frac{\bar{X}^2}{\Sigma(X_i - \bar{X})^2}\right)$$

$$S.e.\,(b_0)\ S_{bo} = S_{y \cdot x}\sqrt{\left(\frac{1}{n} + \frac{\bar{X}^2}{\Sigma(X_i - \bar{X})^2}\right)}.$$

Thus, $\frac{b_0 - \beta_0}{\sigma(bo)}$ is a standard normal.

Substituting $S(b_0)$ for $\sigma(b_0)$, the standardized statistic is distributed as $t_{(n-2)}$. Hence,

$$\left|t^*_{b_0}\right| = \frac{b_0 - \beta_0}{S_{(bo)}} = \frac{b_0}{S_{bo}} \text{ is compared with } t_{\alpha/2}\, n - 2 \text{ in the t-Table.}$$

Decision Rule: Reject H_0, if $\left|t^*_{b_0}\right| > t_{0.025}, n - 2$.

Conclusion: Since the $\left|t^*_{b_0}\right| > t_{0.025}, n - 2$, we conclude that b_0 does not pass through the origin, or not equal to the hypothesized value of H_0.

v.   **Confidence Interval for β_0:**

The $(1 - \alpha)$ 100% confidence limits for β_0 are obtained in the same way as β_1 above:

$$P\left(b_0 - S_{bo}\left(t_{\alpha/2}, n - 2\right)\right) \le \beta_0 \le \left(b_0 + S_{bo}\left(t_{\alpha/2}, n - 2\right)\right) = 1 - \alpha$$

$$(1 - \alpha)\ 100\%\ C.I. = b_0 \pm S_{bo}\left(t_{\alpha/2}, n - 2\right).$$

vi.   **Standard Error of $\hat{Y}$k, the Estimated Mean Value.**

The estimated equation: $\hat{Y}_i = \bar{Y} + b_1(X_i - \bar{X})$, , and because Y and b_1 are subject to error, they influence $\hat{Y}$. In fact, $\hat{Y}_i$ are the estimates of the mean (or average) value of Y for a given value of X. If we are using sample data, we can use these averages to construct confidence intervals for the Y-values for a given X (Cangelosi et al. (1976).

Hence, $\hat{Y}_k$ reflects the variation due to the sampling errors of Y and . The b_1 measure of this variation is called the standard error of $\hat{Y}_k$ obtained from the regression equation:

$\hat{Y}_k = \bar{Y} + b_1 (X_k - \bar{X})$ — when a specific value of X is used in the equation. Thus, the variance of the predicted mean value of Y, $\hat{Y}_k$ at a specific value X_k of X is:

$$V(\hat{Y}_k) = V(\bar{Y}) + (X_k - \bar{X})^2 V(b_1)$$

$$\frac{\sigma^2}{n} + \frac{(X_k - \bar{X})^2 \sigma^2}{\Sigma(X_k - \bar{X})^2}$$

Substituting the unbiased estimator of σ^2 as in β_1, we have:

$$S^2_{\hat{Y}_k} = \frac{MSE}{n} + \frac{(X_k - \bar{X})^2 MSE}{\Sigma(X_i - \bar{X})^2}$$

Since MSE $= S^2_{y.x}$, the variance of $\hat{Y}_k$ is :

$$\text{s.e.} \left(\hat{Y}_k\right) = S^2_{\hat{Y}_k} = \frac{S^2_{y.x}}{n} + \frac{(X_k - \bar{X})^2 S^2_{y.x}}{\Sigma(X_i - \bar{X})^2}$$

$$\text{s.e.} \left(Y_k\right) = S_{\bar{Y}_k} = S_{y.x}\sqrt{\left(\frac{1}{n} + \frac{(X_k - \bar{X})^2}{\Sigma(X_i - \bar{X})^2}\right)}$$

Note: s.e. $(\hat{Y}_k)$ is a minimum value when $X_k = \bar{X}$ and increases as we move away from X in either direction (Draper and Smith, 1966; Neter et al. 1983).

$$\frac{\hat{Y}_k - E(Y_k)}{S(\hat{Y}_k)} \sim t(n-2)$$

vii.   Confidence Interval for the Predicted Mean Value, $\hat{Y}_k$ for a given X_k.

The $(1 - \alpha)100\%$ Confidence Limits for the predicted mean value of $\hat{Y}_k$ for a given X_k can be computed from the standard error above. Thus, the construction of confidence interval for $\hat{Y}_k$ is actually a construction of a range of the possible values for $\hat{Y}_k$ for a given X_k

$$(1 - \alpha)100\% \ C.I. = \hat{Y}_k \pm S(\hat{Y}_k) \ t_{\alpha/2}, n - 2.$$

For each confidence interval, two values are usually computed - - the lower limit and the upper limit. These lower and upper limits will result in a band of lower and upper limits around the estimated regression line which are schematically represented in figure 13.3.3:

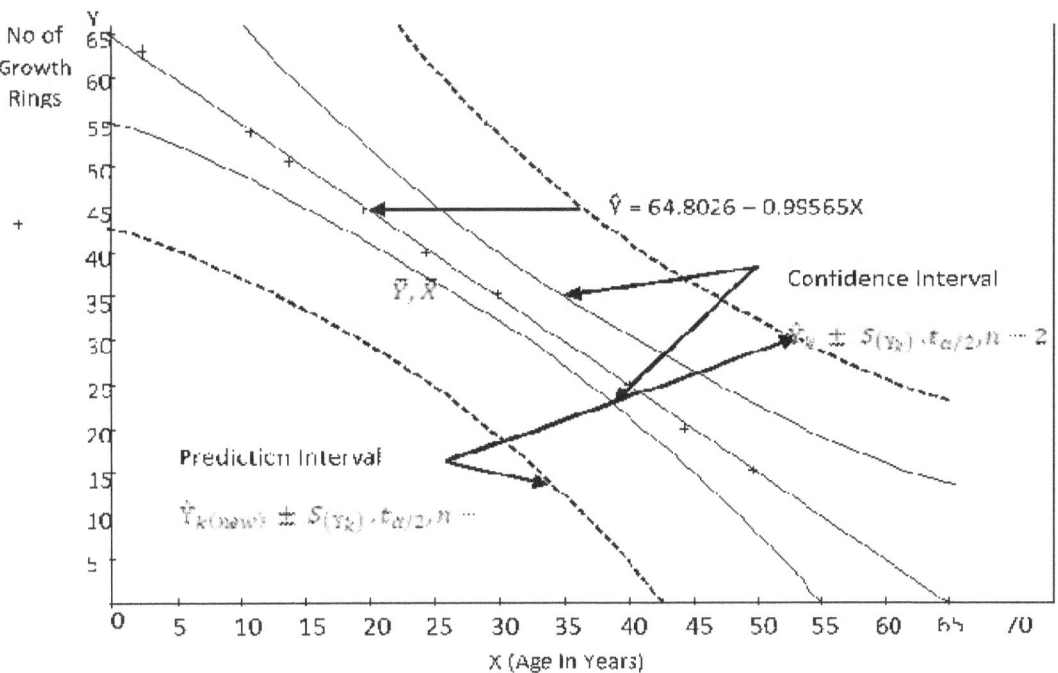

Figure 13.3.3: Confidence Limits and Prediction Limits.

viii.   **Prediction of a New Individual (or Single) Observation, $\hat{Y}_{k(new)}$.**

When a new $\hat{Y}_{k(new)}$ is predicted from the regression equation, by using a new Xnew observation, which is within the range of the data, but never used in the estimated regression, the new $\hat{Y}_{k(new)}$ is treated as a new trial (Neter, et al. (1983):

$$\hat{Y}_{k(new)} = b_0 + b1\ (X_{new} - \bar{X}).$$

While the same regression equation is used by substituting the new, Xnew, to predict the single $\hat{Y}_{k(new)}$ value, the variance of this single predicted value is larger than the mean, $\hat{Y}_k$, value estimated above and is given by Draper and Smith (1966) as:

$$\sigma^2 + V(\hat{Y}_k) = \sigma^2 + \sigma^2 \left[\frac{1}{n} + \frac{(X_{new} - \bar{X})^2}{\sum(X_i - X)^2}\right]$$

$$\sigma^2 \left[1 + \frac{1}{n} + \frac{(X_{new} - \bar{X})^2}{\sum(X_i - X)^2}\right]$$

Substituting the unbiased estimator of σ^2 as before, we have:

$$Var(\hat{Y}_{k(new)}) = S_{y.x}^2 \left[1 + \frac{1}{n} + \frac{(X_{new} - \bar{X})^2}{\sum(X_i - X)^2}\right].$$

$$(1-\alpha)100\%\ C.I.\ for\ \hat{Y}_{k(new)} \hat{Y}_{k(new)} \pm \left(t_{\frac{\alpha}{2}}, n-2\right)\sqrt{S_{y.x}^2\left(1 + \frac{1}{n} + \frac{(X_{new}-\bar{X})^2}{\sum(X_i-X)^2}\right)}$$

300

ix. Prediction of the Mean of "q" New Observations for a given X_k

It is sometimes necessary to predict the mean of "q" new observations on Y, that is $\overline{Y}_k$, of the future, where, as Draper and Smith (1966) observed:

$$\hat{Y}_q \sim N\left(\beta_0 + \beta_1 X_k, \sigma_q^2\right)$$

$$\hat{Y}_k \sim N\left(\beta_0 + \beta_1 X_k, V(\hat{Y}_k)\right) \ and$$

$$\overline{Y}_q - \hat{Y}_k \sim N\left(0, \sigma_q^2 + V(\hat{Y}_k)\right)$$

Hence:

$\frac{\overline{Y}_q - \hat{Y}_k}{s.e}$ is distributed as $t_{(n-2)}$, where s.e. of $\hat{Y}_{q(new)}$ is:

$$s.e.\left(\hat{Y}_{q(new)}\right) = \frac{MSE}{q} + S^2\left(\hat{Y}_k\right)$$

$$= MSE\left(\frac{1}{q} + \frac{1}{n} + \frac{(X_k - \overline{X})^2}{\sum(X_i - X)^2}\right)$$

Thus,

$$S^2\left(\overline{Y}_{q(new)}\right) = S_{y.x}^2\left(\frac{1}{q} + \frac{1}{n} + \frac{(X_k - \overline{X})^2}{\sum(X_i - X)^2}\right).$$

Therefore, $(1 - \alpha)$ 100% C.I. for $\overline{Y}_q$ about $\hat{Y}_k$ is:

$$\hat{Y}_k \pm \left(t_{\frac{\alpha}{2}, n-2}\right)\sqrt{S_{y.x}^2\left[\frac{1}{q} + \frac{1}{n} + \frac{(X_k - \overline{X})^2}{\sum(X_i - X)^2}\right]}$$

x. The Percentage of the Variation in Y Explained by the Regression.

The percentage of the total variation in Y that is explained or accounted for, by the variation of X, is called the coefficient of determination and is measured by R2.

R2 is defined (Draper and Smith, 1966) as the **"proportion of total variation about the mean, $\overline{Y}$, explained by the regression"** and it is frequently expressed as a percentage:

$$R^2 = \frac{\sum \hat{y}^2}{\sum y^2} \ i.e. \frac{Regression\ Sum\ of\ Sqaures}{Total\ Sum\ of\ Squares}$$

$$R^2 = \frac{SS_{Total} - SS_{Error}}{SS_{Total}} = \frac{SS_{Regression}}{SS_{Total}} = 1 - \frac{SS_{Error}}{SS_{Total}}.$$

$$0 \leq R^2 \leq 1.$$

Thus, the larger R^2 is, the more is the total variation of Y reduced by introducing the independent variable X.

If all the observations fall on the fitted regression line, SSE is equal to zero and $R^2 = 1$. However, if the slope of the fitted regression line is $b_1 = 0$, then,

$\hat{Y}_i \equiv \bar{Y}$; $SSE = SS_{Total}$ and $R^2 = 0$ (*Neter et al*, 1983). But in practice R^2, lies between the two extremes - - the closer it is to one (1), the greater is the degree of linear association between X and Y.

The coefficient of determination (R^2) is the total variation in Y that is associated with the regression on X, that is, the degree of linearity between the variable Y and the variable X.

xi. **Correlation Coefficient (R or r) is the square root of the Coefficient of determination**
$$R = \pm\sqrt{R^2}$$
.

However, the correlation coefficient can be directly calculated from the raw data as:

$$R = \frac{\sum(X_i - \bar{X})(Y_i - \bar{Y})}{\sqrt{\sum(X_i - \bar{X})^2 \sum(Y_i - \bar{Y})^2}}$$

$$= \frac{\sum X_i Y_i - \frac{(\sum X_i)(\sum X_i)}{n}}{\sqrt{\left[\sum X_i^2 - \frac{(\sum X_i)^2}{n}\right]\left[\sum Y_i^2 - \frac{(\sum Y_i)^2}{n}\right]}}$$

$$= -1 \leq R \leq 1$$

The sign of R, whether positive or negative, depends (Cangelosi et al., 1976) on the sign of the regression coefficient, b or b1, in the regression equation.

xii. **Coefficient of Non-Determination (or Coefficient of Alienation), k2 = 1 − R2,** is the proportion of the variation in the dependent variable that is not explained by the regression model. It is calculated as the unexplained variation divided by the total variation:

$$k^2 = \frac{\Sigma(Y_i - \hat{Y})^2}{\Sigma(Y_i - \bar{Y})^2}$$

The coefficient of alienation is the square root of the coefficient of non-determination, $k = \sqrt{k^2}$, is an abstract measure of the lack of correlation between Y and X. Unlike R, k is an unsigned absolute rating which can neither be positive nor negative.

xiii. **Testing the Significance of R2.**

The statistical significance of R2 can be tested as:

$H_0 : \rho = 0$ i.e. There is no relationship between X and Y.

$H_0 : \rho \neq 0$ i.e. There is some relationship between X and Y.

$\alpha = 0.05 \ or \ 0.01$.

a) **Compute F-value as:**

$$F^* = \frac{(SS_{REG})/k}{(SS_{ERROR})/(n-k-1)}$$
.

Compare the calculated F-value with the Tabulated F-Values with:

k = number of independent variables in the regression equation.

302

n-k-1= Error degree of freedom.

Conclusion:

The coefficient of determination (R2) is said to be significantly different from zero, if the computed F-value is greater than the Tabular F-value at the α-level of significance.

xiv. **The Assumptions of Regression Analysis.**

The basic assumptions on which Regression analysis is based are:

1. The independent variable, X, is measured without error (i.e. the X's are fixed) and does not vary at random, but is under the control of the investigator.
2. The expected value for the dependent variable, Y, for any given value of X is described by the linear function: $\mu_y = \alpha + \beta X$. That is, the relationship between X and Y variables is linear.
3. For any given value of X, the Y's are independently and normally distributed i.e. $Y = \alpha + \beta X + \varepsilon$. $\varepsilon \sim N(0, \sigma^2)$
4. The error variance (ε's) around the regression line are homoscedastic - - that is, the variance around the regression line is constant and independent of the magnitude of X or Y.

xv. **Uses of Regression Analysis.**
1. To estimate the mean response $\hat{Y}$, for a given X.
2. To make inferences about the regression parameters.
3. To predict a new observation, Y, for a given X.
4. For quality control.

13.3.4. A Worked Example of A Simple Linear Regression.

The following data were obtained by a 300-level Mensuration student in his quest to develop a regression equation to predict or estimate tree volumes from the measurements of diameter at breast height of <u>Gmelina arborea</u> trees in Uniben Plantation in 2009:

Table 13.3.4.1. Dbh in (centimeters) and Volume (cu.meter).

s/n	Dbh (cms)(X_i)	Volume (cu.meters)(Y_i)
1	8	7
2	9	8
3	10	12
4	11	17
5	12	20
6	13	26
7	14	27
8	15	29
9	16	39
10	17	43
Sum:	$\Sigma X_i = 125$	$\Sigma Y_i = 228$
Mean:	$\bar{X} = 12.5$	$\bar{Y} = 22.8$
Sum of Squares:	$\Sigma X_i^2 = 1,645$	$\Sigma Y_i^2 = 6,562$
Sum of products:	$\Sigma X_i Y_i = 3,181$	

i. Procedure of Least Squares Estimates of Regression Parameters.

Before estimating the regression parameters, we must plot the raw data of Y_i versus X_i in a scatter gram to see if there is any linear relationship:

Of course, the scattergram in Figure 13.3.4.1 shows a strongly linear relationship between the tree volume and its diameter at breast height. This indicates that the assumption of a linear function is justified and hence the model: of $Y_i = \beta_0 + \beta X_i + \varepsilon_i$, is adequate.

Figure 13.3.4.1: Scattergram of Tree Volume against Diameter at Breast Height.

ii. Corrected Sums Squares and Cross-Products.

From the uncorrected sums of squares and cross-products in Table 13.3.4.1, we obtain the corrected sums of squares and cross-products as follows:

a) $\sum y_i^2 = \sum Y_i^2 - \frac{(\sum Y_i)^2}{n} = 6{,}562 - \frac{(228)^2}{10} = 6{,}562 - 5{,}198.4 = 1{,}363.6$

b) $\sum x^2 = \sum X_i^2 - \frac{(\sum x_i)^2}{n} = 1{,}645 - \frac{(125)^2}{10} = 1{,}645 - 1{,}562.5 = 82.5$

c) $\sum xy = \sum X Y_i - \frac{(\sum x_i)(\sum Y_i)}{n} = 3{,}181 - \frac{(125)(128)}{10} = 3{,}181 - 2{,}850 = 331.0$

d) **Estimate The Regression Parameters, β_i and β_0**

$b_i = \frac{\Sigma xy}{\Sigma x^2} = \frac{331}{82.5} = 4.012$

$b_0 = \bar{Y} - b_i \bar{X} = 22.8 - (4.012)(12{,}5) = 22.8 - 50.15 = 27.35$

e) **Estimated Regression Equation: $\hat{Y}_i = -27.35 + 4.012 X_i$.**

Estimate SS Regression and SS Residuals.

$$SS_{Re\,gression} = \frac{(\Sigma xy)^2}{\Sigma x^2} = EXPLAINED\ SS = \Sigma \hat{y}^2$$

$$\frac{(331)^2}{82.5} = \frac{109,561}{82.5} = 1,328.012.$$

$$SS_{UNEXPLAINED} = \frac{SS_{DEVIATION}}{n-2} = \frac{\Sigma d_{y.x}^2}{n-2} = S_{y.x}^2$$

$$\Sigma d_{y.x}^2 = \Sigma y^2 - \Sigma \hat{y}^2 = 1,363 - 1,328 = 35.588$$

$$SS_{DEV} = SS_{TOTAL} - SS_{REG}$$

Re-arranging the terms, we have:

$$\Sigma y^2 = \Sigma \hat{y}^2 + \Sigma d_{y.x}^2.$$

$$= \frac{(\Sigma xy)^2}{\Sigma x^2} + \Sigma d_{y.x}^2.$$

$$SS_{RESIDUALS}\ (S_{y.x}^2) = \frac{\Sigma d_{y.x}^2}{n-2} = \frac{35,588}{8} = 4.4485$$

$$S_{y.x} = 2.109146747$$

f) Compute Coefficient of Determination (R^2).

The coefficient of determination (R^2) can be estimated in two ways:

1. Directly from the raw data by calculating the correlation coefficient and squaring it:

$$R = \frac{\Sigma X_i Y_i - \frac{(\Sigma X_i)(\Sigma Y_i)}{n}}{\sqrt{\left[\Sigma X_i^2 - \frac{(\Sigma X_i)^2}{n}\right]\left[\Sigma Y_i^2 - \frac{(\Sigma Y_i)^2}{n}\right]}} \Rightarrow R^2$$

2. From SSREG and SSTOTAL:

$$R^2 = \frac{SS_{REG}}{SS_{TOTAL}} = \frac{\Sigma \hat{y}^2}{\Sigma y^2} = \frac{EXPLAINED\ SS}{TOTAL\ SS} = \text{Proportion of Total variation in Y that is explained by}$$

the Regression equation. That is, the degree of linearity between X and Y.

g) Coefficient of Non-Determination (or Alienation), ($k^2 = 1 - R^2$)

The coefficient of non-determination is: $k^2 = 1 - R^2$

$k^2 = 1 - 0.9739 = 0.026 = $ Proportion of Total variation in Y, that is not explained by the Regression equation.

The Coefficient of alienation $(\sqrt{k^2}) = \sqrt{(1 - k^2)}$, is the square root of the coefficient of non-determination, which is an abstract measure of the lack of correlation between Y and X.

$$k = \sqrt{(1 - 0.9739)} = \sqrt{0.026} = 0.16125$$

h) Making Inferences in Regression.

We must estimate the standard errors of b1 and b0 before we can make any statistical inferences.

(a) Inferences Concerning β_1

0 i.e. The Regression Coefficient is zero.

$H_1: \beta_1 \neq 0$ i.e. The Regression Coefficient is not zero.

$H_0: \beta_1 = \alpha = 0.05$

Decision Rule: Reject H_0, if $\left| t^*_{b_1} \right| > t_{0.025}$, 8.

Variance of the Regression $= s^2_{y \cdot x} = 4.4485$

Variance of $b_1 = s^2_{b_1} = \dfrac{s^2_{y \cdot x}}{\Sigma x^2} = \dfrac{4.4485}{82.5}$

Standard Error of $b_1 = s_{b_1} = \sqrt{s^2_{y \cdot x} / \Sigma x^2}$

$s_{b_1} = \sqrt{\dfrac{4.4485}{82.5}}$.

$\sqrt{0.0539212} = 0.232209388 \approx 0.2322$

$t^*_{b_1} = \dfrac{b_1 - \beta_1}{s_{b_1}} = \dfrac{4.012 - 0}{02322} = 17.2782..$

$t_{0.025}, 8 = 2.306$.

Conclusion: Since $t^*_{b_1} = 17.28 \gg t_{0.025}, 8 = 2.306$ and $t_{0.005}, 8 = 3.355$, reject H_0 and conclude that $\beta_1 \neq 0$, but much greater than zero, both at 5% and 1% level of significance.

A $100(1 - \alpha)\%$ CI for β_1:

Since $\dfrac{b_1 - \beta_1}{s_{b_1}}$ has a t-distribution,

$$P\left[\left(b_1 - s_{b1} \times t_{\frac{\alpha}{2}}, n - 2\right) \leq \beta_1 \leq \left(b_1 + s_{b1} \times t_{\frac{\alpha}{2}}, n - 2\right)\right] = 1 - \alpha..$$

$= b_1 \pm s_{b1} t_{\alpha/2}, n - 2 = 4.012 \pm (0.2322)(2.306)$

$= 4.012 \pm (0.5354532)$

$= 3.47655 \leq \beta_1 \leq 4.54745.$

(b) Inferences Concerning β_0:

$H_0: \beta_0 = 0$ i.e. The Intercept is zero.

$H_1: \beta_0 \neq 0$ i.e. The Intercept is not zero.

306

$\alpha = 0.05$

Decision Rule: Reject H_0, if $|t^*_{b_1}| > t_{0.025,\, 8}$.

$b_0 = \bar{Y} - b_1\bar{X} = 22.8 - (4.012)(12.5) = 22.8 - 50.15 = -27.35$

$Var\,(b_0) = Var(\bar{Y} - b_1\bar{X}) = Var(\bar{Y}) + \bar{X}^2 Var(b_1)$

$$s^2_{b0} = \frac{s^2_{y.x}}{n} + \bar{X}^2 \frac{s^2_{y.x}}{\Sigma x^2}$$

$$= s^2_{y.x}\left(\frac{1}{n} + \frac{\bar{X}^2}{\Sigma x^2}\right)$$

$$\text{S. e. } (b_0) = \sqrt{s^2_{y.x}\left(\frac{1}{n} + \frac{\bar{X}^2}{\Sigma x^2}\right)} = s_{y.x}\sqrt{\left(\frac{1}{n} + \frac{\bar{X}^2}{\Sigma x^2}\right)}$$

$$= \sqrt{4.4485\left(\frac{1}{10} + \frac{(12.5)^2}{82.5}\right)} = \sqrt{4.4485\left(0.10 + \frac{156.25}{82.5}\right)}$$

$$\sqrt{4.4485\,(0.10 + 1.893939394)} = \sqrt{4.4485\,(1.993939394)}$$

$$\sqrt{8.870039394} = 2.978261136$$

$s_{b0} = \approx 2.97826$

$$t^*_{b0} = \frac{b_0 - \beta_0}{s_{b0}} = -\frac{27.35}{2.97826} = -9.1832\,.$$

$t_{0.025,\, 8} = 2.306\,.$

$t_{0.005,\, 8} = 3.355.$

Conclusion: Since, $|\,t^*_{b0}\,| = 9.1832 \gg t_{0.025},8 = 2.306\ and\ t_{0.005},8 = 3.355$, we reject H_0 and conclude that $\beta_0 \neq 0.$, and the regression line does not pass through the origin.

$A\,100\,(1 - \alpha)\%\ C.I.\,for\ \beta_{0:}$

$b_0 \pm s_{b0}\,t_{\alpha/2},n - 2 = -27.35 \pm (2.97826)(2.306)\,.$

$= -27.35 \pm 6.8678675$

$= -27.35 \pm 6.8679$

$\approx -34.2179 \leq \beta_0 \leq -20.482$

i) **Standard Error of An Estimated Mean Value, $\hat{Y}_k$.**

When the regression equation is used to estimate a mean value, $\hat{Y}_k$, when a new $X_k = 15.5$, a value within the range of the data is substituted in :

$= \hat{Y}_k = b_0 + b_1 X_k.$

$$= \bar{Y} - b_1\bar{X} + b_1X_k = \bar{Y} + b_1(X_k - \bar{X}),$$ the standard error of $\hat{Y}_k$ is:

$$Var(\hat{Y}_k) = Var(\bar{Y}) + (X_k - \bar{X})^2\ Var(b_1)$$

$$= \frac{S_{y.x}^2}{n} + \left(\frac{(X_k-\bar{X})^2}{\Sigma X^2}\right) s_{y.x}^2$$

$$s_{\hat{Y}_k}^2 = s_{y.x}^2\left(\frac{1}{n} + \frac{(X_k-\bar{X})^2}{\Sigma x^2}\right)$$

$$s.e.\left(\hat{Y}_k\right) = s_{y.x}\sqrt{\left(\frac{1}{n} + \frac{(X_k-\bar{X})^2}{\Sigma x^2}\right)} = 2.109146747\sqrt{\left[0.10 + \frac{(15.5-12.5)^2}{82.5}\right]}$$

$$= 2.10914675\sqrt{\left(0.10 + \frac{9.0}{82.5}\right)} = 2.10914675\sqrt{0.20909091}$$

$$= (2.10914675)(0.457264594) = 0.964438132$$

$$\approx 0.964438$$

$\therefore$ Estimated Mean Value of $\hat{Y}_k$ for the new $X_k = 15.5$ is:

$$\hat{Y}_k = -27.35 + 4.012\ (15.5)$$

$$= -27.35 + 62.186$$

$$= 34.836\ \text{cu.meters.}$$

A100 $(1 - \alpha)$% C.I. for $\hat{Y}_k$, the Mean Value is:

$$\hat{Y}_k + s.e.\left(\hat{Y}_k\right).t_{\alpha/2}, n - 2$$

$$34.836 \pm (0.964438)(2.306) = 34.836 \pm 0.217774$$

$$34.61823 \le \hat{Y}_k \le 35.0538$$

j) **Prediction of A New Individual (or Single) Value, $\hat{Y}_{k(NEW)}$.**

Using the regression equation to predict a single new $\hat{Y}_k$ value in future, the single $\hat{Y}_{k(NEW)}$ value will still remain as the mean value, 34.836, above. But the standard error of the new single predicted value, $\hat{Y}_{k(NEW)}$, is now different for the same $X_{new} = 15.5$:

The variance, $Var\left(\hat{Y}_{k(NEW)}\right)$, has two components (Neter et al., 1990):

   a. The variance of the sampling distribution of $\hat{Y}_k$ and
   b. The variance of the distribution of Y at $X = X_{k(new)}$

$$Var\left(Y_{k(new)}\right) = Var\left(\hat{Y}_k - Y\right) = Var\left(\hat{Y}_k\right) + Var(Y)$$

$$= \frac{S_{y-x}^2}{n} + \left(\frac{(X_{k(new)}-\bar{X})^2}{\Sigma x^2}\right)s_{y-x}^2 + s_{y.x}^2$$

$$= s_{y.x}^2 + \frac{s_{y.x}^2}{n} + \left(\frac{(X_{k(new)} - \bar{X})^2}{\sum x^2}\right) s_{y-x}^2.$$

$$s_{(Y_{k(new)})}^2 = s_{y.x}^2 \left[1 + \frac{1}{n} + \frac{(X_{k(new)} - \bar{X})^2}{\sum x^2}\right]$$

$$s_{(Y_{k(new)})} = \sqrt{s_{y.x}^2 \left[1 + \frac{1}{n} + \frac{(X_{k(new)} - \bar{X})^2}{\sum x^2}\right]}$$

$$= s_{y.x} \sqrt{1 + \frac{1}{n} + \frac{(X_{k(new)} - \bar{X})^2}{\sum x^2}}$$

Substituting the values above, we have:

$$S_{(Y_{k(new)})} = 2.10914675 \sqrt{1 + 0.10 + 0.10909091}$$

$$= 2.10914675 \sqrt{1.20909091} = (2.10914675)(1.0995867)$$

$$= 2.319189714.$$

A $100(1 - \alpha)\%$ C.I. for $Y_{k(new)}$ is:

$$Y_{k(new)} \pm s(Y_{k(new)}) \times t_{\frac{\alpha}{2}}, n - 2).$$

$$= Y_{k(new)} \pm (2.319189714)(2.306) = Y_{k(new)} \pm 5.34805148$$

$$= 34.836 \pm 5.34805148$$

$$= 29.487949 \le Y_{k(new)} \le 63.437655.$$

It is clearly seen in this result that the prediction limits $(29.488 \le Y_{k(new)} \le 63.438)$, are much wider than the confidence limits $34.618 \le \hat{Y}_k \le 35.054$ in (ix) above.

k) **Prediction of the Mean of 'm' New Observations for a Given Xk.**

Suppose in our data (Table 13.3.4.1), X = 15.5, Y has three values: 30, 31 and 32 cu. meter, a mean of $\bar{Y}_{k(new)} = 31$ cu.ft, which we want to predict for future use, the prediction limits are:

$$\hat{Y}_k \pm s(\bar{Y}_{k(new)}) \times t_{\frac{\alpha}{2}}, n - 2.$$

$$s^2(\bar{Y}_{k(new)}) = s^2(\hat{Y}_k) + \frac{MSE}{m} \quad \left(\text{with } MSE = s_{y.x}^2\right)$$

$$s^2(\bar{Y}_{k(new)}) = s_{y.x}^2 \left[\frac{1}{m} + \frac{1}{n} + \frac{(X_k - \bar{X})^2}{\sum x^2}\right]$$

309

Substituting the values from above:

$$s^2\left(\bar{Y}_{k(new)}\right) = 4.4485\left[\frac{1}{3} + \frac{1}{10} + \frac{(15.5-12.5)^2}{82.5}\right]$$

$$= 4.4485\left[0.3333333 + 0.10 + 0.109090909\right]$$

$$= 4.4485\left(0.542424242\right) = 2.412974242$$

$$s\left(\bar{Y}_{k(new)}\right) = \sqrt{2.412974242} = 1.553375113$$

A 100 $(1-\alpha)$% C.I. for $\bar{Y}_{k(new)}$ is:

$$\bar{Y}_{k(new)} \pm (1.553375113)(2.306) = \bar{Y}_{k(new)} \pm 3.582083$$

$$= 31.0 \pm 3.582083$$

$$27.417917 \le \bar{Y}_{k(new)} \le 34.5821$$

$$\approx 27.418 \le \bar{Y}_{k(new)} \le 34.582$$

Notice that these prediction limits are narrower than those of predicting a single value $(29.488 \le \bar{Y}_{k(new)} \le 63.438)$ because the mean value of the three new values is less variable.

l) **Inferences Concerning R2:**

R2, the coefficient of determination, measures the amount of variation in Y that is accounted for by the variation of X. To test the statistical significance of R2, we test the following hypothesis through the use of analysis of variance approach to Regression Analysis:

$H_0: \rho = 0$ i.e. The X-variable did not explain any variation in Y by its regression.

$H_1: \rho \neq 0$ i.e. The X-variable explained most of the variation in Y by its regression.

$\alpha = 0.05$.

$$F^* = \frac{(SS_{REG})/k}{(SS_{REG})/(n-k-1)} = S_{\hat{Y}}^2 / S_{y \cdot x}^2 \; - \; -$$ compare this with $F_{\alpha}, 1, n-2$ in the F-Table.

Decision Rule: Reject H_0, if $F^* > F - Table$

Table 13.3.4.1a: Anova Table in Regression Analysis.

Source of variation	DF	SS	MS	F*	$F_{0.05}$	$F_{0.01}$
Explained (or regression) $=(\hat{Y}_i - Y)$	k = 1	1,328.012	1,328.012	$\frac{1,328.012}{4.4485} = 306.398^{***}$	5.32	11.26
Unexplained (or error) $(Y_i - \hat{Y}_i)$	n − 2 = 8	35.588	4.4485			
Total$((Y_i - \bar{Y})$	n - 1 = 9	1,363.6				

Conclusion:

Since F* = 306.4 >> F0.05, 1,8 = 5.32 and F0.01, 1,8 = 11.26, we reject H0 and conclude that $\varrho \neq 0$ but is highly significant.

13.3.5. COMPARING THE SLOPES OF TWO INDEPENDENT REGRESSION EQUATIONS.

A final year part-time student, University of Benin experimented on two varieties of Rice in order to compare the slopes of the two independent regression equations and obtained the following data to test the hypotheses that the parameters of the two regressions are equal:

Table 13.3.5: Grain Yields and Tiller Numbers of Two Varieties of Rice.

	Grain Yield(Y_{i1}) (X_{i1}) Rice Variety I	Tiller Number	Grain Yield(Y_{i2}) Rice Variety II	Tiller Number(X_{i2})
1	4862	160	5380	293
2	5244	175	5510	325
3	5128	192	6000	332
4	5052	195	5840	342
5	5298	238	6416	342
6	5410	240	6666	378
7	5234	252	7016	380
8	5608	282	6994	410
	$\sum Y_{i1} = 41{,}836 \qquad \sum X_{i1} = 1{,}734$ $\sum Y_{i1}^2 = 219{,}138{,}992$ $\sum X_{i1}^2 = 388{,}386$ $\bar{Y} = 5{,}229.5 \qquad \bar{X}_1 = 216.75$ $\sum X_{i1}Y_{i1} = 9{,}125{,}084. \bar{Y}_i = 5{,}229.5$		$\sum Y_{i2} = 49{,}822: \bar{Y}_2 = 6{,}2275: \sum X_{i2} = 2{,}802$ $\sum Y_{i2}^2 = 313{,}151{,}004: \sum X_{i2}^2 = 991{,}010$ $\sum X_{i2}Y_{i2} = 17{,}604{,}010: \bar{X}_2 = 350.25$	
	$\sum y_1^2 = 21938992 - \frac{(41836)^2}{8} \bar{X}_1 = 216.75$ $= 219{,}138{,}992 - 218{,}781{,}362 = \mathbf{357{,}630}$		$\sum y_2^2 = 313{,}151{,}004 - \frac{(49822)^2}{8}$ $= 313{,}151{,}004 - 310{,}278{,}960.5$ $= \mathbf{2{,}872{,}043.5}$	
	$\sum x_1^2 = 388{,}386 - \frac{(1734)^2}{8}$ $= 388386 - 375844.5 = 12541.5$		$\sum x_2^2 = 991010 - \frac{(2802)^2}{8}$ $= 991{,}010 - 981{,}400.5 = \mathbf{9{,}609.5}$	

$\sum x_1 y_1 = 9,125,084 - \frac{(41836)(1734)}{8}$ $= 9,125,084 - 9,067,953 = \mathbf{57,131}$	$\sum x_2 y_2 = 17,604,010 - \frac{(4982.2)(2802)}{8}$ $= 17,604,010 - 17,450,155.5 = \mathbf{153,854.5}$
$b_1 = \frac{\sum x_1 y_1}{\sum x_1^2} = \frac{57131}{12541.5} = 4.555356217$	$b_2 = \frac{\sum x_2 y_2}{\sum x_2^2} = \frac{153854.5}{9609.5} = 16.010665.3$
$b_{01} = \bar{Y}_1 - b_1 \bar{X}_1 = 5229.5 - (4.555356217)(216.75)$ $= 5229.5 - 987.3734601 = 4242.12654$	$b_{02} = \bar{Y}_2 - b_1(\bar{X}_2) = 6227.75 - (16.01066653)(350.25)$ $= 6227.75 - 5607.735952 = \mathbf{620.0140479}$
$\hat{Y}_i = b_{01} + b_1 X_{i1} = 4242.127 + 4.5554 X_{ii}$ $\hat{Y}_{i1} = 4242.127 + 4.5554\, X_{i1}$	$\hat{Y}_{i2} = b_{02} + b_{02} X_{i2} = 620.014 + 16.0107 X_{i2}$ $\hat{Y}_{i2} = 620.014 + 16.011\, X_{i2}$

The question is: Are the two regression lines parallel? Are the two intercepts equal?

We need to test these hypotheses to answer the two questions:

1. $H_0: \beta_1 = \beta_2$ i.e. The two regression slopes are equal or parallel.

   $H_1: \beta_1 \neq \beta_2$ i.e. The two regression slopes are not equal or parallel.

2. $H_0: \beta_{01} = \beta_{02}$ i.e. The two intercepts are equal .

   $H_0: \beta_{01} \neq \beta_{02}$ i.e. The two intercepts are not equal.

a) Compute the Variances $\left(S_{y_1 x_1}^2 \text{ and } S_{y_2 x_2}^2\right)$ of the two Regression Lines.

$$1. S_{y_1 x_1}^2 = \frac{\left[\sum y_1^2 - \frac{(\sum x_1 y_1)^2}{\sum x_1^2}\right]}{n-2} = \frac{\left(357630 - \frac{(57131)}{12541.5}\right)}{6} = 357,630 - 260252.0561$$

$$= \frac{97,377.94395}{6} = 16229.65732$$

$$2. S_{y_2 x_2}^2 = \left[\sum y_2^2 - \frac{(\sum x_2 y_2)^2}{\sum x_2^2}\right]/(n-2) = \left(2,872,043.5 - \frac{(153854.5)^2}{9609.5}\right)/6$$

$$= \frac{2,872,043.5 - 2463,313.093}{6} = \frac{408,730.4067}{6} = 68,121.173445.$$

b) Compute The Common Variance By Pooling The Two Variances As:

$$S_{Pooled}^2 = \frac{(n_1 - 2)S_{y_1 x_1}^2 + (n_2 - 2)S_{y_2 x_2}^2}{n_1 + n_2 - 4}$$

But since $n_1 = n_2 = 8, S_P^2 = \frac{(S_{y_1 x_1}^2 + S_{y_2 x_2}^2)}{2} = 16,229.65732 + 68,121.173445$

$$S_P^2 = \frac{84,350.83076}{2} = 42,175.41538$$

1. $H_0: \beta_1 = \beta_2$

$H_1 = \beta_1 \neq \beta_2$

$\alpha = 0.05$ and 0.01. $t_{0.025\,(12)} = 2.179$: $t_{0.005\,(12)} = 3.055$.

Test statistic: $t^* = \dfrac{b_1 - b_2}{\sqrt{S_P^2\left[\dfrac{1}{\Sigma x_1^2} + \dfrac{1}{\Sigma x_2^2}\right]}}$

Decision Rule: Reject H_0, if $|t^*| > t_{0.025}(12) = 2.179$ and $t_{0.005}(12) = 3.055$

$|t^*| = \dfrac{4.555356217 - 16.01066653}{\sqrt{42,175.41538\left(\dfrac{1}{12,541.5} + \dfrac{1}{9,609.5}\right)}}$

$= \dfrac{11.45531031}{\sqrt{(42,175.41538)(1.837989658)}} = \dfrac{11.45531031}{\sqrt{7.751797731}}$

$= \dfrac{11.45531031}{2.784205045} = 4.114391766$

Conclusion: Since $|t^*| = 4.114 > t_{0.025}(12) = 2.179$ and $t_{0.005}(12) = 3.055$ respectively, reject H_0 and conclude that $\beta_1 \neq \beta_2$. Therefore, the two regression lines are not parallel.

2. H_0: $\beta_{01} = \beta_{02}$

H_1: $\beta_{01} \neq \beta_{02}$

$\alpha = 0.05$ and 0.01.

$b_{01} = \bar{Y}_1 - b_1\bar{X}_1$: $b_{02} = \bar{Y}_2 - b_2\bar{X}_2$

$Var(b_{01} - b_{02}) = Var(\bar{Y}_1 - b_1\bar{X}_1) + Var(\bar{Y}_2 - b_2\bar{X}_2)$

$= S_p^2\left[\dfrac{1}{n_1} + \dfrac{\bar{X}_1^2}{\Sigma x_1^2} + \dfrac{1}{n_2} + \dfrac{\bar{X}_2^2}{\Sigma x_2^2}\right]$

$S_P^2 = 42,175.41538\left[\dfrac{1}{8} + \dfrac{(216.75)^2}{12,541.5} + \dfrac{1}{8} + \dfrac{(350.25)^2}{9,609.5}\right]$

$= 42,175.41538\,[0.125 + 3.746008253 + 0.125 + 12.7660193]$

$= 42,175.41538\,(16.76202756) = 706,945.476$

$S_P = \sqrt{706,945.476} = 840.8004965$

$|t^*| = \dfrac{b_{01} - b_{02}}{S_P} = \dfrac{4242.12654 - 620.0140479}{840.8004965} = \dfrac{3.622112492}{840.8004965}$

$= 4.307933341 \approx 4.308$

313

Conclusion:

Since $\mid t^* \mid = 4.308 > t_{0.025}(12) = 2.179 \; and \; t_{0.005}(12) = 3.055$, reject H_0 and conclude that $\beta_{01} \neq \beta_{02}$. i.e the two intercepts are not equal but significantly different.

13.4 .MULTIPLE LINEAR REGRESSION AND CORRELATION

Multiple regression analysis, according to Neter et al. (1983, 1990), is one of the most widely used of all statistical tools. Regression analysis involving more than one independent variable is called **"Multiple Regression Analysis"**. When all independent variables are assumed to affect the dependent variable in a linear fashion and independently of one another, the procedure is called **"Multiple Linear Regression Analysis"**.

The relationship of the dependent variable Y to k independent variables, $X_1, X_2, ..., X_k$, can be expressed as:

$$Y = \beta_0 + \beta_1 X_1 + \beta_2 X_2 + ... + \beta_k X_k.$$

The data required for the application of the multiple linear regression analysis involving k independent variables are (n)(k+1) observations. The (k+1) variables, $Y, X_1, X_2, ..., X_k$, must be measured simultaneously for each of the n units of observation. In addition, there must be enough observations to make n greater than (k+1).

Table 13.4: RAW DATA

Observation No.	Observation Values					
	Y	X_1	X_2	X_3	. . .	X_k
1	Y_1	X_{11}	X_{21}	X_{31}	. . .	X_{k1}
2	Y_2	X_{12}	X_{22}	X_{32}	. . .	X_{k2}
3	Y_3	X_{13}	X_{23}	X_{33}	. . .	X_{k3}
.	.	.	.	.	. . .	.
.	.	.	.	.	. . .	.
.	.	.	.	.	. . .	.
n	Y_n	X_{1n}	X_{2n}	X_{3n}	. . .	X_{kn}

Procedure:

1. Compute the mean and the corrected sums of squares for each of the (k+1) variables, Y, X1,

X2, . . ., Xk, and the correct sums of cross products for all possible pair-combinations of the

(k+1) variables as:

Table 13.4.1: Corrected Sum of Squares and Cross-Products:

Variable	Mean	X_1	X_2	.	.	.	X_K	Y
X_1	$\bar{X}_1$	Σx_1^2	$\Sigma x_1 x_2$	.	.	.	$\Sigma x_1 x_k$	$\Sigma x_1 y$
X_2	$\bar{X}_2$	$\Sigma x_2 x_1$	Σx_2^2	.	.	.	$\Sigma x_2 x_k$	$\Sigma x_2 y$
.	.	.	.	.	.	.	.	.
.	.	.	.	.	.	.	.	.
.	.	.	.	.	.	.	.	.
X_k	$\bar{X}_k$	.	.	.	.	.	Σx_k^2	$\Sigma x_k y$
Y	$\bar{Y}$		.	.	.	.	.	Σy^2

2. Solve for b1, b2, …, bk from the following k simultaneous equations which are generally referred to as **the Normal Equations:**

$$b_1 \Sigma_1^2 + b_2 \Sigma x_1 x_2 + \ldots + b_k \Sigma x_1 x_k = \Sigma x_1 y.$$

$$b_1 \Sigma x_1 x_2 + b_2 \Sigma_2^2 + \ldots + b_k \Sigma x_2 x_k = \Sigma x_2 y.$$

$$b_1 \Sigma x_1 x_k + b_2 \Sigma x_2 x_k + \ldots + b_k \Sigma_k^2 + = \Sigma x_k y.$$

where b1, b2, . . ., bk are estimates of $\beta_1, \beta_2, \ldots \beta_k$ of the multiple linear regression equation, and the values of the sum of squares and sum of cross-products of the (k + 1) variables are those computed in (i). the k-simultaneous equations for the k unknowns can be solved manually or with the aid of computers.

13.4.1.A Worked Example of a Multiple Linear Regression.

We illustrate the procedure for a case where k = 2, using the published data of Gomez and Gomez (1984) on grain yield (Y), plant height (X1) and tiller number (X2) of more than eight rice varieties in Table 13.4.1 below. The multiple linear regression equation with k = 2 can be expressed as:

$$Y = \beta_0 + \beta_2 X_1 + \beta_2 X_2$$

Table 13.4.1a: Computation of a Multiple Linear Regression Equation Relating Plant Height (X1) and Tiller Number (X2) to the Grain Yield of over Eight Rice Varieties.

Variety No.	Grain Yield (Y) kg/ha	Plant Height (X_1) cm	Tiller No/ Hill (X_2)
1	5,755	110.5	14.5
2	5,939	105.4	16.0
3	6,010	118.1	14.6
4	6,545	104.5	18.2
5	6,730	93.6	15.4
6	6,750	84.1	17.6
7	6,899	77.8	17.9
8	7,862	75.6	19.4
Sum	52,488	769.6	133.6
Mean	$\bar{Y}$ 6,561.25	$\bar{X}_1 = 96.2$	$\bar{X}_1 = 16.7$
Sum of squares	$\Sigma y^2 = 3,211,504$	$\Sigma x_1^2 = 1,753.72$	$\Sigma x_2^2 = 23.22$
Sum of Gross-Products	$\Sigma x_1 y = -65,194$	$\Sigma x_2 y = 7,210$	$\Sigma x_2 x_2 = -156.65$

1. **The normal equations are:**

$$b_1 \Sigma x_1^2 + b_2 \Sigma x_1 x_2 = \Sigma x_1 y.$$

$$b_1 \Sigma x_1 x_2 + b_2 \Sigma x_2^2 = \Sigma x_2 y.$$

2. **The solutions for b1 and b2 are:**

$$b_1 = \frac{(\Sigma x_2^2)(\Sigma x_1 y) - (\Sigma x_1 x_2)(\Sigma x_2 y)}{(\Sigma x_1^2)(\Sigma x_2^2) - (\Sigma x_1 x_2)^2}$$

$$b_2 = \frac{(\Sigma x_1^2)(\Sigma x_2 y) - (\Sigma x_1 x_2)(\Sigma x_1 y)}{(\Sigma x_1^2)(\Sigma x_2^2) - (\Sigma x_1 x_2)^2}$$

Substituting the data above, we have:

$$b_1 = \frac{(23.22)(-65.194) - (-156.65)(7,210)}{(1,753.72)(23.22) - (156.65)^2} = -23.75$$

$$b_2 = \frac{(1,753.72)(7,210) - (-156.65)(-65,194)}{(1,753.72)(23.22) - (156.65)^2} = 150.27$$

3. **Compute the estimate of the intercept:**

$\beta_0 = \bar{Y} - b_1 \bar{X}_1 - b_2 \bar{X}_2 \ldots - b_k \bar{X}_k, where \ \bar{Y}, \bar{X}_1, \bar{X}_2, \ldots, \bar{X}_k$ are the means of the (k+1) variables computed in (1) above.

For our example, the estimate of the intercept, β_0 is computed as:

$$b_0 = \bar{Y} - b_1 \bar{X}_1 - b_2 \bar{X}_2$$

$$= 6,561.25 - (-23.75)(96.2) - (150.27)(16.7)$$

$$= 6,561.25 - (-2,284.75) - (2,509.509)$$

$$= 6,561.25 + 2,284.75 - 2,509.509.$$

$$= 6,336.491.$$

4. **The Estimated Multiple Linear Regression is:**

$$\hat{Y} = b_0 \pm b_1 X_1 + b_2 X_2 + \ldots + b_k X_k$$

The estimated multiple linear regression equation relating plant height (X1) and Tiller number (X2) to Rice grain yield (Y) is:

$$\hat{Y} = 6.336.491 \pm 23.75 X_1 + 150.27 X_2$$

5. **Compute:**

(i) **Sum of Squares Due To Regression (SSR):**

$$SSR = \Sigma_{i=1}^{k} (b_1)(\Sigma x_1 y)$$

$$= b_1 \Sigma x_1 y + b_2 \Sigma x_2 y$$

$$= (-23.75)(-65,194) + (150.27)(7,210)$$

$$= 1,548,357.5 + 1,083,446.7.$$

$$= 2,631,804.2.$$

(ii)   Residual Sum of Squares :

$$\Sigma d_{y.x}^2 = SSE = \Sigma y^2 - SSR.$$

$$\Sigma d_{y.x}^2 = SSE = \Sigma y^2 - (b_1 \Sigma x_1 y + b_2 \Sigma x_2 y)$$

$$= 3,211,504 - 2,631,804.5.$$

$$= 579,699.5.$$

$$MSE = \frac{SSE}{n-k-1} = \frac{\Sigma y^2 - b_1 \Sigma x_1 y + b_2 \Sigma x_2 y}{n-k-1} = \frac{579,699.5}{5} = 115,939.9$$

(where k = number of independent variables in the multiple regression equation).

MSE = 115,939.9

$$S_{y.x} = \sqrt{s_{y.x}^2} = \sqrt{MSE} = \sqrt{579,699.5} = 115,939.9.$$

(iii)   **Coefficient of Multiple Determination (R2):**

$$\mathbf{R}^2 = \frac{SSR}{\Sigma y^2} = \frac{SSR}{SS_{TOTAL}}.$$

The coefficient of multiple determination (R2) measures the contribution of the linear function of k independent variables to the variation in Y. It is usually expressed in percentage. Its square root (R) is referred to as the Multiple Correlation Coefficient.

$$\mathbf{R}^2 = \frac{2,631,804.2}{3,211,504} = 0.819493$$

Thus, 81.95% of the total variation in the yields of eight rice varieties can be accounted for by a linear function of plant height and tiller number.

6. **Test the Significance of R2:**

$$H_0: \rho = 0.$$

$$H_1: \rho \neq 0$$

$$\alpha = 0.05; 0.01$$

Decision Rule: Reject H_0, if F* is greater than $F_{0.05}$, 2, 5 = 5.79, $F_{0.01}$, 2,5 = 13.2.

Compute F-value as:

$$F^* = \frac{(SSR)/k}{(SSE)/(n-k-1)} = \frac{(2,631,804.2)/2}{(579,699.5)/(8-2-1)} = \frac{MSR}{MSE}$$

$$\frac{MSR}{MSE} = \frac{1,315,902.25}{115,939.9} = 11.34987 \approx 11.35$$

The coefficient of multiple determination, R2, is said to be significantly different from zero, if the computed F-value (11.35) is greater than the tabular F-value (5.79) at the 5% level of significance but not at the 1% level (13.27).

Thus, the estimated multiple linear regression; $\hat{Y} = 6336.49 - 23.75X1 + 150.27X2$ is only significant at 5% level of significance.

Note:

The size, or magnitude of R2-value provides information on the size of the portion explained by the linear function of the independent variables. The larger the R2 value is, the more important the regression equation is in characterizing Y. If the value of R2 is low (e.g. 0.26), even if the F-value is significant, the estimated regression equation may not be meaningful because 74% of the variation in Y cannot be accounted for by the regression.

13.4.2. The Meaning of the Regression Coefficients.

1. The parameter β_0 is the Y-intercept of the regression plane. If the scope of the model includes $X_1 = 0, X_2 = 0$, β_0 gives the mean response of Y at $X_1 = 0, X_2 = 0$. Otherwise, β_0 does not have any particular meaning as a separate term in the regression model.
2. The parameter β_1 indicates the change in the mean response per unit increase in X_1 when X_2 is held constant. Likewise, β_2 indicates the change in the mean response per unit increase in X_2, when X_1 is held constant.
3. The parameters β_1 and β_2 are frequently called **Partial Regression Coefficients** because they reflect the partial effect of one independent variable when the other independent variable is included in the model and is held constant.

Frequently, the regression function in Multiple Regression is called a **Regression Surface, or a Response Surface.**

13.4.3. Testing the Significance of the Individual Regression Coefficients (β0, β1 and β2):

Multiple regression analysis requires that we test each estimated coefficient $(\beta_0, \beta_1, \beta_2)$ to determine if the particular coefficient is significantly different from zero before we can assume that there is a relationship between the dependent variable and that particular independent variable.

Before we can test the statistical significance of each bi coefficient, we must calculate the standard errors of b0, b1 and b2 from the estimated standard error of the multiple regression $S_{y\cdot x}$.

The variances of the unbiased estimated regression coefficients are obtained (Koutsoyiannis (1973) and Cangelosi et al (1976)) from the following formulas:

$$Var(b_0) = S_{y\cdot x}^2 \left[\frac{1}{n} + \frac{\bar{X}_1^2 \Sigma x_2^2 + \bar{X}_2^2 \Sigma x_1^2 - 2\bar{X}_1 \bar{X}_2 \Sigma x_1 x_2}{\Sigma x_1^2 \Sigma x_2^2 - (\Sigma x_1 x_2)^2}\right]$$

$$S_{b0} = S_{y\cdot x}^2 \sqrt{\left[\frac{1}{n} + \frac{\bar{X}_1^2 \Sigma x_2^2 + \bar{X}_2^2 \Sigma x_1^2 - 2\bar{X}_1 \bar{X}_2 \Sigma x_1 x_2}{\Sigma x_1^2 \Sigma x_2^2 - (\Sigma x_1 x_2)^2}\right]}$$

For our worked example:

$$\bar{X}_1 = 96.2; \ \bar{X}_2 \ 16.7; n = 8$$

$$\Sigma x_1^2 = 1{,}753.72; \ \Sigma x_2^2 = 23.22; \ \Sigma x_1 x_2 = 156.65$$

$$S_{y.x} = 115{,}939.9$$

$$t_{bi}^* = \frac{b_i}{S_{bi}}$$

$$H_0: \beta_0 = 0$$

$$H_1: \beta_1 \neq 0$$

Substituting these values in S_{b0}, we have:

$$S_{b0} = 115{,}939.9 \ S_{b0} = S_{y.x}^2 \sqrt{\left[\frac{1}{8} + \frac{(96.2)^2(23.22) + (16.7)^2(1{,}753.72) - 2(96.7)(16.7)(-156.65)}{(1{,}753.72)(23.22) - (156.65)^2}\right]}$$

$$115{,}939.9 \sqrt{\left(\frac{0.125 + 214{,}888.0968 + 489{,}094.9708 - (-503{,}328.982)}{16{,}182.1559}\right)}$$

$$115{,}939.9 \sqrt{\left(0.125 + \frac{1{,}207{,}312.05}{16{,}182.1559}\right)} = 115{,}939.9 \sqrt{(0.125 + 74.60761453)}$$

$$115{,}939.9 \sqrt{74.73261453} = (115{,}939.9)(8.644802747)$$

$$= 1{,}002{,}277.566.$$

$$t_{b0}^* = \frac{b_0}{S_{b0}} = \frac{6{,}336.491}{1{,}002{,}277.566} = 0.0063221$$

Conclusion:

Since $| t^* | = 0.006 < t0.025, 2.571$, accept H$_0$ and conclude that $\beta_0 = 0$

$$\boldsymbol{Var(b_1)} = \boldsymbol{S_{b1}^2} = \frac{S_{y.x}^2}{\Sigma x_1^2 - \frac{(\Sigma x_1 x_2)^2}{\Sigma x_2^2}}$$

$$S_{b1} = \frac{115{,}939.9}{\sqrt{\left[\Sigma x_1^2 - \frac{(\Sigma x_1 x_2)^2}{\Sigma x_2^2}\right]}}$$

Substituting the values in S_{b1} we have:

$$S_{b1} = \frac{115{,}939.9}{\sqrt{1{,}753.72 - \frac{(-156.65)^2}{23.22}}} = \frac{115{,}939.9}{\sqrt{(1{,}753.72 - 1{,}056.814061)}}$$

$$= \frac{115{,}939.9}{\sqrt{696.9059388}} = \frac{115{,}939.9}{26.39897609} = 4{,}391.833213$$

$$t_{b1}^* = \frac{b_1}{S_{b1}}.$$

$$H_0: \beta_1 = 0.$$

$$H_1: \beta_1 \neq 0.$$

$\alpha = 0.05.\ t_{0.025}, 5 = 2.571;\ t_{0.005} = 4.032$

Decision Rule: Reject H_0, if $|\ t^*\ | > t_{0.025}, 5 = 2.571\ and\ t_{0.005}, = 4.032$

$$t_{bi}^* = -\frac{23.75}{4,391.833} = -0.0054077648 \approx -0.0054$$

Conclusion:

Since $|\ t_{bi}^*\ | = 0.0054 \ll t_{0.025}, = 2.571,$ accept H_0 *that* β_1 is actually zero.

$$\boldsymbol{Var\ (b_2) = S_{b_2}^2} = \frac{S_{y.x}^2}{\Sigma x_1^2 - \frac{(\Sigma x_1 x_2)^2}{\Sigma x_1^2}}$$

Substituting the values in S_{b_2}, we have:

$$S_{b_2} = \frac{115,939.9}{\sqrt{\left(23.22 - \frac{(-156.65)^2}{1,758.72}\right)}} = \frac{115,939.9}{\sqrt{23.22 - 13.99266844}} = \frac{115,939.9}{\sqrt{9.227331558}}$$

$$= \frac{115,939.9}{3.03765231} = 38,167.60055$$

$$H_0: \beta_2 = 0.$$

$$H_1: \beta_2 \neq 0.$$

$\alpha = 0.05.\ t_{0.025}, 5 = 2.571;\ t_{0.005} = 4.032$

$$t_{b_2}^* = \frac{b_2}{S_{b_2}} = \frac{150.27}{38,167.60055} = 0.00393711$$

Any model which is not of the form of simple or multiple regressions as in previous sections, is called a non-linear model, that is, non-linear in the parameters.

Conclusion:

Since $|\ t_{b_2}^*\ | = 0.13 < t_{0.025}, 5 = 2.571,$ accept H_0 *and conclude that* $\beta_2 = 0$.

13.5. SIMPLE NON-LINEAR (CURVILINEAR) MODELS.

Gomez and Gomez (1984) pointed out that the functional relationship between two variables is non-linear, if the rate of change in Y associated with a unit change in X is not constant over a specified range of X-values. A non-linear relationship among variables is common in biological organisms, especially if the range of values is wide.

For example, the pattern of plant growth over time (Gomez and Gomez, 1984), which usually starts slowly, increases to a fast rate intermediate growth stages, and slows toward the end of the life cycle.

When the relationship among the variables under consideration is not linear, the linear regression procedures are inadequate and a researcher must turn to non-linear regression analysis procedures.

There are numerous functional forms that can describe a linear relationship between two variables, and the choice of the appropriate regression and correlation technique depends on the functional form involved.

The non-linear form can be linearized through transformation of variables, or through the creation of new variables. The linearization technique (or Taylor series) is simple and the linear regression procedures are directly applicable; the technique has wide applicability because most of the non-linear relationships found in agricultural research can be linearized through variable transformation or variable creation (Gomez and Gomez, 1984). Thus, models that can be transformed into linear form are said to be intrinsically linear, while models that are impossible to convert into linear forms are said to be intrinsically non-linear (Draper and Smith, 1966; Chatterjee and Price, 1977).

13.5.1. Some Common Non-Linear (or Curvilinear) Forms.

Many of the common non-linear forms generally involve one dependent and one independent variables, and therefore can be represented (Daniel and Wood, 1971) by the equation to be fitted:

$$Y = b_0 + b_1X_1 .$$

It is usually instructive to plot the raw data to see whether, according to Daniel and Wood (1971), the data fall on a straight line, show evidence of curvature, or indicate some other form of ill fit and then try to identify some form of equation that will fit the data better.

For the purposes of fitting equations, Daniel and Wood (1971) and Wahua (1999) have asserted that there two types:

1. Those whose constants are linear, or can be linearized by simple transformations; and
2. Those whose constants are non-linear and cannot be easily linearized.

Wahua (1999), Gomez and Gomez (1984) and Daniel and Wood (1971 identified the following simple non-linear curves and how to linearize them:

1. Sigmoid curves (Logistic curves).
2. Power curves.
3. Logarithmic curves.
4. Exponential curves.
5. Asymptotic curves.
6. Polynomial curves.

The plots in the figures which we shall see in these sections are made from equations which were linearizable by appropriate transformations of Y and X. To identify the type of curve and what logarithmic transformation to use, Wahua (1999), Mead, Curnow and Hasted (1993) and Daniel and Wood (1971) have suggested the following procedure:

13.5.2. Procedure for Identifying the Non-Linear Form and Its Transformation.

1. Plot the original Y-values against those of X on an ordinary graph paper and examine the resulting distribution of points and relate the shape to a familiar curve.
2. Write down the equation and its corresponding transformation (if known).

3. You may plot ln Y versus ln X (i.e natural log of Y and X) on ordinary graph paper. A fairly linear trend would be indicated if your transformation and equation fit the data.

Alternatively, you may plot Y against X on a complete log paper. The trend should also be linear.

13.5.2.1. Types of Power Curves.

In power curves, Y is obtained by raising X to a particular power. The particular power to which X is raised can be positive, negative, a whole number or a fraction:

Curves	Equation	Log transformation
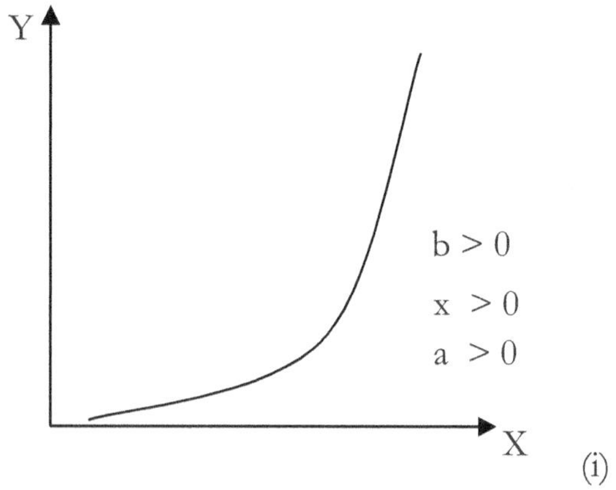 Y = aXb	lnY = lna + blnX
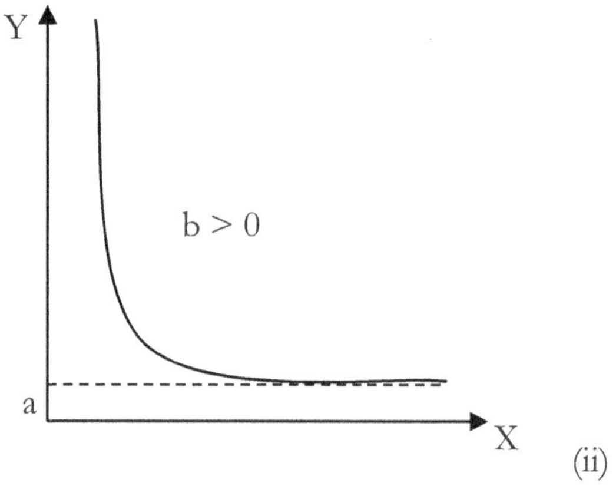 $Y = aX^{-b}$ OR $Y = aX^{\frac{1}{b}}$	$lnY = lna - blnX$ $lnY = lna + \frac{1}{b}lnX$
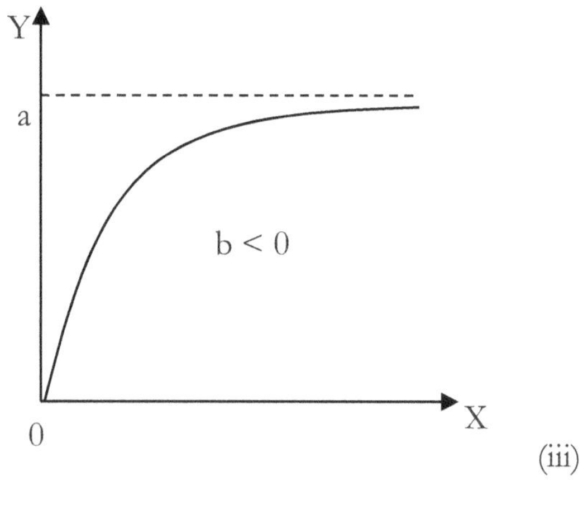 $Y = ae^{b/x}$	$lnY = lna + b / x$

Figure 13.5.2.1: Types of Power Curves.

Curves like Figure 13.5.2.1 (ii) are called "inverse polynomial models" which were first used for yield-density relationships for agricultural crops in the form:

$$\frac{1}{w} = \alpha + \beta_\rho$$

where w = the yield per plant.

ϱ = the plant density per unit area.

These curves are observed when the number of heads, or tillers per plant for cereal crops like rice, wheat and sorghum, is plotted against plant population; also percent light transmission of crop canopies, when plotted against plant population, yields similar curves.

Example 13.5.1

In a study of the relationship between the number of tillers per plant and plant populations in guinea-corn, the following data were obtained. Fit a power curve to the data.

Table 13.5.1: Tiller number per plant and plant population per area

	Tillers/plant (Y)		Plant Population/Area(X)
1	3.6		1
2	2.9		3
3	2.5		5
4	1.8		10
5	1.6		15
6	0.75		20
7	0.68		25
8	0.90		30
9	0.98		35

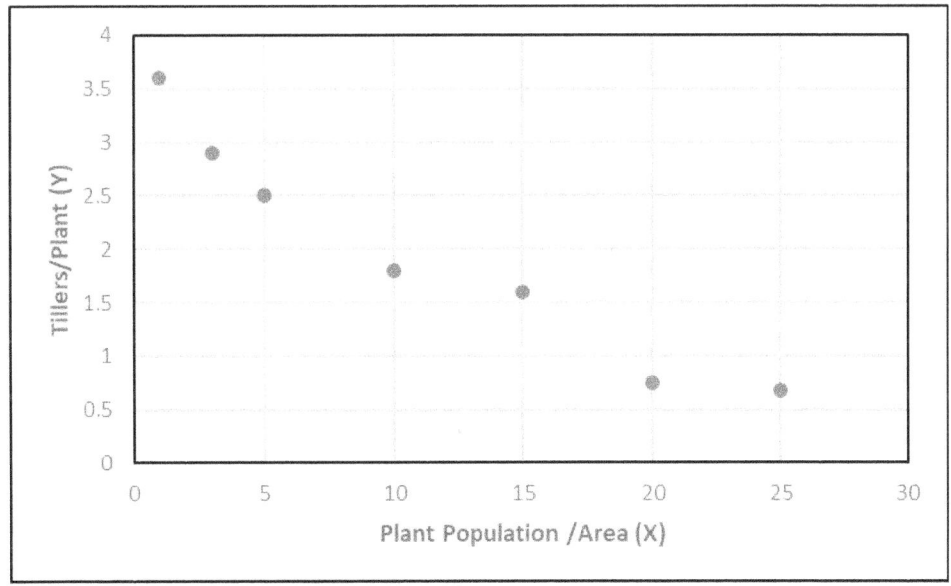

Fig. 13.5.1: Graph of Tiller No Against Plant Population per Unit Area.

The model we need to try is: Y = aX-b which is transformed as lnY = lna-blnX.

Transformed Data and Regression as:

S/N	In Y	In X
1	1.280934	0
2	1.064711	1.098612
3	1.609438	0.916291
4	0.587787	2.302585
5	0.470004	2.708050
6	-0.287682	2.995732
7	-0.105361	3.218876
8	-0.105361	3.401197
9	-0.020203	3.555348
	$\Sigma \ln Y = 3.520817832$	$\Sigma \ln X = 20.88983904$

$$\Sigma y^2 = \Sigma (\ln Y_i)^2 - \frac{(\Sigma \ln Y_i)^2}{n}$$

$$= 7,041635664 - \frac{(3.520817832)^2}{9}$$

$$= 7,041635664 - 1.377350912$$

$$= 5.664284752$$

$$\Sigma x^2 = 58.22370896 - \frac{(20.88983904)^2}{9}$$

$$58.22370896 - 48.4872639$$

$$= 9.736445058$$

$$\Sigma xy' = 2.737241906 - \frac{(3.520817832)(20.88983900)}{9}$$

$$= 2.737241906 - 8.172146422$$

$$\bar{Y} = 0.391201981 \quad \bar{X} = 2.321093227 = -5.434904516$$

$$b = \frac{\Sigma xy'}{\Sigma xy^2} = -\frac{5.434904516}{9.736445058} = -0.558202145$$

$$= -0.5582.$$

$$a = \bar{Y} - b\bar{X} = 0.391201981 - (-0.558202145)(2.321093227)$$

$$= 0.39101981 - (-1.295639218)$$

$$= 1.686841199$$

Anti-log a = e1.686841199 = 5.402388653. =2.71831.686841199 = 5.402388653

The estimated non-linear model is:

$$\hat{Y} = 5.40289 \, X^{0.5582}$$

$$SS_{REG} \left(\Sigma \hat{y}^2\right) = \frac{(\Sigma xy)^2}{\Sigma x^2} = \frac{(-5.434904516)^2}{9.736445058} = \frac{29.5381871}{9.736445058}$$

$$= 3.033775359$$

$$R^2 \frac{SS_{REG}}{SS_{TOTAL}} = \frac{\Sigma \hat{y}^2}{\Sigma y^2} = \frac{3.033775359}{5.664284752} = 0.535597253$$

Hence, our power curve is In $\hat{Y}$ = 1.6868-05582InX

13.5.3.1. Exponential Models (Curves).

In the exponential curve, (Wahua, 1999), X is the power to which the slope, b, is raised. Thus, an exponential equation is one in which the exponent of a constant is a variable of the following general form:

Y = abX.

Which can be transformed: lnY = lna + Xlnb - - which is linear when lnY is plotted against original X values in a semi-log paper.

Some Common Exponential Curves.

CURVES	EQUATION	TRANSFORMATION
(a)	$Y = ab^{x}$ $a > 0$ $b < 1$	InY = Ina + X Inb
(b)	$Y = ab^{-x}$ OR $Y = ae^{-bx}$	InY = Ina - X Inb InY = Ina-bX

Exponential Growth Law:

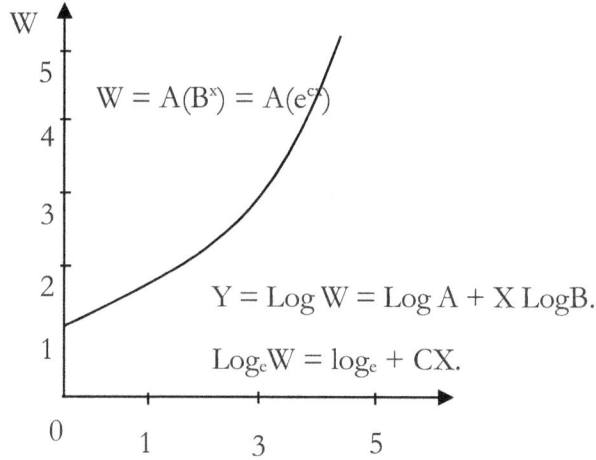

$W = A(B^x) = A(e^{cx})$

$Y = \log W = \log A + X \log B.$

$\log_e W = \log_e + CX.$

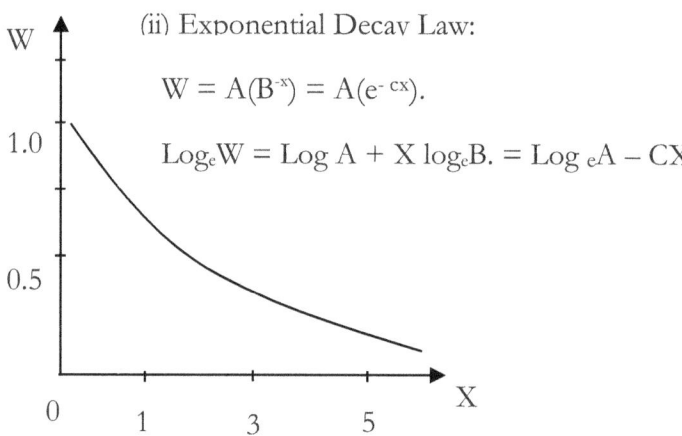

(ii) Exponential Decay Law:

$W = A(B^{-x}) = A(e^{-cx}).$

$\log_e W = \log A + X \log_e B. = \log_e A - CX$

Figure 13.5.3: Some Common Exponential Curves

The exponential decay law has been represented by other models: $Y = Y_o e^{-kt}$, which assumes that the rate of decay is proportional to the current value of Y, and k represents the proportionality factor and Y_0 represents the initial value of the variable before decay begins. Such exponential decay models have been used in agriculture to investigate the decay of chemicals in soils or animals, and survival of medical and other biological populations of risk (Mead et al. 1993).

13.5.3.1. Numerical Example 13.5.3.1:

In a study of the relationship between the weight of the leaves of pumpkin and the weight of the stems at the time of harvesting and the following data were observed;

Table 13.5.2: Stem Weight (X) and Leaf Weight (Y) of Pumpkin Plants.

Stem Weight (x)(kg)	Leaf Weight (Y)(kg)
0.5	7.5
0.6	7.2
1.8	4.2
2.5	2.9
2.8	2.7
3.0	1.0
5.5	0.9

327

Plotting the scatter diagram of the raw data, we have:

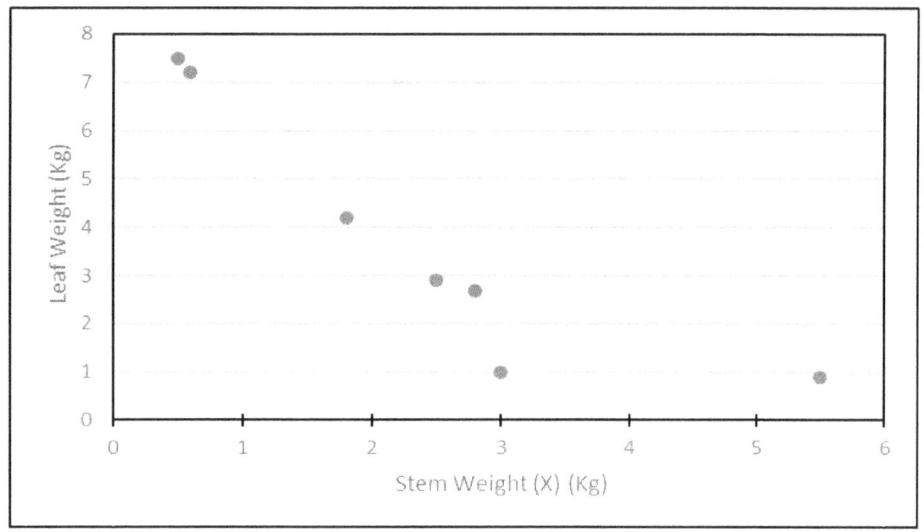

Figure 13.5.3.1: Scattergram of Weight of pumpkin leaves against Weight of the stems.

From the scattergram, we may assume that the relationship is exponential of the form:

Y aebX, which can be linearized as:

ln Y = ln a + bX.

Thus, ln Y is regressed on X as:

	X	Y	In Y
1	0.5	7.5	2.014903
2	0.6	7.2	1.974081
3	1.8	4.2	1.435035
4	2.5	2.9	1.064711
5	2.8	2.7	0.993252
6	3.0	1.0	0.00
7	5.5	0.9	-0.105361
$\sum X = 16.7$		26.4	7.376671
Mean = 2.3357		3.7714	1.0538101
$\Sigma X_i^2 = 57.19$		$143.24 = \Sigma y^2$	$14.75334 = \Sigma y'^2$
$\Sigma X_i ln Y_1 =$	9.638451		

$$\Sigma x^2 = \Sigma X_i^2 - \frac{(\Sigma X_i)}{n} = 57.19 - \frac{(16.7)^2}{7} = 57.19 - 39.841429$$

$$= 17.348571$$

$$\Sigma y^2 = \Sigma Y_i^2 - \frac{(\Sigma Y_i)^2}{n} = 143.24 - \frac{(26.4)^2}{7} = 143.24 - 99.565714$$

$$= 43.674286$$

Transform Y:

$$\Sigma lnY_i = 7.376671 \quad \Sigma y^2 = 14.75334 - \frac{(7.376671)^2}{7} = 14.75334 - 7.773611$$

$$\Sigma (ln\ Y_i)^2 = 14.75334 \quad = 6.9797293$$

$$\Sigma xy' = 9.638451 - \frac{(16.7)(7.376671)}{7} = 9.638451 - 17.5986294$$

$$= -7.9601784$$

$$b = \frac{\Sigma xy'}{\Sigma x^2} = -\frac{7.9601784}{17.348571} = -0.458838$$

$$a' = \overline{Y}' - b\overline{X} = 1.0538101 - (-0.458838)(2.3857)$$

$$= 1.0538101 - (-1.09464982)$$

$$= 2.14845992$$

Anti-log of $2.14845992 = e^{2.14845992} = 8.571647 = 2.71832.14845992$

$$\hat{Y} = 8.5716e^{-0.4588X}$$

$$\Sigma\hat{y}^2 = SS_{REG} = \frac{(\Sigma xy')^2}{\Sigma x^2} = \frac{(-7.9601784)^2}{17.348571} = 3.652429941$$

$$R^2 = SS_{REG}/SS_{TOTAL} = \frac{3.652429941}{14.75334} = 0.247566$$

$$\approx 24.76\%$$

From the $R^2 = 24.76\%$, the model is not very good in explaining the relationship between the leaf weight and stem weight of pumpkin probably because green weights were used. If plant dry weights or biomass were used (Wahua, 1999), a better fit would have been obtained.

13.5.3.2 Numerical Example 13.5.3.2:

Using the data on light transmission ratio (Y) and leaf-area index (X), we examine the data below to see the appropriate non-linear form.

Table 13.5.3.2: Light Transmission Ratio (Y) versus Leaf-Area Index (X).

	Light Transmission Ratio (Y)	Leaf-Area Index (X)	Y' = ln Y
1	75.0	0.50	4.317488
2	72.0	0.60	4.276666
3	42.0	1.80	3.737670
4	29.0	2.50	3.367296
5	27.0	2.80	3.295837
6	10.0	5.45	2.302585
7	9.0	5.60	2.1972246
8	5.0	7.20	1.609438
9	2.0	8.75	0.693147
10	2.0	9.60	0.693147
11	1.0	10.40	0.00000

12	0.9	12.00	-0.105361
ΣY_i 274.9		$\Sigma X_i = 67.20$	26.38513798
$\overline{Y} = 22.908$		$\overline{X} = 5.60$	2.198761498

4. **Compute:** $= 86.79396639 - 58.01462552$

$= 28.77934087$

$$\Sigma x^2 = \Sigma X_i^2 - \frac{(\Sigma X_i)^2}{n} = 551.725 - \frac{(67.2)^2}{12}$$

$= 551.725 - 376.32 = 175.405$

$$\Sigma xy' = \Sigma X_i Y'_i - \frac{(\Sigma X_i)(\Sigma Y'_i)}{n} \quad 76.99555578 - \frac{(67.20)(26.38513798)}{12}$$

$= 76.99555578 - 147.7567727 = -70.76121691$

$$b = \frac{\Sigma xy'}{\Sigma x^2} = -\frac{70.76121691}{175.405} = -0.40341619$$

$$a' = \overline{Y}' - b\overline{X} = 2.198761498 - (-0.40341619)(5.60)$$

$= 2.198761498 - (-2.259130667)$

$= 4.457892165.$

Procedure:

1. Plot the scattergram of the raw data to determine the form of the non-linear model:

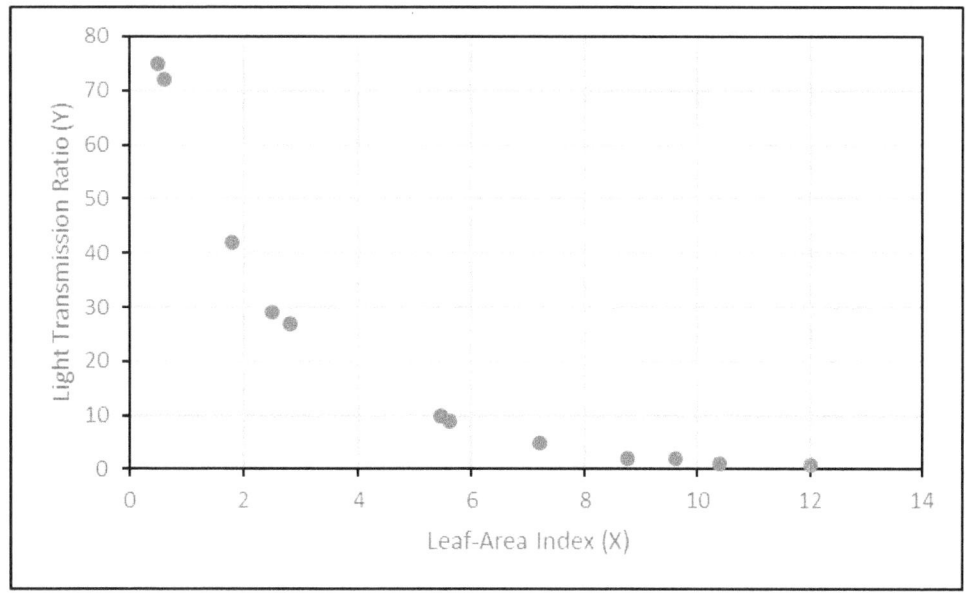

Figure 13.5.3.2: Scattergram of Light Transmission Ratio versus leaf Area Index.

The scattergram looks more like Figure 13.5.3b which is of the form:

Y = ae-bx.

2. Linearize the prescribed non-linear functional form through a proper transformation of one or both variables in $Y = \alpha e^{\beta x}$ as:

$$Y' = \alpha' + \beta X$$

Where $Y' = ln\ Y\ and\ a'\ lna$ (ln is loge or natural logarithm).

3. Compute the transformed values for all units of observation of each variable that is transformed in (3) above for the sums, sums of squares, sum of cross-products of the two variables Y' and X.

   Compute the estimates of α' and β as:

   $$b = \frac{\Sigma xy'}{\Sigma x^2} = \frac{-70.7612691}{175.405} = -0.40341619$$

   $$a' = \bar{Y}' - b\bar{X} = 4.457892165$$

   $$\approx 4.457892\ .$$

4. Using the estimates of the regression parameters of the linearized form obtained in (3) above, derive an appropriate estimate of the original regression based on the specific transformation used in (1).

   To derive the estimate of the original non-linear regression, its regression parameter α needs to be computed as the anti-log of α' as:

   $$a = \text{anti-log of } \alpha' = \text{Anti-log of } 4.457892,$$

   where anti-log of $4.457892 = e^{4.457892}$ $(e = 2.7183)$

   $$= 2.7183^{4.47892} = 86.30795745 \approx 86.308\ .$$

   Therefore, the required estimate of the regression equation is obtained as:

   $$\hat{Y} = 86.308e^{-0.40342}$$

   $$SS_{REG} = \Sigma\hat{y}^2 = \Sigma xy' = \frac{(-70.76121691)^2}{175.405} = 28.54622057$$

   $$\Sigma xy' = 28.77934087$$

   $$R^2 = \frac{\Sigma\hat{y}^2}{\Sigma xy'^2} = \frac{28.54622057}{28.77934087} = 0.991899734$$

   $$= 0.9919.$$

13.5.3.2. ASYMPTOTIC EXPONENTIAL MODELS.

The asymptotic exponential models have been extensively used in agriculture, medicine and other biological sciences. The models have been used for populations of micro-organisms in laboratories, or insects treated which chemicals (Mead et al., 1993). In agricultural investigations, the decay of chemicals in the soil, or in animals, in which the value of the response variable increases or decreases rapidly as X increases at a slower rate to infinity, as Y tends to zero, or a maximum value of X (Wahua, 1999).

For example, Mead et al. (1993) reported a model:

Y = A(1 – e-kx) whose terms represent a tendency to an asymptote which represents a gradual approach of yield to an upper limit, A, imposed by the environment, as shown in the diagram:

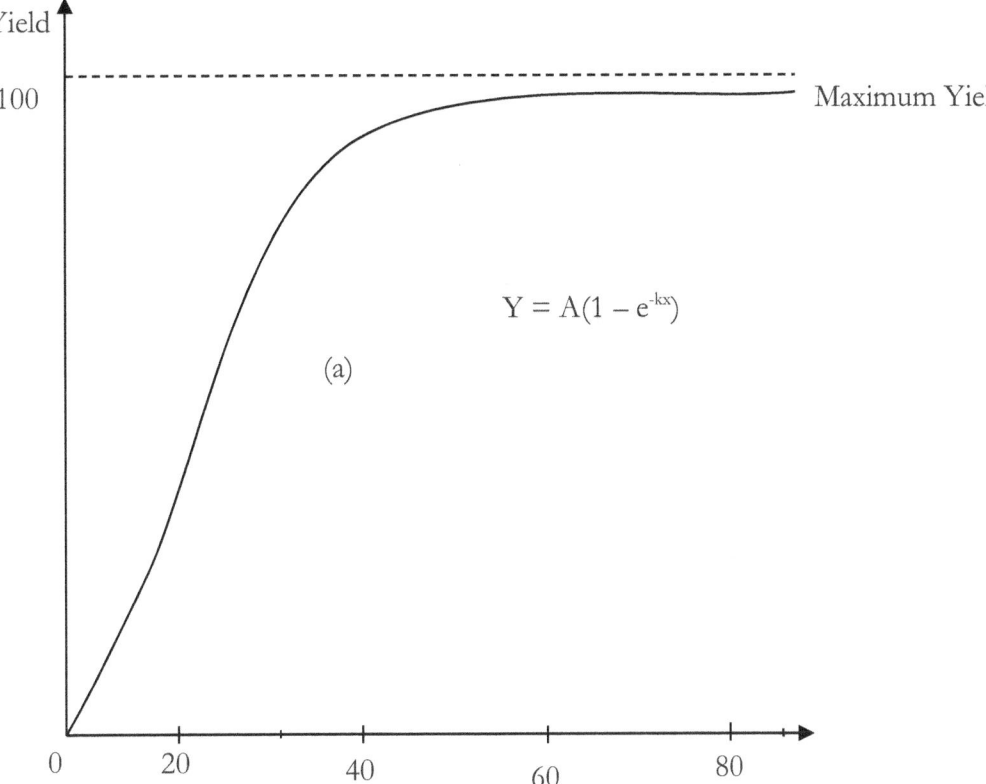

$$Y = A(1 - e^{-kx})$$

(a)

Yield

100

0   20   40   60   80

Amounts of fertilizer added to the soil.

Figure 13.5.3.2a: Asymptotic Response of yield to an upper limit of yield

A more general form of this model is given as:

$$Y = A\left(1 - e^{-k(t-t_0)}\right)$$

where: Y =agronomic yield.

A =asymptotic level which Y approaches as t increases.

k= rate of approach .

t0= time at which Y begins to increase from zero.

The Mitscherlich response function often used for describing fertilizer response is written as:

$$Y = A\left[1 - e^{-b(x+c)}\right],$$

where A = asymptotic level of yield.

c = amount of fertilizer already available in soil;

b = rate at which yield tends to its asymptotic level,

x = amount of fertilizer applied.

Another simple asymptotic exponential curve is the decay of some variable with time (see Fig. 13.5.3 c).

(ii). The model: $Y = Y_o e^{-kt} = A(1 - e^{kt})$.

where:

Y = current value of Y.

Y_0 = initial value of Y before decay begins.

k = proportionality factor.

In fact, this model assumes that the rate of decay, k, is proportional to the current size or value of Y. Snedecor and Cochran (1967) described the model as "Exponential Decay Law" as: $W = A (B^{-x}) = (Ae^{-cx})$

where:

W = yield of crop grown in pots.

X = amount of fertilizer added to the soil in the pots.

A = the asymptote or the initial value of W before declining to zero.

The model: $Y = Y_{oe}^{-kt} = A(1 - e^{-kt})$ is represented by the diagram

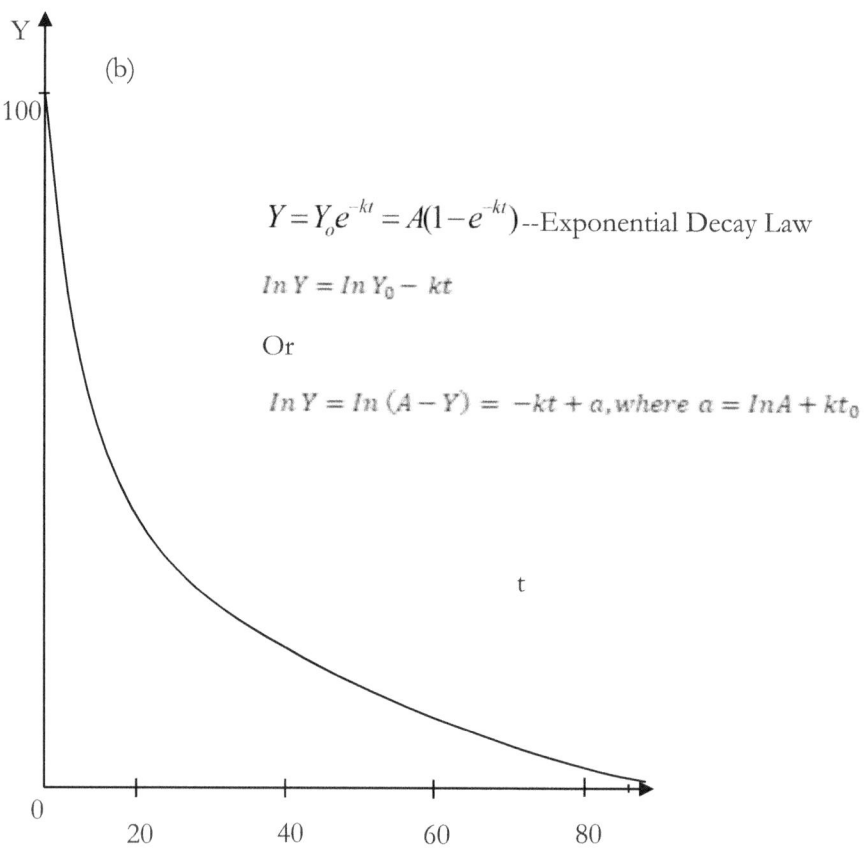

$$Y = Y_o e^{-kt} = A(1 - e^{-kt})\text{--Exponential Decay Law}$$

$$In\, Y = In\, Y_0 - kt$$

Or

$$In\, Y = In\, (A - Y) = -kt + a, where\, a = InA + kt_0$$

Figure 13.5.3.2b: Asymptotic Decay Curve over time

333

13.5.3.3. ASYMPTOTIC RECIPROCAL MODELS (RECTANGULAR HYPERBOLAS).

A third type, of exponential models, are the Asymptotic Reciprocal Models which are commonly based on reciprocal relationships generally referred to as Rectangular Hyperbolas (Daniel and Wood, 1971; Mead et al., 1993 and Wahua, 1999).

One of such a model commonly used in chemical kinetic and simple enzyme relationships is often expressed as:

a. $Y = \dfrac{Ax}{k+x}$ where k is the Michaelis-Menten contant. OR

b. $\dfrac{1}{Y} = \dfrac{1}{A}\left[1 + \dfrac{1}{k(x-x_0)}\right]$ which is used to relate photosynthesis to light intensity, as illustrated in the following diagrams:

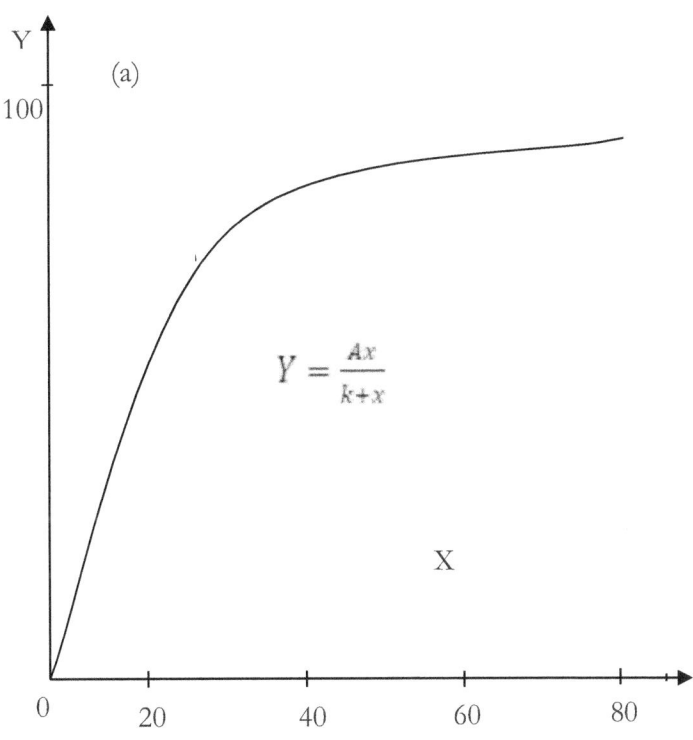

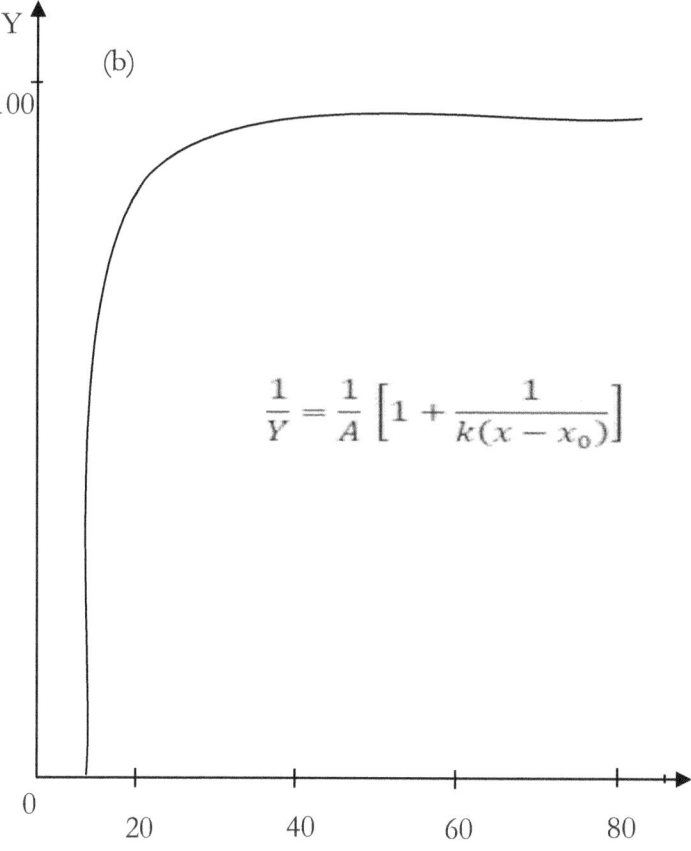

$$\frac{1}{Y} = \frac{1}{A}\left[1 + \frac{1}{k(x - x_0)}\right]$$

Figure 13.5.3.3: Reciprocal Asymptotic Models (a) and (b).

13.5.3.4.LOGISTIC MODELS FOR GROWTH CURVES.

In many biological investigations of the growth of organisms in relation to time, the growth of whole plants, parts of a plant or the growth of animals have been extensively studied. The quantitative growth of any biological organism, according to Mead et al (1993), follows four distinctive stages:

i.   **Early Growth:**

The early growth from a very small initial size is relatively very rapid and can be expressed quantitatively by saying that the rate of growth is proportional to the size of the organisms. Then dy/dt = ky. This form of growth (Mead et a., 1993) is often called "exponential growth" and the size of the organism is described by the exponential function;

$$Y = Y_o e^{\lambda t}.$$

ii.   **Second Stage of Growth:**

The second stage of growth is relatively less because more of the energy of the organism is devoted to maintaining its current size. The growth of the organism can be approximated by the linear relationship:

$$Y = \alpha + \beta t.$$

iii.   **Third Stage of Growth:**

The organism's rate of growth diminishes further in the third stage of growth, as a balance or equilibrium between the energy of the organism and the maintenance requirements is approached. At this stage the size of the organism is asymptotic, approaching an upper limit on size. According to Mead et al. (1993), these three stages of growth can be thought of together in terms of the rate of growth given as:

$$\frac{dy}{dt} = ky \left[\frac{A-y}{A} \right]$$

When Y is small, the rate of growth of Y is approximately proportional to Y, since the expression in brackets is close to 1. As Y increases, the relative growth rate declines, and as Y approaches A, dy/dt approaches zero.

iv.   **Fourth Stages of Growth:**

The fourth stage of growth is the anti-thesis of growth through senescence. The equation for Y implied by this form of growth rate is the logistic curve given as:

$$Y = \frac{A}{1+e^{(a-kt)}},$$

where:

Y= Agronomic Yield.

A = Maximum level of Yield possible.

k= Rate of approach.

t= Time at which Y begins to increase.

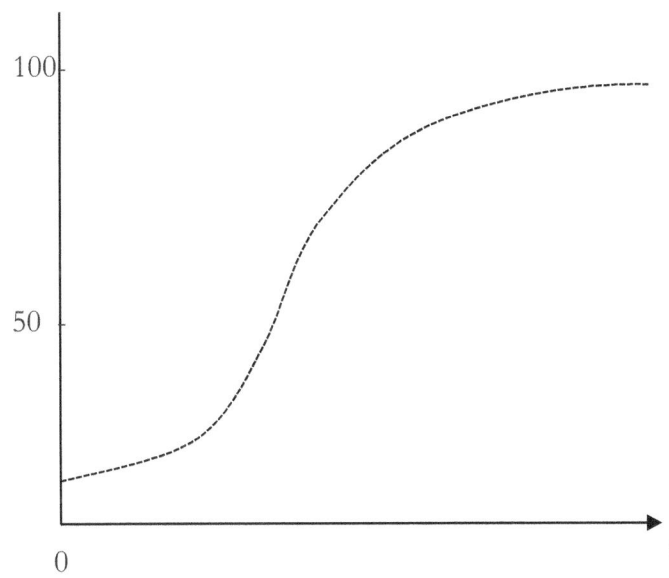

Figure 13.5.3.4: Logistic Growth Function of Size as a Function of Time.

In fitting the Logistic Model, Chatter Jee and Price (1977) used the relationship between the dosage, X_i of a pesticide and the proportion of insects that died after exposure to a dose (X_i) of the pesticide $P_i = \frac{r_i}{n_i}$. The authors obtained a non-linear function which accurately represented the relationship between dose, X_i, and the proportion, P_i of dying as: $P_i = \frac{e^{\beta_0 + \beta_1 X_i}}{1+e^{\beta_0+\beta_1 X_i}}$. (Sometimes called Logit Analysis).

This relationship is called the logistic response function, which is monotonic and bounded.

between 0 and 1, as shown in Figure 13.5.3.5:

336

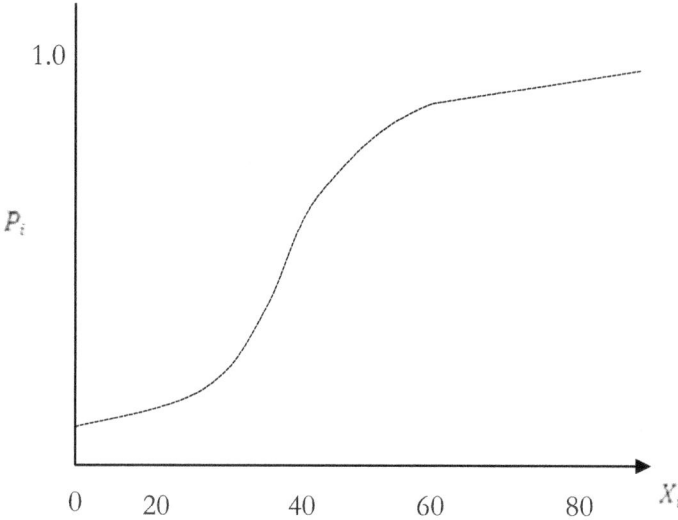

Figure 13.5.3.5: Logistic Response Function of chemical Dose and Proportion of Dead Insects

It can be seen that the two Logistic Response Functions:

$$Y = \frac{A}{1 + e^{(a - kt)}}.$$

$$P_i = \frac{e^{\beta_0 + \beta_1 X_i}}{1 + e^{\beta_0 + \beta_1 X_i}},$$

produce the same curve.

13.6 .SIMPLE POLYNOMIAL REGRESSION MODEL

Polynomial regression models can contain one, two, or more than two independent variables. In addition, each independent variable can be present in various powers.

13.6.1.ONE INDEPENDENT VARIABLE POLYNOMIAL –SECOND ORDER MODEL.

The regression model:

$Y_i = \beta_0 + \beta_1 x_i + \beta_2 x_i^2 + \varepsilon_i$ $(where: x_i = X_i - \bar{X})$ is called a Second Order Model with one independent variable because the single independent variable appears to the first and second powers

It should be noted that the independent variable is expressed as a deviation around its mean, $\bar{X}$, and i^{th} the observation is denoted by X_i. The reason for using deviations around the mean in polynomial regression models is that X, X2, and higher-power terms often will be highly correlated, and can cause serious computational difficulties, when the X'X matrix is inverted for estimating regression coefficients. Expressing the independent variable as a deviation from its mean reduces the multicollinearity substantially.

The regression coefficients in polynomial regression are frequently to reflect the pattern of the exponents:

$$Y_i = \beta_0 + \beta_1 x_i + \beta_{11} x_i^2 + \varepsilon_i.$$

The response function for the regression model is:

$E[Y] = \beta_0 + \beta_1 x + \beta_{11} x^2$, which is a parabola and is frequently called a Quadratic Response Function.

The regression coefficient β_0 represents the mean response of Y when i.e. $x = 0$ when $X = \bar{X}$. The regression coefficient, β_1, is often called the Linear Effect Coefficient, while β_{11} is called

337

The Quadratic Effect Coefficient.

With second-order Models where p = 2, Z1 = X, Z2 = X2 and $\beta_2 = \beta_{11}$, which involves one independent variable, Draper and Smith (1966) also suggested a model of the form:

$$Y = \beta_0 + \beta_1 X + \beta_{11} X_i^2 + \varepsilon.$$

13.6.2. CREATION OF NEW VARIABLES.

Some non-linear relationship between two variables can be linearized, according to Gomez and Gomez (1984), through the creation of one or more variables such that they can account for the non-linear component of the original function. These techniques are most commonly applied to the k-th degree polynomial:

$$Y = \alpha + \beta_1 X + \beta_2 X^2 + \beta_3 X^3 + \cdots + \beta_k X^k$$

Such an equation can be linearized by creating k new variables: $Z_1 \cdots, Z_k$ to form a multiple linear equation of the form:

$$Y = \alpha + \beta_1 Z_1 + \beta_2 Z_2 + \ldots + \beta_k Z_k.$$

where $Z_1 = X, Z_2 = X^2, \ldots, and\ Z_k = X^k$

With the linearized form resulting from the creation of new variables, the procedure for multiple linear regression and correlation analysis can be directly applied.

We illustrate this technique by using the data in the Table 13.6.1 below reported by Gomez and Gomez (1984) where the relationship between yield (Y) and nitrogen rate (X) is assumed to be quadratic (Second-degree polynomial):

$$Y = \alpha + \beta_1 X + \beta_2 X^2$$

Procedure:

Step 1: Linearize the prescribed non-linear functional form through the creation of an appropriate set of new variables. The linearized form of the second-degree polynomial above is:

$$Y = \alpha + \beta_1 Z_1 + \beta_2 Z_2,$$

where the two newly created variables Z_1 and Z_2 are defined as:

$$Z_1 = X; Z_2 = X^2.$$

Step 2: Compute the values of the newly created variables for all n units of observation. The Z_2 values are computed by squaring the corresponding values of the original variable X.

Step 3: Apply the appropriate multiple linear regression technique to the linearized form derived in Step 1, using the data derived in Step 2.

   i.   Compute the means, sums, sums of squares and sums of cross-products for the three variables Y, Z_1 and Z_2 .

   ii.   Compute the estimates of the three parameters: $\alpha, \beta_1\ and\ \beta_2$.

Table 13.6.1: Computation of an Estimated Quadratic Regression Equation

Pair Number	Grain Yield kg/ha (Y)	Nitrogen Rate kg/ha(Z_1 = X)	$Z_2 = X^2$
1	4,878	0	0
2	5,506	30	900
3	6,083	60	3,600
4	6,291	90	8,100
5	6,361	120	14,400
Sum	29,119	300	27,000
Mean	$\bar{Y} = 5{,}823.8$	$\bar{Z}_1 = 60$	$\bar{Z}_2 = 5{,}400$
Sums of squares:	$\Sigma y^2 = 1{,}569{,}579$	$\Sigma z_1^2 = 9{,}000$	$\Sigma z_2^2 = 140{,}940{,}000$
Sums of Gross-products:	$\Sigma z_1 y = 112{,}530$	$\Sigma z_2 y = 12{,}167{,}100$	$\Sigma z_1 z_2 = 1{,}080{,}000$

Compute the parameter estimates:

$$\hat{Y} = a + b_1 X + b_2 X^2, \text{ or its linearized form: } \hat{Y} = a + b_1 Z_1 + b_2 Z_2$$

$$b_1 = \frac{(\Sigma z_2^2)(\Sigma z_1 y) - (\Sigma z_1 z_2)(\Sigma z_2 y)}{(\Sigma z_1^2)(\Sigma z_2^2) - (\Sigma z_1 z_2)} = \frac{(10{,}940{,}000)(112{,}530) - (1{,}080{,}000)(12{,}167{,}100)}{(9000)(14{,}940{,}000) - (1{,}080{,}000)^2}$$

$$= \frac{(1585\,997\,820\,000\,000) - (314\,046\,800\,000)}{(1268\,460\,000\,000) - (1166\,400\,000\,000)} = \frac{2719\,510\,200\,000}{102\,060\,000\,000}$$

$$= 26.64619048 \approx 26.646$$

$$b_2 = \frac{(\Sigma z_1^2)(\Sigma z_2 y) - (\Sigma z_1 z_2)(\Sigma z_1 y)}{(\Sigma z_1^2)(\Sigma z_2^2) - (\Sigma z_1 z_2)} = \frac{(9000)(12{,}167{,}100) - (1{,}080{,}000)(112{,}530)}{(9000)(14{,}940{,}000) - (1{,}080{,}000)^2}$$

$$= \frac{109\,503\,900\,000 - 121\,532\,400\,000}{102\,060\,000\,000} = \frac{12\,028\,500\,000}{102\,060\,000\,000} = -0.117857142$$

$$= -0.11786$$

$$a = \bar{Y} - b_1 \bar{Z}_1 - b_2 \bar{Z}_2 = 5{,}823.8 - (26.646)(60) - (-0.11786)(5{,}400)$$

$$= 5{,}823.8 - 1{,}598.76 + 636.444$$

$$= 5{,}823.8 + 636.444 - 1{,}598.76$$

$$= 6{,}460.244 - 1{,}598.76$$

$$= 4{,}861.484$$

Thus, the second-degree polynomial regression equation describing the yield response of rice to the rates of nitrogen applied, within the range of 0 – 120kg N per hectare, is estimated as:

$$\hat{Y} = 4{,}861.484 + 26.646X - 0.11786X^2, \text{ for } 0 \leq X \leq 120.$$

iii.   Compute the coefficient of determination, R^2, as:

$$R^2 = \frac{b_1\sum z_1 y + b_{12}\sum z_2 y}{\sum y^2}$$

$$R^2 = \frac{b_1 + b_2\sum z_2 y}{\sum y^2}.$$

$$= \frac{(26.646)(112,530) + (-0.11786)(12,167,100)}{1,569,579}$$

$$= \frac{2,998,474.38 - 1,434,014.406}{1,569,579} = \frac{1,564,459.974}{1,569,579}$$

$$= 0.996738599 \approx 0.9967$$

iv.   Compute the F-value as:

$$F = \frac{(n-k-1)(b_1\sum z_1 y + b_2\sum z_2 y)}{k(\sum y^2 - b_1\sum z_1 y - b_2\sum z_2 y)}$$

$$= \frac{(5-2-1)(b_1\sum z_1 y + b_2\sum z_2 y)}{(2)(1,569,579 - 1,564,459.974)} = \frac{3,128,919.948}{10,238.052} = 305.6167$$

$$\approx 305.62$$

Conclusion:

Because the computed F-value exceeds the tabular F-value with V1 = n-k-1 = 2; V2 = k = 2 degrees of freedom at the 1% level of significance of 99.0, the estimated quadratic regression equation is significant at the 1% level.

The results indicate that the yield response for the rice variety to nitrogen fertilization can be adequately described by the quadratic equation.

The computed R2 – value of 0.9967 indicates that 99.67% of the total variation in the mean yields was explained by the quadratic regression equation estimated.

14 SIMPLE LINEAR CORRELATION ANALYSIS

The simple Linear Correlation Analysis deals with the estimation and test of significance of the sample linear correlation coefficient, r, which is a measure of the degree of linear association between two variables X and Y.

In correlation, we are concerned largely whether two variables are inter-dependent, or co-vary i.e. vary together. We do not express one as a function of the other - - no distinction between independent and dependent variables.

There are numerous correlation coefficients in statistics. The most common of these is called the **Product-Moment Correlation Coefficient**, which is also called **Pearson's Correlation Coefficient.**

Suppose we want to estimate the correlation between Y1 and Y2, then, we have to obtain the sum of products, which is a measure of co-variation, from which the formula for the correlation coefficient is finally obtained.

The sum of products of Y1 and Y2 is given by: $\Sigma y_1 y_2$ and their co-variance is given by:

$$S_{12} = [\Sigma y_1 y_2] / (n - 1)$$

A measure of the association should be independent of the original scale of measurement so that we can compare the degree of association in one pair of variables with that of another. To accomplish this, we divide the covariance by the standard deviations of variables Y1 and Y2, thus resulting in a standardized deviate:

$$r_{Y_1 Y_2} = \frac{\Sigma y_1 y_2}{(n-1) s y_1 s y_2}$$

This is the formula for the Product-Moment Correlation Coefficient, $r_{Y_1 Y_2}$, between variables Y1 and Y2.

This formula can be simplified as: $r_{12} = \frac{\Sigma y_1 y_2}{(n-1) s_1 s_2} = \frac{s_{12}}{s_1 s_2}$, which can be re-written as $r_{12} = \frac{\Sigma y_1 y_2}{\sqrt{\Sigma y_1^2 \Sigma y_2^2}}$, which is preferable for computation.

14.1. Computation of the Product-Moment Correlation Coefficient.

The computation of a product-moment correlation coefficient is quite simple. The basic quantities needed are the same as those required for the computation of the regression coefficient.

Example 14.1: A Fishery Biologist investigated the relationship between the body weight and gill weight of Tilapia species and obtained the following data:

	Gill weight (mm) Y_1	Body weight (gm) Y_2
1	159	14.40
2	179	15.20
3	100	11.30
4	45	2.50
5	384	22.70
6	230	14.90
7	100	1.41
8	320	15.81
9	80	4.19
10	220	15.39
11	320	17.25
12	210	9.52
Sum = $\Sigma y_{1i} = 2{,}347$		$\Sigma y_{1i} = 144.57$
Sum of squares = $\Sigma y_{1i}^2 = 583{,}403$		$\Sigma y_{2i}^2 = 2{,}204.1853$
Sum of Products = $\Sigma y_{1i} y_{2i} = 34{,}837.10$		

1. Sum of squares of $Y_1 = \Sigma y_1^2 = \Sigma Y_{1i}^2 - \dfrac{(\Sigma Y_{1i})^2}{n} = 583{,}403 - \dfrac{(2347)^2}{12}$

   $= 583{,}403 - 459{,}034.0833 = 124{,}368.9167$

2. Sum of squares of $Y_2 = \Sigma y_2^2 = \Sigma Y_{2i}^2 - \dfrac{(\Sigma Y_{2i})^2}{n}$

   $= 2204.1853 - \dfrac{(144.57)^2}{12} = 2{,}204.1853 - 1{,}741.707075 = 462.478225$

3. Sum of Products $= \Sigma y_1 y_2 - \Sigma Y_{1i} Y_2 - \Sigma Y_{1i} Y_{2i} - \dfrac{(\Sigma Y_{1i})(\Sigma Y_{2i})}{n}$

   $= 34{,}837.10 - \dfrac{(23.47)(144.57)}{12}$

   $= 34{,}837.10 - 28{,}275.4825.$

   $= 6{,}561.6175.$

4. Product-Moment Correlation Coefficient:

   $$r_{12} = \frac{\Sigma y_1 y_2}{\sqrt{\Sigma y_1^2 \, \Sigma y_2^2}}$$

   $$= \frac{6{,}561.6175}{\sqrt{(124{,}368.9167)(462.4782)}} = \frac{6{,}561.6175}{7{,}584.056483} = 0.865185737$$

   ≈ 0.8652

Note: The interesting feature of this example is that it is Part-Whole Correlation. Although gill weight is only a small fraction of body weight, it is still a part of it, and the correlation is, therefore, between two variables, one of which is a part of another. Therefore, it should be obvious that gill weight and body weight from the same specimen should be positively correlated.

14.1.1. Properties of r:

The correlation coefficient is a measure of the degree of closeness of the linear relationship between two variables. Some properties of r should be noted:

1. r is a pure number without units or dimension, because of the scales in which X_1 and X_2 are measured.
2. r always lies between -1 and + 1. Positive values of r indicate a tendency of X_1 and X_2 to increase together. When r is negative, large values of X_1 are associated with small values of X_2.
3. r is zero when there is no correlation between X_1 and X_2.
4. The slope of the axis is determined by the scales of measurement adopted for the two axes of the graph and is therefore not a reliable indicator of the magnitude of r.

Larger correlations, either positive or negative, are fairly obvious from the graphs. It is not easy to make visual evaluation, if the absolute value of r is less than 0.5; even the direction of inclination of the ellipse may elude you, if r is between -0.3 and + 0.3.

In regression analysis:

$$\Sigma \hat{y}^2 = (\Sigma xy)^2 / \Sigma x^2$$

$$r^2 = \frac{\Sigma \hat{y}^2}{\Sigma y^2}$$

$\Sigma \hat{y}^2 = r^2 \Sigma y^2 \equiv$ the explained SS. The unexplained SS is:

$$\Sigma d_{1.2}^2 = (1 - r^2) \Sigma y^2$$

The quantity $1 - r^2$ is sometimes known as the Coefficient of Non-Determination, which expresses the proportion of the variance of a variable that has not been explained by another variable.

The square root of the coefficient, $\sqrt{1 - r^2}$, is known as the Coefficient Of Alienation, measuring the lack of association between variables Y_1 and Y_2.

14.1.2. Significance Test in Correlation

The most common significance test is whether a sample correlation coefficient could have come from a population with a parametric correlation coefficient of zero.

Therefore, the null hypothesis is:

$H_0 : \rho = 0$. This implies that the two variables are uncorrelated

$H_1 : \rho \neq 0$. This implies that the two variables are correlated.

If the sample comes from a bivariate normal distribution and $\varrho = 0$, so it cannot be used to test the hypothesis that ϱ is a specific value other than zero. The hypothesis is tested as a t-test with $n - 2$ degrees of freedom:

$$t = (r - 0) / \sqrt{\frac{(1-r^2)}{n-2}} = r \sqrt{(n - 2) / (1 - r^2)}$$

343

Using the data from above:

r = 0.8652

n = 12

$t = 0.8652 \sqrt{(12 - 2) / (1 - 1 - 0.8652)}$

$08652\sqrt{10/0.25142896}$

$= 0.8652\sqrt{39.7725} = 0.8652(6.3065) = 5.4564$

$t_{0.05}, 10 = 0.576$

$t_{0.01}, 10 = 0.708$

Therefore, reject H_0.

Note: The t-table can be used only for testing the null hypothesis, $\varrho = 0$. It is unsuited for testing other null hypotheses, such as:

$\rho = 0.5, or\ \rho_1 = \rho_2$ or making confidence statements about ϱ.

14.2 WHEN n > 50, MAKE USE OF THE Z – TRANSFORMATION.

When $\varrho \neq 0$, the distribution of sample values of r is markedly asymmetrical, and, although a standard error has been found for r in such cases, it should not be applied, unless the sample is very large (n > 500).

To overcome this difficulty, we transform r to a function z, developed by Fisher (1921):

$$z = \frac{1}{2} In \left(\frac{1+r}{1-r}\right) = \frac{1}{2}[log_e(1+r) - log_e(1-r)]$$

When $r = 0,\ z = 0, since \frac{1}{2}ln1 = 0$.

As r approaches 1, (1 + r)/(1- r) approaches infinity; consequently, z approaches infinity. Therefore, substantial differences between r and z occur at higher values for r.

While correlation coefficients are distributed in skewed fashion for values of $\varrho \neq 0$, the values of z are approximately normally distributed for any value of its parameter.

The variance of z is:

$\sigma_z^2 = \frac{1}{n-3}$ -- adequate for sample sizes $n \geq 50$ and a tolerable approximation even when $n \geq 25$.

The variance of z is independent of the magnitude of r, but is simply a function of the sample size n.

14.2.1. TESTING TWO CORRELATION COEFFICIENTS.

To test the hypothesis:

$H_0 : \varrho 1 = \varrho 2$

$H_1 : \varrho 1 \neq \varrho 2$

Compute Z_1 and Z_2 as shown above and compute t^* as:

$$t^* = \frac{Z_1 - Z_2}{\sqrt{\frac{1}{n_1 - 3} + \frac{1}{n_2 - 3}}}$$

Then, compare this value with $t_\alpha(\infty)$.

Numerical Example 14.2.1:

An Entomologist found the correlation between body weight and wing length of a dragon-fly species in Ovia North-east of Edo State, Nigeria, to be $r_1 = 0.552$, in a sample of $n_1 = 39$ and $r_2 = 0.665$ in a sample of $n_2 = 20$ in Etsakor Central. Test the hypothesis that $\varrho_1 = \varrho_2$ versus the alternative that they are not equal at $\alpha = 0.05$.

Solution:

Compute z_1 and z_2 as:

$$z_1 = \frac{1}{2} \ln\left(\frac{1+r_1}{1-r_1}\right) = \frac{1}{2}[log_e(1+r_1) - log_e(1-r_1)]$$

$$= \frac{1}{2}[log_e(1+0.552) - log_e(1-0.552)]$$

$$= \frac{1}{2}[log_e 1.551 - log_e 0.448]$$

$$\frac{1}{2}[0.439544421 - (-0.802962046)]$$

$$= 0.5(1.242506468) = 0.621253233$$

$$z_1 \approx 0.6213$$

$$z_2 = \frac{1}{2}[log_e(1-r_2) - log_e(1-r_2)]$$

$$= \frac{1}{2}[log_e(1+0.665) - log_e(1-0.665)]$$

$$= \frac{1}{2}[log_e(1.665) - log_e(0.335)]$$

$$\frac{1}{2}[0.509825123 - (-1.093624747)] = 0.5$$

$$= 0.5(1.60344987)$$

$$= 0.801724935$$

$$t^* = \frac{z_1 - z_2}{\frac{1}{n_1-3} + \frac{1}{n_2-3}} = \frac{0.6213 - 0.8017}{\frac{1}{36} + \frac{1}{17}} = -\frac{0.1804}{\sqrt{0.027777777 + 0.058823529}}$$

$$= -\frac{0.1804}{\sqrt{0.08660137}} = -\frac{0.1804}{0.294281}$$

$$= -6.130195281$$

$$\approx -6.1302$$

$$t_{0.05}, \infty = 1.960$$

Conclusion:

Since $t^* = -6.1302 \ll -1.960$ and we accept H_0 and conclude that $\rho_1 = \rho_2$.

14.3. NON-PARAMETRIC METHOD (OR RANK CORRELATION), OR SPEARMAN CORRELATION.

Quite often, a bivariate population is far from normal. In that event, the computation of r as an estimate of ϱ is no longer valid. Yet, we may still wish to test for the significance of association between two variables.

One method of analyzing such data according to Sokal and Rohlf (1967), is by ranking the variates and calculating a coefficient of rank correlation. This approach belongs to the general family of Non-parametric methods where the ranks of the variables are used, instead of the variables themselves.

There are several rank correlation coefficients. But the rank correlation coefficient due to Spearman, rs, usually denoted by rs, which is the ordinary correlation coefficient, r, between the ranked values of X1 and X2. The easier method of computing rs is given by the formula:

$$r_s = 1 - \frac{6\sum d_i^2}{n(n^2-1)}.$$

Like r, the rank correlation can range in samples from -1 **(Complete discordance)** to $+1$ **(Complete concordance)**.

Numerical Example 14.3.1:

Seven rats were fed on a deficient diet for three weeks after which two Animal Scientists ranked their condition. The rankings by the two scientists are presented below. Using these ranks, calculate the coefficient of the rank correlation and test at $\alpha = 0.05 \ and \ 0.01$ hether the rank correlation is significant.

Rate No.	Ranking by: scientist 1(R_1):	Scientist 2(R_2)	Difference in Ranks (d_1)	d_i^2
1	4	4	0	0
2	1	2	-1	1
3	6	5	1	1
4	5	6	-1	1
5	3	1	2	4
6	2	3	-1	1
7	7	7	0	0
			$\Sigma d_i = 0$	$\Sigma d_i^2 = 8$

$$r_s = 1 - \frac{6\Sigma d_i^2}{n(n^2-1)}$$

Substituting the data we have:

$$r_s = 1 - \frac{6(8)}{7(49-1)} = 1 - \frac{48}{336}$$

$$= 1 - 0.142857142 = 0.85714257$$

$$\approx 0.857$$

Note: Spearman's correlation, r_s, can be computed directly from the differences between the ranks R_1 and R_2 of paired variables 1 and 2 as follows:

$$r_s = 1 - \frac{6\Sigma(R_1 - R_2)^2}{n(n^2 - 1)}$$

Comparing r_s with the tabled values of r_s, shows that it is significant at both 5% and 1% levels of significance. See Table A11 (Snedecor and Cochran, 1967).

Conclusion:

Since r_s calculated $= 0.857 > r_{0.05} = 0.707$ and $r_{0.01} = 0.834$, we conclude that r_s is highly significant.

Applications of Correlation:

i.   To measure the intensity of association observed in any pair of variables.
ii. To test whether the association is greater than could be expected by chance alone.
iii. Such associations lead to reasoning about causal relationships between the variables.

Can be applied in many areas of ecology, systematics and other fields (Animal Science, Entomology, etc.) in which experimental methods are difficult to apply

15. ANALYSIS OF COVARIANCE

Analysis of Covariance is a technique that combines the features of analysis of variance and regression that can be used for designed experiments and observational studies. The basic idea is to augment the analysis of variance model containing the factor effects with one or more additional quantitative variables that are related to the dependent variable. The augmentation is intended to reduce the variance of the error terms in the model.

15.1 DEVELOPMENT OF COVARIANCE MODEL

The single-factor ANOVA model in terms of fixed factor effects is:

$Y_{ij} = \mu + \tau_i + \varepsilon_{ij}$

The covariance model starts with this model and simply adds another or several terms, reflecting the relationship between the concomitant (X_{ij}) and dependent (Y_{ij}) variables by using a linear relation as a first approximation like:

$Y_{ij} = \mu + \tau_i + \gamma X_{ij} + \varepsilon_{ij}$

where:

μ = an overall mean.

τ_i = the fixed treatment effects, subject to the restriction $\Sigma\tau_i = 0$.

γ = the regression coefficient for the relation between Y and X.

X_{ij} = constants.

Y_{ij} = the dependent variable $\sim N(\mu_{ij}, \sigma^2)$

ε_{ij} = independent $N(0, \sigma^2)$,

$i = 1, 2, \ldots r; \quad j = 1, 2, \ldots, n_i.$

If we express the concomitant as a deviation from the overall mean, $\bar{X}..$, the resulting model is the usual covariance model for a single-factor study with fixed factor levels:

$Y_{ij} = \mu + \tau_i + \gamma(X_{ij} - \bar{X}..) + \varepsilon_{ij}.$

Thus, the covariance model corresponds to ANOVA, except the added term $\gamma(X_{ij} - \bar{X}..)$ to reflect the relationship between Y and X. Since X_{ij} are assumed to be constants:

$E(Y_{ij}) = \mu + \tau_i + \gamma(X_{ij} - \bar{X}..).$

$\sigma^2(Y_{ij}) = \sigma^2$

Y_{ij} are independent $N(\mu_{ij}, \sigma^2)$ where: $\mu_{ij} = \mu. + \tau_i + \gamma(X_{ij} - \bar{X}..).$

$\Sigma\tau_i = 0.$

The covariance model may be expanded to include several concomitant variables that are related to the dependent variable.

$Y_{ij} = \mu + \tau_i + \gamma(X1_{ij} - \bar{X}1..) + \gamma(X2_{ij} - \bar{X}2..) + \varepsilon_{ij}.$

It is not only the linearly related Y and X variables that can be used in covariance analysis. The relation between Y and X may be quadratic and covariance analysis can still be used as:

$Y_{ij} = \mu + \tau_i + \beta_1 (X_{ij} - \bar{X}..) + \beta_2 (X_{ij} - \bar{X}..)2 + \varepsilon_{ij}$ – where β_1 and β_2 are regression coefficients.

	TREATMENT 1		. . .	TREATMENT j		. . .	TREATMENT k		
	Y11	X11	. . .	Y1j	X1j	. . .	Y1k	X1k	
	Y21	X21	. . .	Y2j	X2j	. . .	Y2k	X2k	
	.	.		.	.		.	.	
	.	.		.	.		.	.	
	.	.		.	.		.	.	
	Yn1	Xn1	. . .	Ynj	Xnj	. . .	Ynk	Xnk	
SUM	Ty1	Tx1	. . .	Tyj	Txj	. . .	Tyk	Txk	$\Sigma Tyj = Gy$
MEAN	$\bar{Y}_1$	$\bar{X}_1$	. . .	$\bar{Y}j$	$\bar{X}j$	. . .	$\bar{Y}k$	$\bar{X}k$	$\Sigma Txj = Gx$ $\bar{Y}$ $\bar{X}$

$Txx = n\Sigma(\bar{X}j - \bar{X})2$	$Txy = n\Sigma(\bar{X}j - \bar{X})(\bar{Y}j - \bar{Y})$	$Tyy = n\Sigma(\bar{Y}j - \bar{Y})2$
$Exx = \sum_j (X_{ij} - \bar{X}_j)^2$	$Exyj = \Sigma(X_{ij} - \bar{X}_j)(Y_{ij} - \bar{X}_j)$	$Eyyj = \Sigma(Y_{ij} - \bar{Y}_j)^2$
$Exx = \sum_j E_{xxj}$	$Exy = \Sigma E_{xxj}$	$Eyy = \sum_j E_{yyj}$
$Sxx = Txx + Exx = \Sigma\Sigma(X_{ij} - \bar{X})^2$	$Sxy = Txy + Exy$	$Syy = Tyy + Eyy$
	$= \Sigma\Sigma(X_{ij} - \bar{X})(X_{ij} - \bar{Y})$	$= \Sigma\Sigma(Y_{ij} - \bar{Y})^2$

15.1   NOTATION FOR THE ANALYSIS OF COVARIANCE

The covariate is denoted by the symbol X and the variate by the symbol Y. Thus:

Txj = sum of measurements on the covariate under treatment j.

Tyj = sum of measurements on the variate under treatment j.

$\bar{X}j$ = mean of measurements on the covariate under treatment j.

$\bar{Y}j$ = mean of measurements on the variate under treatment j.

$Exxj$ = variation on the covariate under treatment j.

$Eyyj$ = variation on the variate under treatment j.

$Exyj$ = covariation between the variate and the covariate under treatment j.

$Exx = \Sigma Exxj$; $Eyy = \Sigma Eyyj$; $Exy = \Sigma Exyj$.

Txx = between-treatment variation on the covariate.

Tyy = between-treatment variation on the variate.

Txy = between-treatment covariation.

Sxx = overall variation on the covariate = $Txx + Exx$.

Syy = overall variation on the variate = $Tyy + Eyy$.

Sxy = overall covariation = $Txy + Exy$.

15.2 GENERAL PROCEDURE FOR ANALYSIS OF LINEAR COVARIANCE

DESIGN	SOURCES OF VARIANCE	DEGREES OF FREEDOM	SUM OF PRODUCTS			DEGREES OF FREEDOM	ADJUSTED $\Sigma y2$	M.S
			x.y	x.y	y.y			
CRD	TOTAL	$rt - 1$	$\Sigma x2$	Exy	$\Sigma y2$			
	TRT	$t - 1$	Txx	Txy	Tyy			
	ERROR	$t(r - 1)$	Exx	Exy	Eyy	$t(r - 1) - 1$	$Eyy - \dfrac{(Exy)^2}{Exx}$	$S_{y.x}^2$
	TRT + ERROR	$rt - 1$	Sxx	Sxy	Syy	$(rt - 1) - 1$	$Syy - \dfrac{(S_{xy})^2}{S_{xx}}$	
RCB	TOTAL	$rt - 1$	$\Sigma x2$	Σxy	$\Sigma y2$			
	BLOCK	$r - 1$	Rxx	Rxy	Ryy			
	TRT	$t - 1$	Txx	Txy	Tyy			
	ERROR	$(r - 1)(t - 1)$	Exx	Exy	Eyy	$(r-1)(t-1)-1$	$Eyy - \dfrac{(E_{xy})^2}{E_{xx}}$	$S_{y.x}^2$
	TRT + ERROR	$r(t - 1)$	Sxx	Sxy	Syy	$r(t - 1) - 1$	$Syy - \dfrac{(S_{xy})^2}{S_{xx}}$	$S_{y.x}^2$
LATIN SQUARE	TOTAL	$k2 - 1$	$\Sigma x2$	Σxy	$\Sigma y2$			
	COLUMN	$k - 1$	Cxx	Cxy	Cyy			
	ROW	$k - 1$	Rxx	Rxy	Ryy			
	TRT	$k - 1$	Txx	Txy	Tyy			
	ERROR	$(k-1)(k-2)$	Exx	Exy	Eyy	$(k-1)(k-2)-1$		$S_{y.x}^2$
	TRT + ERROR		Sxx	Sxy	Syy	$(k-1)(k-2)-1$		
TREATMENT ADJUSTMENT						$(t - 1)$ OR $(k - 1)$	$Syy - \dfrac{(S_{xy})^2}{S_{xx}}$ $Eyy - \dfrac{(E_{xy})^2}{E_{xx}}$	

Note:

1.　Each column for sum of products is a complete analysis of variance for the variable or item indicated.

2.　Each row gives the elements needed for linear regression.

3.　In the adjustment column, the item, $(Exy)2/Exx$, is the linear regression SS for Y on X. When we subtract it from Eyy, the Residual SS is obtained. The mean square of the Residual is the $S^2_{y,x}$.

A Numerical Example of a Covariance Analysis

In a study conducted to determine if the initial body weight of the pigs affected the final body weight gain, after feeding them with four different rations for one month. Four pigs were randomly selected from each of the five litters and four rations were randomized among the pigs of the same litter. The initial and final body weights were recorded.

i.　With covariance analysis, interpret the results of the experiment from the data below.

ii.　Were there significant differences in the initial weights of the pigs?

iv.　By t-test, determine whether the effects of the rations 1 and 4, on adjusted weight gain, are significantly different.

Table 2:　INITIAL WEIGHT (X) AND FINAL WEIGHT (Y) OF PIGS FED ON DIETS

TRT	BLOCKS										TRT TOTALS	
	I		II		III		IV		V		X	Y
	X	Y	X	Y	X	Y	X	Y	X	Y		
1	2.7	5.5	3.2	5.0	2.8	5.6	3.3	7.5	4.5	8.0	16.5	31.6
2	2.4	7.0	2.1	6.6	3.3	6.9	1.9	6.4	4.8	9.5	14.5	36.4
3.	3.5	8.0	2.8	8.0	2.9	6.5	3.0	6.7	2.6	7.5	14.8	36.9
4	4.5	10.2	2.5	8.8	4.3	9.5	2.7	8.1	3.8	8.9	17.8	45.5
BLOCK	13.1		10.6		13.3		10.9		15.7		63.6	
TOTAL	30.9		28.4		28.5		28.7		33.9		150.4	

COMPUTATIONAL PROCEDURE:

1.　Examine the problem carefully and identify the design and the various variables. We note that the pens or litters are separate experimental blocks making the design an RCB. The Rations were randomly assigned litter by litter – i.e. randomization within Blocks.

Hence:

Litters are the Blocks or Replications = 5.

Rations are the Treatments = 4.

X = Initial Body Weight.

Y = Final Body Weight.

2.   Arrange the data so as to obtain Treatment and Replication totals easily for both X and Y, as shown in Table 1 above. Obtain the Totals.

3.   Check the guiding Table 1 for the items to be calculated for the design, RCB concerned.

4.   Sketch the ANOVA Table and check the degrees of freedom.

5.   Calculate the various items, row by row for the design, as indicated in Table 1.

For RCB, calculate for:

– Total Regression = $\Sigma y2$.

– Regression within Blocks = $\Sigma x2$.

– Regression within Treatments = Σxy.

– Then the Error.

A.   FOR TOTAL REGRESSION (CHECK INDIVIDUAL VALUES OF X AND Y in Table 2):

Total SS for Y = $\Sigma y2$ = 5.52 + 5.02 + . . . + 8.92 $-\dfrac{(150.4)^2}{20}$ = Eyy.

Eyy = 1169.42 – 1131.01 = 38.41.

Total SS for X = Exx = $\Sigma x2$ = 2.72 + 3.22 + . . . + 3.82 $-\dfrac{(63.6)^2}{20}$.

= 215.2 – 202.25 = 12.95.

Total Sum of Cross Products = Exy = Σxy = (2.7) (5.5) + (3.2) (5.0) + . . . + (3.8) (8.9) $-\dfrac{(63.6)(150.4)}{20}$ = 492.40 – 478.27 = 14.19.

B.   REGRESSION FOR THE BLOCK – CHECK BLOCK TOTALS:

BLOCK $\Sigma y2$ = Ryy = $\dfrac{30.9^2 + 28.4^2 + \ldots + 33.9^2}{4} - \dfrac{(150.4)^2}{20}$

$= \dfrac{4{,}546.52}{4} - \dfrac{22{,}620.16}{20}$

= 1,136.63 – 1,131.01 = 5.62.

BLOCK $\Sigma x2$ = Rxx = $\dfrac{13.1^2 + 10.6^2 + \ldots + 15.7^2}{4} - \dfrac{(63.6)^2}{20}$

$$= \frac{826.16}{4} - \frac{4{,}044.96}{20}$$

$$= 206.54 - 202.25 = 4.29.$$

$$\text{BLOCK } \Sigma xy = Rxy = \frac{(13.1)\,(30.9) + (10.6)\,(28.4) + \ldots + (15.7)\,(33.9)}{4} - \frac{(63.6)\,(150.4)}{20}$$

$$= \frac{1{,}929.94}{4} - 478.27 = 482.49 - 478.27 = 4.22.$$

C.   REGRESSION FOR THE TREATMENT – CHECK TREATMENT TOTALS:

$$\text{TREATMENT } \Sigma y2 = Tyy = \frac{31.6^2 + 36.4^2 + \ldots + 45.5^2}{5} - \frac{(150.4)^2}{20}$$

$$= \frac{5{,}755.38}{5} - 1{,}131.01$$

$$= 1{,}151.08 - 1{,}131.01 = 20.07.$$

$$\text{TREATMENT } \Sigma x2 = Txx = \frac{16.5^2 + 14.5^2 + \ldots + 17.8^2}{5} - \frac{(63.6)^2}{20}$$

$$= \frac{1{,}018.38}{5} - 202.25 = 203.68 - 202.25 = 1.43.$$

TREATMENT $\Sigma xy =$

$$Txy = \frac{(16.5)\,(31.6) + (14.5)\,(36.4) + \ldots + (17.8)\,(45.5)}{5} - \frac{(63.6)\,(150.4)}{20}$$

$$= \frac{2{,}405.22}{5} - \frac{9{,}565.44}{20} = 481.04 - 478.27 = 2.77.$$

D.   SUM OF PRODUCTS FOR ERROR:

These are obtained by subtraction:

ERROR $\Sigma y2 = Eyy = \text{Total } yy - Ryy - Tyy$

$$= 38.41 - 5.62 - 20.07 = 12.72.$$

ERROR $\Sigma x2 = Exx = \text{Total } xx - Rxx - Txx$

$$= 12.95 - 4.29 - 1.43 = 7.23.$$

ERROR $\Sigma xy = Exy = \text{Total } xy - Rxy - Txy$

$$= 14.19 - 4.22 - 2.77 = 7.19.$$

E.   ADJUSTMENT OF Y FOR X:

FOR ERROR SS: $Eyy - \dfrac{(E_{xy})^2}{E_{xx}} = 12.72 - \dfrac{(7.19)^2}{7.23} = 12.72 - 7.15 = 5.57.$

RESIDUAL SS = 5.57.

FOR TREATMENT + ERROR:

$$\text{Syy} - \frac{(S_{xy})^2}{S_{xx}} = 32.79 - \frac{(9.97)^2}{8.66} = 32.79 - 11.48 = 21.31.$$

$$\text{i.e. } (20.07 + 12.72) \frac{(2.77 + 7.19)^2}{(1.43 + 7.19)} = 32.79 - \frac{(9.97)^2}{8.66}$$

$$= 21.31.$$

TREATMENT ADJUSTED:

$$\text{Syy} - \frac{(S_{xy})^2}{S_{xx}} - \text{Eyy} - \frac{(E_{xy})^2}{E_{xx}} = 21.31 - 5.57 = 15.74.$$

5. Look at the Sketch ANOVA and fix the values calculated according to the guiding Table 1.

TABLE 3: ANALYSIS OF VARIANCE FOR LINEAR COVARIANCE IN RCB FOR DATA IN TABLE 2

SOURCES OF VARIATION	DF	SUM OF PRODUCTS			Y ADJUSTED FOR X			
		Xx	xy	yy	DF	SS	MS	FCAL
TOTAL	19	12.95	14.19	38.41				
BLOCK	4	4.29	4.22	5.62				
TREAMENT	3	1.43	2.78	20.07				
ERROR	12	7.23	7.19	12.72	11	5.57	20.29	
TRT + ERROR	15	8.66	9.97	32.79	14	21.31		
TRT ADJUSTED					3	15.74	5.25	0.51

7. Interpretation of the Results:

To interpret the Results, we have three basic questions to answer:

Were the initial weights of the animals significantly different? That is, were the Xi's, especially from treatment to treatment, really different?

Were there significant differences among unadjusted treatment (diet) means?

Were there significant differences among adjusted treatment means?

Let us set and test the various hypotheses:

(i) Test for Initial Weights (Xi's):

H0: The initial weights of the animals did not differ i.e. the same.

H1: the initial weights differed.

Let us extract the portion of Table 3 relevant to this test:

SOURCES OF VARIATION	DF	SS	MS	FCAL	FTABLE	
			xx VALUES			
TOTAL	19	12.95				
BLOCK	4	4.29				
TREATMENT	3	1.43	0.48	0.8ns	3.49	
ERROR	12	7.23	0.60			

Comparing F_{cal} and F_{tab} i.e. $F_{cal} = 0.8 << F_{tab} = 3.49$, we see that the initial weights of the animals did not differ significantly at 5% probability. This means that any observed effects of the treatments (diets) could not be due to the differences in the initial weights.

Let us now test the Treatment Effects:

(ii)   Test for Unadjusted Treatment (Diet) Means:

H0: Unadjusted diet means did not differ.

H1: Unadjusted diet means differed.

Let us extract the portion of Table 3 relevant to this test: i.e. Data on final weights:

SOURCES OF VARIATION	DF	SS	MEAN	FCAL	FTABLE	
			VALUES			
TOTAL	19	38.41				
BLOCK	4	5.62				
TREATMENT	3	20.07	6.69	6.31**	3.49	
ERROR	12	12.72	1.06			

$F_{cal} = 6.31 >> F_{tab} = 3.49$. Since the Unadjusted Treatment Means differed significantly and the initial weights of the animals did not differ, one could rely on the observed means without any adjustments.

(iii)   Test for Adjusted Treatment Means:

Let us look at Table 3 again and extract the data from the Adjustment section:

H0: Adjusted diet means did not differ.

H1: Adjusted diet means differed.

SOURCES OF VARIATION	DF	SS	MS	FCAL	FTABLE	
TREATMENT ADJUSTED	3	15.74	5.25	10.29**	3.59	
ERROR	11	5.57	0.51			

Since the Adjusted Treatment Means differed significantly and the initial weights (X) of the animals did not differ, we conclude that any differences in the observed treatment means were really due to the various diets, not to the initial weights.

8. **OTHER POSSIBLE SITUATIONS**

Let us look at other situations that could arise:

1. **Significant unadjusted but no significant Adjusted Treatment Means:**

Should this situation arise, it simply means that any observed differences among the Treatment means are due to differences in the X's i.e. initial weights in this case. Of course, the Treatment means should not be used for decision making without adjustment.

2. **Significant Adjusted but not Unadjusted Treatment Means:**

Obviously, this indicates that the co-variate really had significant effects on the Treatment Means. That is, the X variable really masked the Treatment effects. Adjustment is necessary in this case.

9. **ADJUSTMENT OF TREATMENT MEANS**

In linear regression, the slope is the regression coefficient, defined as the change in the dependent variable (Y) per unit change in the fixed variable (X). It is the same slope, by.x, that we shall use in adjusting the Treatment Means, according to the differences in the covariate, X values.

by.x = Exy/Exx

From Table 3 above, Exy = 7.19; Exx = 7.23

$$by.x = \frac{E_{xy}}{E_{xx}} = \frac{7.19}{7.23} = 0.994.$$

The Equation for adjusted treatment mean is:

$$\bar{y}_i = \hat{Y}_i - b_{y.x}(\bar{X}_i - \bar{X}_{..})$$

where:

$\bar{y}_i$ = each adjusted treatment mean.

$\bar{X}_i$ = each mean of the covariate, X.

$\bar{X}_{..}$ = grand mean of the covariate, X.

Let us go back to Table 2 above and extract information from the Treatment Totals and use it to construct Table 4 as:

Table 4: ADJUSTMENT OF TREATMENT MEANS FROM TABLE 2

4.2.2	X VARIABLES				Y VARIABLES		
TREATMENT	TOTALS	$\bar{X}_i$	$\bar{X}_i - \bar{X}_{..}$ (d_i)	$b_{y.x}(\bar{X}_i - \bar{X}_{..})$ (S_i)	TOTALS	$\bar{Y}_i$	$\bar{Y}_i - S_i$
1	16.5	3.30	0.12	0.994(0.12) = 0.119	31.6	6.32	6.201
2	14.5	2.90	-0.28	0.994(-0.28) = -0.278	36.4	7.28	7.002
3	14.8	2.96	-0.22	0.994(-0.22) = -0.219	36.9	7.38	7.599
4	17.8	3.56	0.38	0.994(0.38) = 0.378	45.5	9.10	8.722

TOTALS	63.6	12.72	0	0	150.4	30.08	29.524
MEANS		3.18	0	0		7.52	7.381

Note:

(i) $\Sigma di = 0$

(ii) $\Sigma \bar{Y}_i = (\bar{Y}_i - S_i) = \Sigma \bar{Y}_i$

Test to compare adjusted means for Diet 1 and 4:

$$t = \frac{\bar{Y}_1 - \bar{Y}_4}{S_d}$$

where $S_d = \sqrt{S_{y.x}^2 \left[\frac{2}{r} + \frac{(\bar{X}_1 - \bar{X}_4)^2}{E_{xx}}\right]}$

$S_{y.x}^2 = 0.51$ from Table 3.

$r = 5$ (number of observations per mean).

$E_{xx} = 7.23$.

From Table 4:

$\bar{X}_1 = 3.30$

$\bar{X}_4 = 3.56$

$\bar{Y}_1 = 6.201$

$\bar{Y}_4 = 8.722$

Substituting into the formula for S_d, we have:

$$S_d = \sqrt{0.51 \left[\frac{2}{5} + \frac{(3.30 - 3.56)^2}{7.23}\right]} \qquad = \sqrt{0.51 \left[0.4 + \frac{(-0.26)^2}{7.23}\right]}$$

$$= \sqrt{0.51 \left[0.4 + \frac{(0.0676)^2}{7.23}\right]} \qquad = \sqrt{0.51 \left[0.4 + 0.00935\right]}$$

$$= \sqrt{0.51 \left[0.40935\right]} \qquad = \sqrt{0.20877} \quad = 0.4569$$

$$t_{cal} = \frac{\bar{Y}_1 - \bar{Y}_4}{0.4569} = \frac{6.201 - 8.722}{0.4569} = \frac{-2.521}{0.4569} = -5.5176$$

$t_{0.025} (11) = 2.201$

Inference: the adjusted treatment means for diet 1 and 4 did not differ significantly.

REFERENCES

Alika, J. E. (1997): Statistics and Research Methods. Ambik Press. Benin-City. 269p.

Byrkit, Donald R. (1975): Elements of Statistics. An Introduction to Probability and Statistical Inference, 2nd Edition. D. Van Nostrand Company, New York. Cincinnati. Toronto. London. 431p.

Cangelosi, V. E., Taylor, P.H. And Rice, P.F.(1976): Basic Statistics: A Real World Approach. West Publishing Co., St Paul. New York. Boston. 432p

Chambers, John M. (1977): Computational Methods for Data Analysis. A Wiley Publication in Applied Statistics. John Wiley and Sons. New York. Chichester. Brisbane. Toronto. 268p.

Chatterjee, Samprit and Price, Bertram (1977): Regression Analysis by Example. John Wiley and Sons. New York. London. Sidney. Toronto. 228p.

Cochran, W. G. And Cox, G. M. (1957): Experimental Designs. Second Edition. A Wiley Publication in Applied Statistics. John Wiley and Sons, Inc. London. Sidney.611p

Cochran, William G. (1963): Sampling Techniques. Second Edition. A Wiley Publication in Applied Statistics. John Wiley and Sons, Inc. New York. London. Sidney. 413p.

Cox, D. R. (1958): Planning Of Experiments. A Wiley Publication in Applied Statistics. John Wiley and Sons, Inc. New York. London. Sidney. 308p.

Daniel, Cuthbert and Wood, Fred S. (1971): Fitting Equations to Data. Computer Analysis of Multifactor Data for Scientists and Engineers. Wiley-Interscience. A Division of John Wiley and Sons, Inc., New York, Sidney. Toronto.

Dixon, Wilfrid J. and Massey Jr., Frank J. (1969): Introduction to Statistical Analysis. Third Edition. McGraw-Hill Book Company, New York. San Francisco. Toronto. London. Sidney. 638p.

Draper, N. R. and Smith H. (1966): Applied Regression Analysis. John Wiley and Sons, Inc. New York. London. Sidney. 407p.

Freund, John E. (1971): Mathematical Statistics. Second Edition. Prentice Hall, Inc., Englewood Cliffs, New Jersey. 463p.

Gomez, Kwanchai A. And Gomez, Arturo A. (1984): Statistical Procedures for Agricultural Research. Second Edition. An International Rice Research Institute Book. Wiley-Interscience Publication. John Wiley and Sons, New York. Chichester. Brisbane. Toronto. Singapore. 299p.

Hicks, Charles R. (1973): Fundamental Concepts in the Design of Experiments. Holt, Rinehart and Winston, New York. Chicago. London. Toronto. Sidney. 344p.

Koutsoyiannis A. (1977): Theory of Econometrics. An Introductory Exposition of Econometric Methods. Second Edition. ELBS with Macmillan. Educational Low-Priced Books Scheme Funded By the British Government. 681p.

Little, Thomas M. And Hills, Jackson F. (1972) Statistical Methods in Agricultural Extension University of California, Riverside and Davis. 90 University Hall, Berkeley, California 94720. 242pp.

Mendenhall, William (1975): Introduction to Probability and Statistics. Fourth Edition. Duxbury Press. A Division of Wadsworth Publishing Company, Inc. North Scituate, Massachusetts. 460p.

Mead, R., Curnow and Hasted, A. M. (1993): Statistical Methods in Agriculture and Experimental Biology. Second Edition. Chapman and Hall. London. Glasgow. New York. Tokyo. 415p.

Mendenhall William (1968): Introduction to Linear Models and the Design and Analysis Of Experiments. Duxbury Press, a Division of Wadsworth Publishing Company, Inc., Belmont, California. 465p.

Meyer, Stuart L. (1975): Data Analysis for Scientists and Engineers. John Wiley and Sons, Inc. New York. London. Sidney. Toronto. 513p.

Morrison, Donald F. (1976): Multivariate Statistical Methods. Second Edition. McGraw-Hill Book Company, New York. London. Sidney. Toronto. 415p.

Neter, John, Wasserman, William and Kutner, Michael H. (1983): Applied Linear Regression Models. Richard D. Irwin, Inc. Homewood, Illinois. 547p.

Neter, John, Wasserman, William and Kutner, Michael H. (1990): Applied Linear Models: Regression, Analysis Of Variance and Experimental Designs. Third Edition. Irwin, Burr Ridge, Illinois. Boston. Sidney. Massachusetts. 1181 P.

Ross, Sheldon M. (2005): Introductory Statistics. Second Edition. Elsevier Academic Press. Amsterdam. Boston. London. New York. Oxford. Paris. Sidney. Tokyo. 809p.

Schaeffer, R. L. And Mendenhall,W. (1975): Introduction To Probability: Theory And Applications. Duxbury Press. A Division of Wadsworth Publishing Company, Inc. North Scituate, Massachusetts. 289p.

Sheldon, M. Ross (1972): Introduction to Probability Models. Academic Press, New York, San Francisco, London. A Subsidiary of Harcourt Brace Jovanick Publishers. 272p.

Sokal, Robert R. And Rohlf, F. James (1969): Biometry. The Principles and Practice of Statistics in Biological Research. W.H. Freeman and Company. San Francisco. 757p.

Steel, Robert G.D. And Torrie, James H. (1981): Principles and Procedures of Statistics. A Biomedical Approach. Second Edition International Student Edition. McGraw-Hill International Book Company, Auckland, London. 620p.

Wahua, T.A.T. (1999): Applied Statistics for Scientific Studies. Afrika-Link Books (A Division of Afrika-Link Communications Limited). Owerri. Ibadan. Abuja. 356p.

Walpole, R.E. And Myers, R.H. 1972: Probability and Statistics for Engineers and Statisticians. Macmillan Publishing Company. Inc, USA. 506pp

Walpole, Ronald E. (1974): Introduction to Statistics. Macmillan Publishing Co., Inc New York; Collier Macmillan Publishers, London. 340p.

Winer B. J. (1971): Statistical Principles in Experimental Design. Second Edition, McGraw-Hill Book Company, New York. San Francisco. Sidney. Toronto. 907p.

Winkler Robert L. And Hays, William L. (1975): Statistics: Probability, Inference and Design. Second Edition. Holt, Rinehart and Winston. New York. Chicago. London. Sidney. 889p.

INDEX

ABOUT THE AUTHOR

DR. WILLIAM WILLOWS MODUGU was born on June 8, 1948, in Ikpeshi, Akoko-Edo Local Government Area, Edo State. He attended Saint Paul's Anglican Primary School, Ikpeshi (1955-1960). He attended Christ the King Anglican Modern School, Auchi (1961-1963). He later attended Our Lady of Fatima Grammar School, Auchi (1964-1967), where he obtained a Division One in the West African School Certificate (WASC) in 1967.

From there he proceeded to Immaculate Conception College, Benin City, for his Higher School Certificate (HSC) (1968-1969) and went back to Our Lady of Fatima to teach Biology and Mathematics, in January 1970 to September 1970.

In his quest for education, he was admitted to the Department of Forestry, University of Ibadan, Ibadan, in September 1970 and later graduated with a Second Class (Upper Division) in June 1973. After the mandatory one-year National Youth Service, he was employed at the Federal Department of Forestry, Ibadan, where he served for four years before proceeding to the University Of Washington, Seattle, U.S.A, for his Master's Degree (1978-1979) in Forest Biometrics, under Professor K. J. Turnbull.

Finally, Dr. W. W. Modugu, acquired his Ph.D. at the College of Environmental Science And Forestry (SUNY-ESF), State University of New York, Syracuse, U.S.A (1981-1985). Upon his return to Nigeria, he worked briefly as the Agricultural Specialist, American Embassy, Lagos (1986-1988). From the American Embassy, Lagos, he was employed as the Project Monitoring and Evaluation Expert of the IBRD Assisted Afforestation Programme Co-ordinating Unit (APCU), Kano, from 1991 to 1999. At the completion of the Project Implementation in 1999, Dr. Modugu was employed at the Department of Forestry and Wildlife, University of Benin, Benin City, to teach both Undergraduate and Post-Graduate Statistics and Forest Mensuration from December 1999, until his retirement in 2013, after fifteen years. Dr. Modugu still teaches Statistics and Forest Mensuration, as a Contract Staff, at the Department of Forestry and Wildlife, Faculty of Agriculture, University of Benin, Benin-City.

Dr. W. W. Modugu is popularly addressed by his colleagues as "WW.Com" as well as the "Youth Leader". Dr. Modugu is married to Mrs. Margaret Oyemhomhe Modugu and they are blessed with seven children.

APPENDIX
Table 1: Individual Terms of the Binomial Distribution

The $(x + 1)^{st}$ term in the expansion of the binomial $[\theta+(1-\theta)]^n$ is given by

$$f(x; n, \theta) = \binom{n}{x} \theta^x (1-\theta)^{n-x}, \quad x = 0, 1, 2, \ldots, n.$$

This is the probability of exactly x successes in n independent binomial trials with probability of success on a single trial equal to θ. This table contains the individual terms of $f(x;n,\theta)$ for specified choices of x, n, and θ.

For $\theta > 0.5$, the value of $\binom{n}{x} \theta^x (1-\theta)^{n-x}$ is found by using the table entry for $\binom{n}{x} \theta^x (1-\theta)^{n-x} \theta^x$

Source: William H. Beyer, ed. *Handbook of Tables for Probability and Statistics.* Cleveland: The Chemical Rubber Company, 1966. Reprinted by permission of the publisher.

n	x	.05	.10	.15	.20	.25	.30	.35	.40	.45	.50
1	0	.9500	.9000	.8500	.8000	.7500	.7000	.6500	.6000	.5500	.5000
	1	.0500	.1000	.1500	.2000	.2500	.3000	.3500	.4000	.4500	.5000
2	0	.9025	.8100	.7225	.6400	.5625	.4900	.4225	.3600	.3025	.2500
	1	.0950	.1800	.2550	.3200	.3750	.4200	.4550	.4800	.4950	.5000
	2	.0025	.0100	.0225	.0400	.0625	.0900	.1225	.1600	.2025	.2500
3	0	.8574	.7290	.6141	.5120	.4219	.3430	.2746	.2160	.1664	.1250
	1	.1354	.2430	.3251	.3840	.4219	.4410	.4436	.4320	.4084	.3750
	2	.0071	.0270	.0574	.0960	.1406	.1890	.2389	.2880	.3341	.3750
	3	.0001	.0010	.0034	.0080	.0156	.0270	.0429	.0640	.0911	.1250
4	0	.8145	.6561	.5220	.4096	.3164	.2401	.1785	.1296	.0915	.0625
	1	.1715	.2916	.3685	.4096	.4219	.4116	.3845	.3456	.2995	.2500
	2	.0135	.0486	.0975	.1536	.2109	.2646	.3105	.3456	.3675	.3750
	3	.0005	.0036	.0115	.0256	.0469	.0756	.1115	.1536	.2005	.2500
	4	.0000	.0001	.0005	.0016	.0039	.0081	.0150	.0256	.0410	.0625
5	0	.7738	.5905	.4437	.3277	.2373	.1681	.1160	.0778	.0503	.0312
	1	.2036	.3280	.3915	.4096	.3955	.3602	.3124	.2592	.2059	.1562
	2	.0214	.0729	.1382	.2048	.2637	.3087	.3364	.3456	.3369	.3125
	3	.0011	.0081	.0244	.0512	.0879	.1323	.1811	.2304	.2757	.3125
	4	.0000	.0004	.0022	.0064	.0146	.0284	.0488	.0768	.1128	.1562
	5	.0000	.0000	.0001	.0003	.0010	.0024	.0053	.0102	.0185	.0312
6	0	.7351	.5314	.3771	.2621	.1780	.1176	.0754	.0467	.0277	.0156
	1	.2321	.3543	.3993	.3932	.3560	.3025	.2437	.1866	.1359	.0938
	2	.0305	.0984	.1762	.2458	.2966	.3241	.3280	.3110	.2780	.2344
	3	.0021	.0146	.0415	.0819	.1318	.1852	.2355	.2765	.3032	.3125
	4	.0001	.0012	.0055	.0154	.0330	.0595	.0951	.1382	.1861	.2344
	5	.0000	.0001	.0004	.0015	.0044	.0102	.0205	.0369	.0609	.0938
	6	.0000	.0000	.0000	.0001	.0002	.0007	.0018	.0041	.0083	.0156
7	0	.6983	.4783	.3206	.2097	.1335	.0824	.0490	.0280	.0152	.0078
	1	.2573	.3720	.3960	.3670	.3115	.2471	.1848	.1306	.0872	.0547
	2	.0406	.1240	.2097	.2753	.3115	.3177	.2985	.2613	.2140	.1641
	3	.0036	.0230	.0617	.1147	.1730	.2269	.2679	.2903	.2918	.2734
	4	.0002	.0026	.0109	.0287	.0577	.0972	.1442	.1935	.2388	.2734
	5	.0000	.0002	.0012	.0043	.0115	.0250	.0466	.0774	.1172	.1641
	6	.0000	.0000	.0001	.0004	.0013	.0036	.0084	.0172	.0320	.0547
	7	.0000	.0000	.0000	.0000	.0001	.0002	.0006	.0016	.0037	.0078
8	0	.6634	.4305	.2725	.1678	.1001	.0576	.0319	.0168	.0084	.0039
	1	.2793	.3826	.3847	.3355	.2670	.1977	.1373	.0896	.0548	.0312
	2	.0515	.1488	.2376	.2936	.3115	.2965	.2587	.2090	.1569	.1094
	3	.0054	.0331	.0839	.1468	.2076	.2541	.2786	.2787	.2568	.2188
	4	.0004	.0046	.0185	.0459	.0865	.1361	.1875	.2322	.2627	.2734
	5	.0000	.0004	.0026	.0092	.0231	.0467	.0808	.1239	.1719	.2188
	6	.0000	.0000	.0002	.0011	.0038	.0100	.0217	.0413	.0703	.1094
	7	.0000	.0000	.0000	.0001	.0004	.0012	.0033	.0079	.0164	.0312
	8	.0000	.0000	.0000	.0000	.0000	.0001	.0002	.0007	.0017	.0039

Table 1:
Terms of the
Distribution.
interpolations
θ will in
accurate at
decimal places

n	x	.05	.10	.15	.20	.25	.30	.35	.40	.45	.50	
9	0	.6302	.3874	.2316	.1342	.0751	.0404	.0207	.0101	.0046	.0020	Individual Binomial *Linear with respect to general be most to two
	1	.2985	.3874	.3679	.3020	.2253	.1556	.1004	.0605	.0339	.0176	
	2	.0629	.1722	.2597	.3020	.3003	.2668	.2162	.1612	.1110	.0703	
	3	.0077	.0446	.1069	.1762	.2336	.2668	.2716	.2508	.2119	.1641	
	4	.0006	.0074	.0283	.0661	.1168	.1715	.2194	.2508	.2600	.2461	
	5	.0000	.0008	.0050	.0165	.0389	.0735	.1181	.1672	.2128	.2461	
	6	.0000	.0001	.0006	.0028	.0087	.0210	.0424	.0743	.1160	.1641	
	7	.0000	.0000	.0000	.0003	.0012	.0039	.0098	.0212	.0407	.0703	
	8	.0000	.0000	.0000	.0000	.0001	.0004	.0013	.0035	.0083	.0176	
	9	.0000	.0000	.0000	.0000	.0000	.0000	.0001	.0003	.0008	.0020	
10	0	.5987	.3487	.1969	.1074	.0563	.0282	.0135	.0060	.0025	.0010	Individual Binomial
	1	.3151	.3874	.3474	.2684	.1877	.1211	.0725	.0403	.0207	.0098	
	2	.0746	.1937	.2759	.3020	.2816	.2335	.1757	.1209	.0763	.0439	
	3	.0105	.0574	.1298	.2013	.2503	.2668	.2522	.2150	.1665	.1172	
	4	.0010	.0112	.0401	.0881	.1460	.2001	.2377	.2508	.2384	.2051	
	5	.0001	.0015	.0085	.0264	.0584	.1029	.1536	.2007	.2340	.2461	
	6	.0000	.0001	.0012	.0055	.0162	.0368	.0689	.1115	.1596	.2051	
	7	.0000	.0000	.0001	.0008	.0031	.0090	.0212	.0425	.0746	.1172	
	8	.0000	.0000	.0000	.0001	.0004	.0014	.0043	.0106	.0229	.0439	
	9	.0000	.0000	.0000	.0000	.0000	.0001	.0005	.0016	.0042	.0098	
	10	.0000	.0000	.0000	.0000	.0000	.0000	.0000	.0001	.0003	.0010	
11	0	.5688	.3138	.1673	.0859	.0422	.0198	.0088	.0036	.0014	.0004	
	1	.3293	.3835	.3248	.2362	.1549	.0932	.0518	.0266	.0125	.0055	
	2	.0867	.2131	.2866	.2953	.2581	.1998	.1395	.0887	.0513	.0269	
	3	.0137	.0710	.1517	.2215	.2581	.2568	.2254	.1774	.1259	.0806	
	4	.0014	.0158	.0536	.1107	.1721	.2201	.2428	.2365	.2060	.1611	
	5	.0001	.0025	.0132	.0388	.0803	.1321	.1830	.2207	.2360	.2256	
	6	.0000	.0003	.0023	.0097	.0268	.0566	.0985	.1471	.1931	.2256	
	7	.0000	.0000	.0003	.0017	.0064	.0173	.0379	.0701	.1128	.1611	
	8	.0000	.0000	.0000	.0002	.0011	.0037	.0102	.0234	.0462	.0806	
	9	.0000	.0000	.0000	.0000	.0001	.0005	.0018	.0052	.0126	.0269	
	10	.0000	.0000	.0000	.0000	.0000	.0000	.0002	.0007	.0021	.0054	
	11	.0000	.0000	.0000	.0000	.0000	.0000	.0000	.0000	.0002	.0005	
12	0	.5404	.2824	.1422	.0687	.0317	.0138	.0057	.0022	.0008	.0002	
	1	.3413	.3766	.3012	.2062	.1267	.0712	.0368	.0174	.0075	.0029	
	2	.0988	.2301	.2924	.2835	.2323	.1678	.1088	.0639	.0339	.0161	
	3	.0173	.0852	.1720	.2362	.2581	.2397	.1954	.1419	.0923	.0537	
	4	.0021	.0213	.0683	.1329	.1936	.2311	.2367	.2128	.1700	.1208	
	5	.0002	.0038	.0193	.0532	.1032	.1585	.2039	.2270	.2225	.1934	
	6	.0000	.0005	.0040	.0155	.0401	.0792	.1281	.1766	.2124	.2256	
	7	.0000	.0000	.0006	.0033	.0115	.0291	.0591	.1009	.1489	.1934	
	8	.0000	.0000	.0001	.0005	.0024	.0078	.0199	.0420	.0762	.1208	
	9	.0000	.0000	.0000	.0001	.0004	.0015	.0048	.0125	.0277	.0537	
	10	.0000	.0000	.0000	.0000	.0000	.0002	.0008	.0025	.0068	.0161	
	11	.0000	.0000	.0000	.0000	.0000	.0000	.0001	.0003	.0010	.0029	
	12	.0000	.0000	.0000	.0000	.0000	.0000	.0000	.0000	.0001	.0002	

Table 1:
Terms of the
Distribution:

Table 1: Individual Terms of the Binomial Distribution

n	x	.05	.10	.15	.20	θ.25	.30	.35	.40	.45	.50
13	0	.5133	.2542	.1209	.0550	.0238	.0097	.0037	.0013	.0004	.0001
	1	.3512	.3672	.2774	.1787	.1029	.0540	.0259	.0113	.0045	.0016
	2	.1109	.2448	.2937	.2680	.2059	.1388	.0836	.0453	.0220	.0095
	3	.0214	.0997	.1900	.2457	.2517	.2181	.1651	.1107	.0660	.0349
	4	.0028	.0277	.0838	.1535	.2097	.2337	.2222	.1845	.1350	.0873
	5	.0003	.0055	.0266	.0691	.1258	.1803	.2154	.2214	.1989	.1571
	6	.0000	.0008	.0063	.0230	.0559	.1030	.1546	.1968	.2169	.2095
	7	.0000	.0001	.0011	.0058	.0186	.0442	.0833	.1312	.1775	.2095
	8	.0000	.0000	.0001	.0011	.0047	.0142	.0336	.0656	.1089	.1571
	9	.0000	.0000	.0000	.0001	.0009	.0034	.0101	.0243	.0495	.0873
	10	.0000	.0000	.0000	.0000	.0001	.0006	.0022	.0065	.0162	.0349
	11	.0000	.0000	.0000	.0000	.0000	.0001	.0003	.0012	.0036	.0095
	12	.0000	.0000	.0000	.0000	.0000	.0000	.0000	.0001	.0005	.0016
	13	.0000	.0000	.0000	.0000	.0000	.0000	.0000	.0000	.0000	.0001
14	0	.4877	.2288	.1028	.0440	.0178	.0068	.0024	.0008	.0002	.0001
	1	.3593	.3559	.2539	.1539	.0832	.0407	.0181	.0073	.0027	.0009
	2	.1229	.2570	.2912	.2501	.1802	.1134	.0634	.0317	.0141	.0056
	3	.0259	.1142	.2056	.2501	.2402	.1943	.1366	.0845	.0462	.0222
	4	.0037	.0349	.0998	.1720	.2202	.2290	.2022	.1549	.1040	.0611
	5	.0004	.0078	.0352	.0860	.1468	.1963	.2178	.2066	.1701	.1222
	6	.0000	.0013	.0093	.0322	.0734	.1262	.1759	.2066	.2088	.1833
	7	.0000	.0002	.0019	.0092	.0280	.0618	.1082	.1574	.1952	.2095
	8	.0000	.0000	.0003	.0020	.0082	.0232	.0510	.0918	.1398	.1833
	9	.0000	.0000	.0000	.0003	.0018	.0066	.0183	.0408	.0762	.1222
	10	.0000	.0000	.0000	.0000	.0003	.0014	.0049	.0136	.0312	.0611
	11	.0000	.0000	.0000	.0000	.0000	.0002	.00.10	.0033	.0093	.0222
	12	.0000	.0000	.0000	.0000	.0000	.0000	.0001	.0005	.0019	.0056
	13	.0000	.0000	.0000	.0000	.0000	.0000	.0000	.0001	.0002	.0009
	14	.0000	.0000	.0000	.0000	.0000	.0000	.0000	.0000	.0000	.0001
15	0	.4633	.2059	.0874	.0352	.0134	.0047	.0016	.0005	.0001	.0000
	1	.3658	.3432	.2312	.1319	.0668	.0305	.0126	.0047	.0016	.0005

2	.1348	.2669	.2856	.2309	.1559	.0916	.0476	.0219	.0090	.0032
3	.0307	.1285	.2184	.2501	.2252	..1700	.1110	.0634	.0318	.0139
4	.0049	.0428	.1156	.1876	.2252	.2186	.1792	.1268	.0780	.0417
5	.0006	.0105	.0449	.1032	.1651	.2061	.2123	.1859	.1404	.0916
6	.0000	.0019	.0132	.0430	.0917	.1472	.1906	.2066	.1914	.1527
7	.0000	.0003	.0030	.0138	.0393	.0811	.1319	.1771	.2013	.1964
8	.0000	.0000	.0005	.0035	.0131	.0348	.0710	.1181	.1647	.1964
9	.0000	.0000	.0001	.0007	.0034	.0116	.0298	.0612	.1048	.1527
10	.0000	.0000	.0000	.0001	.0007	.0030	.0096	.0245	.0515	.0916
11	.0000	.0000	.0000	.0000	.0001	.0006	.0024	.0074	.0191	.0417
12	.0000	.0000	.0000	.0000	.0000	.0001	.0004	.0016	.0052	.0139
13	.0000	.0000	.0000	.0000	.0000	.0000	.0001	.0003	.0010	.0032
14	.0000	.0000	.0000	.0000	.0000	.0000	.0000	.0000	.0001	.0005
15	.0000	.0000	.0000	.0000	.0000	.0000	.0000	.0000	.0000	.0000

Table 1: Individual Terms of the Binomial Distribution

n	X	.05	.10	.15	.20	.25	.30	.35	.40	.45	.50
16	0	.4401	.1853	.0743	.0281	.0100	.0033	.0010	.0003	.0001	.0000
	1	.3706	.3294	.2097	.1126	.0535	.0228	.0087	.0030	.0009	.0002
	2	.1463	.2745	.2775	.2111	.1336	.0732	.0353	.0150	.0056	.0018
	3	.0359	.1423	.2285	.2463	.2079	.1465	.0888	.0468	.0215	.0085
	4	.0061	.0514	.1311	.2001	.2252	.2040	.1553	.1014	.0572	.0278
	5	.0008	.0137	.0555	.1201	.1802	.2099	.2008	.1623	.1123	.0667
	6	.0001	.0028	.0180	.0550	.1101	.1649	.1982	.1983	.1684	.1222
	7	.0000	.0004	.0045	.0197	.0524	.1010	.1524	.1889	.1969	.1746
	8	.0000	.0001	.0009	.0055	.0197	.0487	.0923	.1417	.1812	.1964
	9	.0000	.0000	.0001	.0012	.0058	.0185	.0442	.0840	.1318	.1746
	10	.0000	.0000	.0000	.0002	.0014	.0056	.0167	.0392	.0755	.1222
	11	.0000	.0000	.0000	.0000	.0002	.0013	.0049	.0142	.0337	.0667
	12	.0000	.0000	.0000	.0000	.0000	.0002	.0011	.0040	.0115	.0278
	13	.0000	.0000	.0000	.0000	.0000	.0000	.0002	.0008	.0029	.0085
	14	.0000	.0000	.0000	.0000	.0000	.0000	.0000	.0001	.0005	.0018
	15	.0000	.0000	.0000	.0000	.0000	.0000	.0000	.0000	.0001	.0002
	16	.0000	.0000	.0000	.0000	.0000	.0000	.0000	.0000	.0000	.0000
17	0	.4181	.1668	.0631	.0225	.0075	.0023	.0007	.0002	.0000	.0000
	1	.3741	.3150	.1893	.0957	.0426	.0169	.0060	.0019	.0005	.0001
	2	.1575	.2800	.2673	.1914	.1136	.0581	.0260	.0102	.0035	.0010
	3	.0415	.1556	.2359	.2393	.1893	.1245	.0701	.0341	.0144	.0052
	1	.9076	.0605	.1457	.2093	.2209	.1868	.1320	.0796	.0411	.0182
	5	.0010	.0175	.0668	.1361	.1914	.2081	.1849	.1379	.0875	.0472
	6	.0001	.0039	.0236	.0680	.1276	.1784	.1991	.1839	.1432	.0944
	7	.0000	.0007	.0065	.0267	.0668	.1201	.1685	.1927	.1841	.1484
	8	.0000	.0001	.0014	.0084	.0279	.0644	.1134	.1606	.1883	.1855
	9	.0000	.0000	.0003	.0021	.0093	.0276	.0611	.1070	.1540	.1855
	10	.0000	.0000	.0000	.0004	.0025	.0095	.0263	.0571	.1008	.1484
	11	.0000	.0000	.0000	.0001	.0005	.0026	.0090	.0242	.0525	.0944

	12	.0000	.0000	.0000	.0000	.0001	.0006	.0024	.0081	.0215	.0472
	13	.0000	.0000	.0000	.0000	.0000	.0001	.0005	.0021	.0068	.0182
	14	.0000	.0000	.0000	.0000	.0000	.0000	.0001	.0004	.0016	.0052
	15	.0000	.0000	.0000	.0000	.0000	.0000	.0000	.0001	.0003	.0010
	16	.0000	.0000	.0000	.0000	.0000	.0000	.0000	.0000	.0000	.0001
	17	.0000	.0000	.0000	.0000	.0000	.0000	.0000	.0000	.0000	.0000
18	0	.3972	.1501	.0536	.0180	.0056 '	.0016	.0004	.0001	.0000	.0000
	1	.3763	.3002	.1704	.0811.	.0338	.0126	.0042	.0012	.0003	.0001
	2	.1683	.2835	.2556	.1723	.0958	.0458	.0190	.0069	,.0022	.0006
	3	.0473	.1680	.2406	.2297	.1704	.1046	.0547	.0246	.0095	.0031
	4	.0093	.0700	.1592	.2153	.2130	.1681	.1104	.0614	.0291	.0117
	5	.0014	.0218	.0787	.1507	.1988	.2017	.1664	.1146	.0666	.0327
	6	.0002	.0052	.0301	.0816	.1436	.1873	.1941	.1655	.1181	.0708
	7	.0000	.0010	.0091	.0350	.0820	.1376	.1792	.1892	.1657	.1214
	8	.0000	.0002	.0022	.0120	.0376	.0811	.1327	.1734	.1864	.1669
	9	.0000	.0000	.0004	.0033	.0139	.0386	.0794	.1284	.1694	.1855
	10	.0000	.0000	.0001	.0008	.0042	.0149	.0385	.0771	.1248	.1669
	11	.0000	.0000	.0000	.0001	.0010	.0046	.0151	.0374	.0742	.1214

n	X	.05	.10	.15	.20	.25	.30	.35	.40	.45	.50
	12	.0000	.0000	.0000	.0000	.0002	.0012	.0047	.0145	.0354	.0708
	13	.0000	.0000	.0000	.0000	.0000	.0002	.0012	.0045	.0134	.0327
	14	.0000	.0000	.0000	.0000	.0000	.0000	.0002*	.0011	.0039	.0117
	15	.0000	.0000	.0000	.0000	.0000	.0000	.0000	.0002	.0009	.0031
	16	.0000	.0000	.0000	.0000	.0000	.0000	.0000	.0000	.0001	.0006
	17	.0000	.0000	.0000	.0000	.0000	.0000	.0000	.0000	.0000	.0001
	18	.0000	.0000	.0000	.0000	.0000	.0000	.0000	.0000	.0000	.0000
19	0	.3774	.1351	.0456	.0144	.0042	.0011	.0003	.0001	,.0000	.0000
	1	.3774	.2852,	.1529	.0685	.0268	.0093	.0029	.0008	.0002	.0000
	2	.1787	.2852	.2428	.1540	.0803	.0358	.0138	.0046	.0013	.0003
	3	.0533	.1796	.2428	.2182	.1517	.0869	.0422	.0175	.0062	.0018
	4	.0112	.0798	.1714	.2182	.2023	.1491	.0909	.0467	.0203	.0074

n	x										
	5	.0018	.0266	.0907	.1636	.2023	.1916	.1468	.0933	.0497	.0222
	6	.0002	.0069	.0374	.0955	.1574	.1916	.1844	.1451	.0949	.0518
	7	.0000	.0014	.0122	.0443	.0974	.1525	.1844	.1797	.1443	.0961
	8	.0000	.0002	.0032	.0166	.0487	.0981	.1489	.1797	.1771	.1442
	9	.0000	.0000	.0007	.0051	.0198	.0514	.0980	.1464	.1771	.1762
	10	.0000	.0000	.0001	.0013	.0066	.0220	.0528	.0976	.1449	.1762
	11 *	.0000	.0000	.0000	.0003	.0018	.0077	.0233	.0532	.0970	.1442
	12	.0000	.0000	.0000	.0000	.0004	.0022	.0083	.0237	.0529	.0961
	13	.0000	.0000	.0000	.0000	.0001	.0005	.0024	.0085	.0233	.0518
	14	.0000	.0000	.0000	.0000	.0000	.0001	.0006	.0024	.0082	.0222
	15	.0000	.0000	.0000	.0000	.0000	.0000	.0001	.0005	.0022	.0074
	16	.0000	.0000	.0000	.0000	.0000	.0000	.0000	.0001	.0005	.0018
	17	.0000	.0000	.0000	.0000	.0000	.0000	.0000	.0000	.0001	.0003
	18	.0000	.0000	.0000	.0000	.0000	.0000	.0000	.0000	.0000	.0000
	19	.0000	.0000	.0000	.0000	.0000	.0000	.0000	.0000	.0000	.0000
20	0	.3585	.1216	.0388	.0115	.0032	.0008	.0002	.0000	.0000	.0000
	1	.3774	.2702	.1368-	.0576	.0211	.0068	.0020	.0005	.0001	.0000
	2	.1887	.2852	.2293	.1369	.0669	.0278	.0100	.0031	.0008	.0002
	3	.0596	.1901	.2428	.2054	.1339	.0716	.0323	.0123	.0040	.0011
	4	.0133	.0898	..1821	.2182	.1897	.1304	.0738	.0350	.0139	.0046
	5	.0022	.0319	.1028	.1746	.2023	.1789	.1272	.0746	.0365	.0148
	6	.0003	.0089	.0454	.1091	.1686	.1916	.1712	.1244	.0746	.0370
	7	.0000	.0020	.0160	.0545	.1124	.1643	.1844	.1659	.1221	.0739
	8	.0000	.0004	.0046	.0222	.0609	.1144	.1614	.1797	.1623	.1201
	9	.0000	.0001	.0011	.0074	.0271	.0654	.1158	.1597	.1771	.1602
	10	.0000	.0000	.0002	.0020	.0099	.0308	.0686	.1171	.1593	.1762
	11	.0000	.0000	.0000	.0005	.0030	.0120	.0336	.0710	.1185	.1602
	12	.0000	.0000	.0000	.0001	.0008	.0039	.0136	.0355	.0727	.1201
	13	.0000	.0000	.0000	.0000	.0002	.0010	.0045	.0146	.0366	.0739
	14	.0000	.0000	.0000	.0000	.0000	.0002	.0012	.0049	.0150	.0370
	15	.0000	.0000	.0000	.0000	.0000	.0000	.0003	.0013	.0049	.0148

16	.0000	.0000	.0000	.0000	.0000	.0000	.0000	.0003	.0013	.0046
17	.0000	.0000	.0000	.0000	.0000	.0000	.0000	.0000	.0002	.0011
18	.0000	.0000	.0000	.0000	.0000	.0000	.0000	.0000	.0000	.0002
19	.0000	.0000	.0000	.0000	.0000	.0000	.0000	.0000	.0000	.0000
20	.0000	.0000	.0000	.0000	.0000	.0000	.0000	.0000	.0000	.0000

Table 2: Individual Terms of the Poisson Distribution

The Poisson probability function is given by

$$f(x:\lambda) = \frac{\lambda^x e^{-x}}{x!} \quad \lambda > 0, x = 0,1,2 \ldots$$

This table contains the individual terms of $f(x:\lambda)$ for specified values of x and λ.

Source: William H. Beyer, ed, *Handbook of Tables for Probability and Statistics.* Cleveland: The Chemical Rubber Company, 1966. Reprinted by permission of the publisher.

1/

Table 2: Individual Terms of the Poisson Distribution

_X_	0.1	0.2	0.3	0.4	0.5	0.6	0.7	0.8	0.9	1.0
0	.9048	.8187	.7408	.6703	.6065	.5488	.4966	.4493	.4066	.3679
1	.0905	.1637	.2222	.2681	.3033	.3293	.3476	.3595	.3659	.3679
2	.0045	.0164	.0333	.0536	.0758	.0988	.1217	.1438	.1647	.1839
3	.0002	.0011	.0033	.0072	.0126	.0198	.0284	.0383	.0494	.0613
4	.0000	.0001	.0003	.0007	.0016	.0030	.0050	.0077	.0111	.0153
5	.0000	.0000	.0000	.0001	.0002	.0004	.0007	.0012	.0020	.0031
6	.0000	.0000	.0000	.0000	.0000	.0000	.0001	.0002	.0003	.0005
7	.0000	.0000	.0000	.0000	.0000	.0000	.0000	.0000	.0000	.0001

X	1.1	1.2	1.3	1.4	1.5	1.6	1.7	1.8	1.9	2.0
0	.3329	.3012	.2725	.2466	.2231	.2019	.1827	.1653	.1496	.1353
1	.3662	.3614	.3543	.3452	.3347	.3230	.3106	.2975	.2842	.2707
2	.2014	.2169	.2303	.2417	.2510	.2584	.2640	.2678	.2700	.2707
3	.0738	.0867	.0998	.1128	.1255	.1378	.1496	.1607	.1710	.1804
4	.0203	.0260	.0324	.0395	.0471	.0551	.0636	.0723	.0812	.0902
5	.0045	.0062	.0084	.0111	.0141	.0176	.0216	.0260	.0309	.0361
6	.0008	.0012	.0018	.0026	.0035	.0047	.0061	.0078	.0098	.0120
7	.0001	.0002	.0003	.0005	.0008	.0011	.0015	.0020	.0027	.0034
8	.0000	.0000	.0001	.0001	.0001	.0002	.0003	.0005	.0006	.0009
9	.0000	.0000	.0000	.0000	.0000	.0000	.0001	.0001	.0001	.0002

X	2.1	2.2	2.3	2.4	2.5	2.6	2.7	2.8	2.9	3.0

0	.1225	.1108	.1003	.0907	.0821	.0743	.0672	.0608	.0550	.0498
1	.2572	.2438	.2306	.2177	.2052	.1931	.1815	.1703	.1596	.1494
2	.2700	.2681	.2652	.2613	.2565	.2510	.2450	.2384	.2314	.2240
3	.1890	.1966	.2033	.2090	.2138	.2176	.2205	.2225	.2237	.2240
4	.0992	.1082	.1169	.1254	.1336	.1414	.1488	.1557	.1622	.1680
5	.0417	.0476	.0538	.0602	.0668	.0735	.0804	.0872	.0940	.1008
6	.0146	.0174	.0206	.0241	.0278	.0319	.0362	.0407	.0455	.0504
7	.0044	.0055	.0068	.0083	.0099	.0118	.0139	.0163	.0188	.0216
8	.0011	.0015	.0019	.0025	.0031	.0038	.0047	.0057	.0068	.0081
9	.0003	.0004	.0005	.0007	.0009	.0011	.0014	.0018	.0022	.0027
10	.0001	.0001	.0001	.0002	.0002	.0003	.0004	.0005	.0006	.0008
11	.0000	.0000	.0000	.0000	.0000	.0001	.0001	.0001	.0002	.0002
12	.0000	.0000	.0000	.0000	.0000	.0000	.0000	.0000	.0000	.0001

X	3.1	3.2	3.3	3.4	λ 3.5	3.6	3.7	3.8	3.9	4.0
0	.0450	.0408	.0369	.0334	.0302	.0273	.0247	.0224	.0202	.0183
1	.1397	.1304	.1217	.1135	.1057	.0984	.0915	.0850	.0789	.0733
2	.2165	.2087	.2008	.1929	.1850	.1771	.1692	.1615	.1539	.1465
3	.2237	.2226	.2209	.2186	.2158	2125	.2087	.2046	.2001	.1954
4	.1734	.1781	.1823	.1858	.1888	.1912	.1931	.1944	.1951	.1954
5	.1075	.1140	.1203	.1264	.1322	.1377	.1429	.1477	.1522	.1563
6	.0555	.0608	.0662	.0716	.0771	.0826	.0881	.0936	.0989	.1042
7	.0246	.0278	.0312	.0348	.0385	.0425	.0466	.0508	.0551	.0595
8	.0095	.0111	.0129	.0148	.0169	.0191	.0215	.0241	.0269	.0298
9	.0033	.0040	.0047	.0056	.0066	.0076	.0089	.0102	.0116	.0132

Table 2: Individual Terms of the Poisson Distribution

	λ									
	3.1	3.2	3.3	3.4	3.5	3.6	3.7	3.8	3.9	4.0
10	.0010	.0013	.0016	.0019	.0023	.0028	.0033	.0039	.0045	.0053
11	.0003	.0004	.0005	.0006	.0007	.0009	.0011	.0013	.0016	.0019
12	.0001	.0001	.0001	.0002	.0002	.0003	.0003	.0004	.0005	.0006
13	.0000	.0000	.0000	.0000	.0001	.0001	.0001	.0001	.0002	.0002
14	.0000	.0000	.0000	.0000	.0000	.0000	.0000	.0000	.0000	.0001

	λ									
X	4.1	4.2	4.3	4.4	4.5	4.6	4.7	4.8	4.9	5.0
0	.0166	.0150	.0136	.0123	.0111	.0101	.0091	.0082	.0074	.0067
1	.0679	.0630	.0583	.0540	.0500	.0462	.0427	.0395	.0365	.0337
2	.1393	.1323	.1254	.1188	.1125	.1063	.1005	.0948	.0894	.0842
3	.1904	.1852	.1798	.1743	.1687	.1631	.1574	.1517	.1460	.1404
4	.1951	.1944	.1933	.1917	.1898	.1875	.1849	.1820	.1789	.1755
5	.1600	.1633	.1662	.1687	.1708	.1725	.1738	.1747	.1753	.1755
6	.1093	.1143	.1191	.1237	.1281	.1323	.1362	.1398	.1432	.1462
7	.0640	.0686	.0732	.0778	.0824	.0869	.0914	.0959	.1002	.1044
8	.0328	.0360	.0393	.0428	.0463	.0500	.0537	.0575	.0614	.0653
9	.0150	.0168	.0188	.0209	.0232	.0255	.0280	.0307	.0334	.0363
10	.0061	.0071	.0081	.0092	.0104	.0118	.0132	.0147	.0164	.0181
11	.0023	.0027	.0032	.0037	.0043	.0049	.0056	.0064	.0073	.0082
12	.0008	.0009	.0011	.0014	.0016	.0019	.0022	.0026	.0030	.0034
13	.0002	.0003	.0004	.0005	.0006	.0007	.0008	.0009	.0011	.0013
14	.0001	.0001	.0001	.0001	.0002	.0002	.0003	.0003	.0004	0005
15	.0000	.0000	0000	0000	.0001	.0001	.0001	.0001	.0001	.0002

	λ									
X	5.1	5.2	5.3	5.4	5.5	5.6	5.7	5.8	5.9	6.0
0	.0061	.0055	.0050	.0045	.0041	.0037	.0033	.0030	.0027	.0025
1	.0311	.0287	.0265	.0244	.0225	.0207	.0191	.0176	.0162	.0149
2	.0793	.0746	.0701	.0659	.0618	.0580	.0544	.0509	.0477	.0446
3	.1348	.1293	.1239	.1185	.1133	.1082	.1033	.0985	.0938	.0892
4	.1719	.1681	.1641	.1600	.1558	.1515	.1472	.1428	.1383	.1339
5	.1753	.1748	.1740	.1728	.1714	.1697	.1678	.1656	.1632	.1606
6	.1490	.1515	.1537	.1555	.1571	.1584	.1594	.1601	.1605	.1606
7	.1086	.1125	.1163	.1200	.1234	.1267	.1298	.1326	.1353	.1377
8	.0692	.0731	.0771	.0810	.0849	.0887	.0925	.0962	.0998	.1033
9	.0392	.0423	.0454	.0486	.0519	.0552	.0586	.0620	.0654	.0688
10	.0200	.0220	.0241	.0262	.0285	.0309	.0334	.0359	.0386	.0413
11	.0093	.0104	.0116	.0129	.0143	.0157	.0173	.0190	.0207	.0225
12	.0039	.0045	.0051	.0058	.0065	.0073	.0082	.0092	.0102	.0113
13	.0015	.0018	.0021	.0024	.0028	.0032	.0036	.0041	0046	.0052
14	.0006	.0007	.0008	.0009	.0011	.0013	.0015	.0017	.0019	.0022
15	.0002	.0002	.0003	.0003	.0004	.0005	.0006	.0007	.0008	.0009
16	.0001	.0001	.0001	.0001	.0001	.0002	.0002	.0002	.0003	.0003
17	.0000	.0000	.0000	.0000	.0000	.0000	.0001	.0001	.0001	.0001

Table 2: Individual Terms of the Poisson Distribution

x	6.1	6.2	6.3	6.4	6.5	6.6	6.7	6.8	6.9	7.0
0	.0022	.0020	.0018	.0017	.0015	.0014	.0012	.0011	.0010	.0009
1	.0137	.0126	.0116	.0106	.0098	.0090	.0082	.0076	.0070	.0064
2	.0417	.0390	.0364	.0340	.0318	.0296	.0276	.0258	.0240	.0223
3	.0848	.0806	.0765	.0726	.0688	.0652	.0617	.0584	.0552	.0521
4	.1294	.1249	.1205	.1162	.1118	.1076	.1034	.0992	.0952	.0912
6	.1679	.154ft	.1519	.1487	.1454	.1420	.1385	.1349	.1314	.1277
6	.1605	.1601	.1595	.1586	.1575	.1562	.1546	.1529	.1511	.1490
7	.1399	.1418	.1435	.1450	.1462	.1472	.1480	.1486	.1489	.1490
8	.1066	.1099	.1130	.1160	.1188	.1215	.1240	.1263	.1284	.1304
9	.0723	.0757	.0791	.0825	.0858	.0891	.0923	.0954	.0985	.1014
10	.0441	.0469	.0498	.0528	.0558	.0588	.0618	.0649	.0679	.0710
11	.0245	.0265	.0285	.0307	.0330	.0353	.0377	.0401	.0426	.0452
12	.0124	.0137	.0150	.0164	.0179	.0194	.0210	.0227	.0245	.0264
13	.0058	.0065	.0073	.0081	.0089	.0098	.0108	.0119	.0130	.0142
14	.0025	.0029	.0033	.0037	.0041	.0046	.0052	.0058	.0064	.0071
15	.0010	.0012	.0014	.0016	.0018	.0020	.0023	.0026	.0029	.0033
16	.0004	.0005	.0005	.0006	.0007	.0008	.0010	.0011	.0013	.0014
17	.0001	.0002	.0002	.0002	.0003	.0003	.0004	.0004	.0005	.0006
18	.0000	,0001	.0001	.0001	.0001	.0001	.0001	.0002	.0002	.0002
19	.0000	.0000	.0000	.0000	.0000	.0000	.0000	.0001	.0001	.0001

x	7.1	7.2	7.3	7.4	7.5	7.6	7.7	7.8	7.9	8.0
0	.0008	.0007	.0007	.0006	.0006	.0005	.0005	.0004	.0004	.0003
1	.0059	.0054	.0049	.0045	.0041	.0038	.0035	.0032	.0029	.0027
2	.0208	.0194	.0180	.0167	.0156	.0145	.0134	.0125	.0116	.0107
3	.0492	.0464	.0438	.0413	.0389	.0366	.0345	.0324	.0305	.0286
4	.0874	.0836	.0799	.0764	.0729	.0696	.0663	.0632	.0602	.0573
5	.1241	.1204	.1167	.1130	.1094	.1057	.1021	.0986	.0961	.0916
6	.1468	.1445	.1420	.1394	.1367	.1339	.1311	.1282	.1252	.1221
7	.1489	.1486	.1481	.1474	.1465	.1454	.1442	.1428	.1413	.1396

8	.1321	.1337	.1351	.1363	.1373	.1382	.1388	.1392	.1395	.1396
9	.1042	.1070	.1096	.1121	.1144	.1167	.1187	.1207	.1224	.1241
10	.0740	.0770	.0800	.0829	.0858	.0887	.0914	.0941	.0967	.0993
11	.0478	.0504	.0531	.0558	.0585	.0613	.0640	.0667	.0695	.0722
12	.0283	.0303	.0323	.0344	.0366	.0388	.0411	.0434	.0457	.0481
13	.0154	.0168	.0181	.0196	.0211	.0227	.0243	.0260	.0278	.0296
14	.0078	.0086	.0095	.0104	.0113	.0123	.0134	.0145	.0157	.0169
15	.0037	.0041	.0046	.0051	.0057	0062	.0069	.0075	.0083	.0090
16	.0016	.0019	.0021	.0024	.0026	.0030	.0033	.0037	.0041	.0045
17	.0007	.0008	.0009	.0010	.0012	.0013	.0015	.0017	.0019	.0021
18	.0003	.0003	.0004	.0004	.0005	.0006	.0006	.0007	.0008	.0009
19	.0001	.0001	.0001	.0002	.0002	.0002	.0003	.0003	.0003	.0004
20	.0000	.0000	.0001	.0001	.0001	.0001	.0001	.0001	.0001	.0002
21	.0000	.0000	.0000	.0000	.0000	.0000	.0000	.0000	.0001	.0001

Table 2: Individual Terms of the Poisson Distribution

8.1	8.2	8.3	8.4	8.5	8.6	8.7	7.8	8.9	9.0
0003	.0003	.0002	.0002	.0002	.0002	-.0002	.0002	.0001	.0001
0025	.0023	.0021	.0019	.0017	.0016	.0014	.0013	.0012	.0011
0100	.0092	.0086	.0079	.0074	.0068	.0063	.0058	.0054	.0050
0269	.0252	.0237	.0222	.0208	.0195	.0183	.0171	.0160	.0150
0544	.0517	.0491	.0466	.0443	.0420	.0398	.0377	.0357	.0337
.0882	.0849	.0816	.0784	.0752	.0722	.0692	.0663	.0635	.0607
1191	.1160	.1128	.1097	.1066	.1034	.1003	.0972	.0941	.0911
.1378	.1358	.1338	.1317	.1294	.1271	.1247	.1222	1197	.1171
.1395	.1392	.1388	.1382	.1375	.1366	.1356	.1344	.1332	.1318
.1256	.1269	.1280	.1290	.1299	.1306	.1311	.1315	.1317	.1318
.1017	.1040	.1063	.1084	.1104	.1123	.1140	.1157	.1172	.1186
.0749	.0776	.0802	.0828	.0853	.0878	.0902	.0925	.0948	.0970
.0505	.0530	.0555	.0579	.0604	.0629	.0654	.0679	.0703	.0728
.0315	.0334	.0354	.0374	.0395	.0416	.0438	.0459	.0481	.0504
.0182	.0196	.0210	.0225	.0240	.0256	.0272	.0289	.0306	.0324
.0098	.0107	.0116	.0126	.0136	.0147	.0158	.0169	.0182	.0194
.0050	.0055	.0060	.0066	.0072	.0079	.0086	.0093	.0101	.0109
.0024	.0026	.0029	.0033	.0036	.0040	.0044	.0048	.0053	.0058
.0011	.0012	.0014	.0015	.0017	.0019	.0021	.0024	.0026	.0029
.0005	.0005	.0006	.0007	.0008	.0009	.0010	.0011	.0012	.0014
.0002	.0002	.0002	.0003	.0003	.0004	.0004	.0005	.0005	.0006
.0001	.0001	.0001	.0001	.0001	.0002	.0002	.0002	.0002	.0003
.0000	.0000	.0000	.0000	.0001	.0001	.0001	.0001	.0001	.0001

9.1	9.2	9.3	9.4	λ	9.6	9.7	9.8	9.9	10
.0001	.0001	.0001	.0001	.0001	.0001	.0001	.0001	.0001	.0000
0010	0009	0009	0008	0007	0007	0006	0005	0005	0005
0046	0043	0040	0037	0034	0031	0029	0027	0025	0023
0140	0131	0123	0115	0107	0100	0093	0087	0081	0076
.0319	.0302	.0285	.0269	.0254	.0240	.0226	.0213	.0201	.0189
.0581	.0555	.0530	.0506	.0483	.0460	.0439	.0418	.0398	.0378
0881	0851	0822	0793	0764	0736	0709	0682	0656	0631
1145	1118	1091	1064	1037	1010	0982	0955	0928	0901
1302	1286	1269	1251	1232	1212	1191	1170	1148	1126
.1317	.1315	.1311	.1306	.1300	.1293	.1284	.1274	.1263	.1251
.1198	.1210	.1219	.1228	.1235	.1241	.1245	.1249	.1250	.1251
0991	1012	1031	1049	1067	1083	1098	1112	1125	1137
0752	0776	0799	0822	0844	0866	0888	0908	0928	0948
0526	0549	0572	0594	0617	0640	0662	0685	0707	0729
.0342	.0361	.0380	.0399	.0419	.0439	.0459	.0479	.0500	.0521
.0208	.0221	.0235	.0250	.0265	.0281	.0297	.0313	.0330	.0347
0118	0127	0137	0147	0157	0168	0180	0192	0204	0217
0063	0069	0075	0081	0088	0095	0103	0111	0119	0128
0032	0035	0039	0042	0046	0051	0055	0060	0065	0071
0015	0017	0019	0021	0023	0026	0028	0031	0034	0037

X	λ									
	9.1	9.2	9.3	9.4	9.5	9.6	9.7	9.8	9.9	10
20	.0007	.0008	.0009	.0010	.0011	.0012	.0014	.0015	.0017	.0019
21	.0003	.0003	.0004	0004	.0005	.0006	.0006	.0007	.0008	.0009
22	.0001	.0001	.0002	.0002	.0002	.0002	.0003	.0003	.0004	.0004
23	.0000	.0001	.0001	.0001	.0001	.0001	.0001	.0001	.0002	.0002
24	.0000	.0000	.0000	.0000	.0000	.0000	.0000	.0001	.0001	.0001

X	λ									
	11	12	13	14	15	16	17	18	19	20
0	.0000	.0000	.0000	.0000	.0000	.0000	.0000	.0000	.0000	.0000
1	.0002	.0001	.0000	.0000	.0000	.0000	.0000	.0000	.0000	.0000
2	.0010	.0004	.0002	.0001	.0000	.0000	.0000	.0000	.0000	.0000
3	.0037	.0018	.0008	.0004	.0002	.0001	.0000	.0000	.0000	.0000
4	.0102	.0053	.0027	.0013	.0006	.0003	.0001	.0001	.0000	.0000
5	.0224	.0127	.0070	.0037	.0019	.0010	.0005	.0002	.0001	.0001
6	.0411	.0255	.0152	.0087	.0048	.0026	.0014	.0007	.0004	.0002
7	.0646	.0437	.0281	.0174	.0104	.0060	.0034	.0018	.0010	.0005
8	.0888	.0655	.0457	.0304	.0194	.0120	.0072	.0042	.0024	.0013
9	.1085	.0874	.0661	.0473	.0324	.0213	.0135	.0083	.0050	.0029
10	.1194	.1048	.0859	.0663	.0486	.0341	.0230	.0150	.0095	.0058
11	.1194	.1144	.1015	.0844	.0663	.0496	.0355	.0245	.0164	.0106
12	.1094	.1144	.1099	.0984	.0829	.0661	.0504	.0368	.0259	.0176
13	.0926	.1056	.1099	.1060	.0956	.0814	.0658	.0509	.0378	.0271
14	.0728	.0905	.1021	.1060	.1024	.0930	.0800	.0655	.0514	.0387
15	.0534	.0724	.0885	.0989	.1024	.0992	.0906	.0786	.0650	.0516
16	.0367	.0543	.0719	.0866	.0960	.0992	.0963	.0884	.0772	.0646
17	.0237	.0383	.0550	.0713	.0847	.0934	.0963	.0936	.0863	.0760
18	.0145	.0256	.0397	.0554	.0706	.0830-	.0909	.0936	.0911	.0844
19	.0084	.0161	.0272	.0409	.0557	.0699	.0814	.0887	.0911	.0888
20	.0046	.0097	.0177	.0286	.0418	.0559	.0692	.0798	.0866	.0888
21	.0024	.0055	.0109	.0191	.0299	.0426	.0560	.0684	.0783	.0846

22	.0012	.0030	.0065	.0121	.0204	.0310	.0433	.0560	.0676	.0769
23	.0006	.0016	.0037	.0074	.0133	.0216	.0320	.0438	.0559	.0669
24	.0003	.0008	.0020	.0043	.0083	.0144	.0226	.0328	.0442	.0557
25	.0001	.0004	.0010	.0024	.0050	.0092	.0154	.0237	.0336	.0446
26	.0000	.0002	.0005	.0013	.0029	.0057	.0101	.0164	.0246	.0343
27	.0000	.0001	.0002	.0007	.0016	.0034	.0063	.0109	.0173	.0254
28	.0000	.0000	.0001	.0003	.0009	.0019	.0038	.0070	.0117	.0181
29	.0000	.0000	.0001	.0002	.0004	.0011	.0023	.0044	.0077	.0125
30	.0000	.0000	.0000	.0001	.0002	.0006	.0013	.0026	.0049	.0083
31	.0000	.0000	.0000	.0000	.0001	.0003	.0007	.0015	.0030	.0054
32	.0000	.0000	.0000	.0000	.0001	.0001	.0004	.0009	.0018	.0034
33	.0000	.0000	.0000	.0000	.0000	.0001	.0002	,0005	.0010	.0020
34	.0000	.0000	.0000	.0000	.0000	.0000	.0001	.0002	.0006	.0012
35	.0000	.0000	.0000	.0000	.0000	.0000	.0000	.0001	.0003	.0007
36	.0000	.0000	.0000	.0000	.0000	.0000	.0000	.0001	.0002	.0004
37	.0000	.0000	.0000	.0000	.0000	.0000	.0000	.0000	.0001	.0002
38	.0000	.0000	.0000	.0000	.0000	.0000	.0000	.0000	.0000	.0001
39	.0000	.0000	.0000	.0000	.0000	.0000	.0000	.0000	.0000	.0001

	00–04	05–09	10–14	15–19	20–24	25–29	30–34	35–39	40–44	45–49
50	64249	63664	39652	40646	97306	31741	07294	84149	46797	82487
51	26538	44249	04050	48174	65570	44072	40192	51153	11397	58212
52	05845	00512	78630	55328	18116	69296	91705	86224	29503	57071
53	74897	68373	67359	51014	33510	83048	17056	72506	82949	54600
54	20872	54570	35017	88132	25730	22626	86723	91691	13191	77212
55	31432	96156	89177	75541	81355	24480	77243	76690	42507	84362
56	66890	61505	01240	00660	05873	13568	76082	79172	57913	93448
57	41894	57790	79970	33106	86904	48119	52503	24130	72824	21627
58	11303	87118	81471	52936	08555	28420	49416	44448	04269	27029
59	54374	57325	16947	45356	78371	10563	97191	53798	12693	27928
60	64852	34421	61046	90849	13966	39810	42699	21753	76192	10508
61	16309	20384	09491	91588	97720	89846	30376	76970	23063	35894
62	42587	37065	24526	72602	57589	98131	37292	05967	26002	51945
63	40177	98590	97161	41682	84533	67588	62036	49967	01990	72308
64	82309	76128	93965	26743	24141	04838	40254	26065	07938	76236
65	79788	68243	59732	04257	27084	14743	17520	95401	55811	76099
66	40538	79000	89559	25026	42274	23489	34502	75508	06059	86682
67	64016	73598	18609	73150	62463	33102	45205	87440	96767	67042
68	49767	12691	17903	93871	99721	79109	09425	26904	07419	76013
69	76974	55108	29795	08404	82684	00497	51126	79935	57450	55671
70	23854	08480	85983	96025	50117	64610	99425	62291	86943	21541
71	68973	70551	25098	78033	98573	79848	31778	29555	61446	23037
72	36444	93600	65350	14971	25325	00427	52073	64280	18847	24768
73	03003	87800	07391	11594	21196	00781	32550	57158	58887	73041
74	17540	26188	36647	78386	04558	61463	57842	90382	77019	24210
75	38916	55809	47982	41968	69760	79422	80154	91486	19180	15100
76	64288	19843	69122	42502	48508	28820	59933	72998	99942	10515
77	86809	51564	38040	39418	49915	19000	58050	16899	79952	57849
78	99800	99566	14742	05028	30033	94889	53381	23656	75787	59223
79	92345	31890	95712	08279	91794	94068	49337	88674	35355	12267
80	90363	65162	32245	82279	79256	80834	06088	99462	56705	06118
81	64437	32242	48431	04835	39070	59702	31508	60935	22390	52246
82	91714	53662	28373	34333	55791	74758	51144	18827	10704	76803
83	20902	17646	31391	31459	33315	03444	55743	74701	58851	27427
84	12217	86007	70371	52281	14510	76094	96579	54853	78339	20839
85	45177	02863	42307	53571	22532	74921	17735	42201	80540	54721
86	28325	90814	08804	52746	47913	54577	47525	77705	95330	21866
87	29019	28776	56116	54791	64604	08815	46049	71186	34650	14994
88	84979	81353	56219	67062	26146	82567	33122	14124	46240	92973
89	50371	26347	48513	63915	11158	25563	91915	18431	92978	11591
90	53422	06825	69711	67950	64716	18003	49581	45378	99878	61130
91	67453	35651	89316	41620	32048	70225	47597	33137	31443	51445
92	07294	85353	74819	23445	68237	07202	99515	62282	53809	26685
93	79544	00302	45338	16015	66613	88968	14595	63836	77716	79596
94	64144	85442	82060	46471	24162	39500	87351	36637	42833	71875
95	90919	11883	58318	00042	52402	28210	34075	33272	00840	73268
96	06670	57353	86275	92276	77591	46924	60839	55437	03183	13191
97	36634	93976	52062	83678	41256	60948	18685	48992	19462	96062
98	75101	72891	85745	67106	26010	62107	60885	37503	55461	71213
99	05112	71222	72654	51583	05228	62056	57390	42746	39272	96659

z	0.00	0.01	0.02	0.03	0.04	0.05	0.06	0.07	0.08	0.09
0.0	0.000	0.010	0.020	0.030	0.040	0.050	0.060	0.070	0.080	0.090
.1	.100	.110	.119	.129	.139	.149	.159	.168	.178	.187
.2	.197	.207	.216	.226	.236	.245	.254	.264	.273	.282
.3	.291	.300	.310	.319	.327	.336	.345	.354	.363	.371
.4	.380	.389	.397	.405	.414	.422	.430	.438	.446	.454
.5	.462	.470	.478	.485	.493	.500	.508	.515	.523	.530
.6	.537	.544	.551	.558	.565	.572	.578	.585	.592	.598
.7	.604	.611	.617	.623	.629	.635	.641	.647	.653	.658
.8	.664	.670	.675	.680	.686	.691	.696	.701	.706	.711
.9	.716	.721	.726	.731	.735	.740	.744	.749	.753	.757
1.0	.762	.766	.770	.774	.778	.782	.786	.790	.793	.797
1.1	.800	.804	.808	.811	.814	.818	.821	.824	.828	.831
1.2	.834	.837	.840	.843	.846	.848	.851	.854	.856	.859
1.3	.862	.864	.867	.869	.872	.874	.876	.879	.881	.883
1.4	.885	.888	.890	.892	.894	.896	.898	.900	.902	.903
1.5	.905	.907	.909	.910	.912	.914	.915	.917	.919	.920
1.6	.922	.923	.925	.926	.928	.929	.930	.932	.933	.934
1.7	.935	.937	.938	.939	.940	.941	.942	.944	.945	.946
1.8	.947	.948	.949	.950	.951	.952	.953	.954	.954	.955
1.9	.956	.957	.958	.959	.960	.960	.961	.962	.963	.963
2.0	.964	.965	.965	.966	.967	.967	.968	.969	.969	.970
2.1	.970	.971	.972	.972	.973	.973	.974	.974	.975	.975
2.2	.976	.976	.977	.977	.978	.978	.978	.979	.979	.980
2.3	.980	.980	.981	.981	.982	.982	.982	.983	.983	.983
2.4	.984	.984	.984	.985	.985	.985	.986	.986	.986	.986
2.5	.987	.987	.987	.987	.988	.988	.988	.988	.989	.989
2.6	.989	.989	.989	.990	.990	.990	.990	.990	.991	.991
2.7	.991	.991	.991	.992	.992	.992	.992	.992	.992	.992
2.8	.993	.993	.993	.993	.993	.993	.993	.994	.994	.994
2.9	.994	.994	.994	.994	.994	.995	.995	.995	.995	.995

TABLE OF $z = \frac{1}{2} \log_e (1 + r)/(1 - r)$ TO TRANSFORM THE CORRELATION COEFFICIENT

r	0.00	0.01	0.02	0.03	0.04	0.05	0.06	0.07	0.08	0.09
.0	0.000	0.010	0.020	.0.030	0.040	0.050	0.060	0.070	0.080	0.090
.1	.100	.110	.121	.131	.141	.151	.161	.172	.182	.192
.2	.203	.213	.224	.234	.245	.255	.266	.277	.288	.299
.3	.310	.321	.332	.343	.354	.365	.377	.388	.400	.412
.4	.424	.436	.448	.460	.472	.485	.497	.510	.523	.536
.5	.549	.563	.576	.590	.604	.618	.633	.648	.662	.678
.6	.693	.709	.725	.741	.758	.775	.793	.811	.829	.848
.7	.867	.887	.908	.929	.950	.973	.996	1.020	1.045	1.071
.8	1.099	1.127	1.157	1.188	1.221	1.256	1.293	1.333	1.376	1.422

r	0.000	0.001	0.002	0.003	0.004	0.005	0.006	0.007	0.008	0.009
.90	1.472	1.478	1.483	1.488	1.494	1.499	1.505	1.510	1.516	1.522
.91	1.528	1.533	1.539	1.545	1.551	1.557	1.564	1.570	1.576	1.583
.92	1.589	1.596	1.602	1.609	1.616	1.623	1.630	1.637	1.644	1.651
.93	1.658	1.666	1.673	1.681	1.689	1.697	1.705	1.713	1.721	1.730
.94	1.738	1.747	1.756	1.764	1.774	1.783	1.792	1.802	1.812	1.822
.95	1.832	1.842	1.853	1.863	1.874	1.886	1.897	1.909	1.921	1.933
.96	1.946	1.959	1.972	1.986	2.000	2.014	2.029	2.044	2.060	2.076
.97	2.092	2.109	2.127	2.146	2.165	2.185	2.205	2.227	2.249	2.273
.98	2.298	2.323	2.351	2.380	2.410	2.443	2.477	2.515	2.555	2.599
.99	2.646	2.700	2.759	2.826	2.903	2.994	3.106	3.250	3.453	3.800

CORRELATION COEFFICIENTS AT THE 5% AND 1% LEVELS OF SIGNIFICANCE

Degrees of Freedom	5%	1%	Degrees of Freedom	5%	1%
1	.997	1.000	24	.388	.496
2	.950	.990	25	.381	.487
3	.878	.959	26	.374	.478
4	.811	.917	27	.367	.470
5	.754	.874	28	.361	.463
6	.707	.834	29	.355	.456
7	.666	.798	30	.349	.449
8	.632	.765	35	.325	.418
9	.602	.735	40	.304	.393
10	.576	.708	45	.288	.372
11	.553	.684	50	.273	.354
12	.532	.661	60	.250	.325
13	.514	.641	70	.232	.302
14	.497	.623	80	.217	.283
15	.482	.606	90	.205	.267
16	.468	.590	100	.195	.254
17	.456	.575	125	.174	.228
18	.444	.561	150	.159	.208
19	.433	.549	200	.138	.181
20	.423	.537	300	.113	.148
21	.413	.526	400	.098	.128
22	.404	.515	500	.088	.115
23	.396	.505	1,000	.062	.081

df	level	2	3	4	5	6	7	8	9	10	11	12	13	14	15
13	.95	3.06	3.73	4.15	4.45	4.69	4.88	5.05	5.19	5.32	5.43	5.53	5.63	5.71	5.79
	.99	4.26	4.96	5.40	5.73	5.98	6.19	6.37	6.53	6.67	6.79	6.90	7.01	7.10	7.19
14	.95	3.03	3.70	4.11	4.41	4.64	4.83	4.99	5.13	5.25	5.36	5.46	5.55	5.64	5.72
	.99	4.21	4.89	5.32	5.63	5.88	6.08	6.26	6.41	6.54	6.66	6.77	6.87	6.96	7.05
16	.95	3.00	3.65	4.05	4.33	4.56	4.74	4.90	5.03	5.15	5.26	5.35	5.44	5.52	5.59
	.99	4.13	4.78	5.19	5.49	5.72	5.92	6.08	6.22	6.35	6.46	6.56	6.66	6.74	6.82
18	.95	2.97	3.61	4.00	4.28	4.49	4.67	4.82	4.96	5.07	5.17	5.27	5.35	5.43	5.50
	.99	4.07	4.70	5.09	5.38	5.60	5.79	5.94	6.08	6.20	6.31	6.41	6.50	6.58	6.65
20	.95	2.95	3.58	3.96	4.23	4.45	4.62	4.77	4.90	5.01	5.11	5.20	5.28	5.36	5.43
	.99	4.02	4.64	5.02	5.29	5.51	5.69	5.84	5.97	6.09	6.19	6.29	6.37	6.45	6.52
24	.95	2.92	3.53	3.90	4.17	4.37	4.54	4.68	4.81	4.92	5.01	5.10	5.18	5.25	5.32
	.99	3.96	4.54	4.91	5.17	5.37	5.54	5.69	5.81	5.92	6.02	6.11	6.19	6.26	6.33
30	.95	2.89	3.49	3.84	4.10	4.30	4.46	4.60	4.72	4.83	4.92	5.00	5.08	5.15	5.21
	.99	3.89	4.45	4.80	5.05	5.24	5.40	5.54	5.56	5.76	5.85	5.93	6.01	6.08	6.14
40	.95	2.86	3.44	3.79	4.04	4.23	4.39	4.52	4.63	4.74	4.82	4.91	4.98	5.05	5.11
	.99	3.82	4.37	4.70	4.93	5.11	5.27	5.39	5.50	5.60	5.69	5.77	5.84	5.90	5.96
60	.95	2.83	3.40	3.74	3.98	4.16	4.31	4.44	4.55	4.65	4.73	4.81	4.88	4.94	5.00
	.99	3.76	4.28	4.60	4.82	4.99	5.13	5.25	5.36	5.45	5.53	5.60	5.67	5.73	5.79
120	.95	2.80	3.36	3.69	3.92	4.10	4.24	4.36	4.48	4.56	4.64	4.72	4.78	4.84	4.90
	.99	3.70	4.20	4.50	4.71	4.87	5.01	5.12	5.21	5.30	5.38	5.44	5.51	5.56	5.61
∞	.95	2.77	3.31	3.63	3.86	4.03	4.17	4.29	4.39	4.47	4.55	4.62	4.68	4.74	4.80
	.99	3.64	4.12	4.40	4.60	4.76	4.88	4.99	5.08	5.16	5.23	5.29	5.35	5.40	5.45

r = number of steps between ordered means

df for $s_{\bar{x}}$	$1-\alpha$	2	3	4	5	6	7	8	9	10	11	12	13	14	15
1	.95	18.0	27.0	32.8	37.1	40.4	43.1	45.4	47.4	49.1	50.6	52.0	53.2	54.3	55.4
	.99	90.0	135	164	186	202	216	227	237	246	253	260	266	272	277
2	.95	6.09	8.3	9.8	10.9	11.7	12.4	13.0	13.5	14.0	14.4	14.7	15.1	15.4	15.7
	.99	14.0	19.0	22.3	24.7	26.6	28.2	29.5	30.7	31.7	32.6	33.4	34.1	34.8	35.4
3	.95	4.50	5.91	6.82	7.50	8.04	8.48	8.85	9.18	9.46	9.72	9.95	10.2	10.4	10.5
	.99	8.26	10.6	12.2	13.3	14.2	15.0	15.6	16.2	16.7	17.1	17.5	17.9	18.2	18.5
4	.95	3.93	5.04	5.76	6.29	6.71	7.05	7.35	7.60	7.83	8.03	8.21	8.37	8.52	8.66
	.99	6.51	8.12	9.17	9.96	10.6	11.1	11.5	11.9	12.3	12.6	12.8	13.1	13.3	13.5
5	.95	3.64	4.60	5.22	5.67	6.03	6.33	6.58	6.80	6.99	7.17	7.32	7.47	7.60	7.72
	.99	5.70	6.97	7.80	8.42	8.91	9.32	9.67	9.97	10.2	10.5	10.7	10.9	11.1	11.2
6	.95	3.46	4.34	4.90	5.31	5.63	5.89	6.12	6.32	6.49	6.65	6.79	6.92	7.03	7.14
	.99	5.24	6.33	7.03	7.56	7.97	8.32	8.61	8.87	9.10	9.30	9.49	9.65	9.81	9.95
7	.95	3.34	4.16	4.69	5.06	5.36	5.61	5.82	6.00	6.16	6.30	6.43	6.55	6.66	6.76
	.99	4.95	5.92	6.54	7.01	7.37	7.68	7.94	8.17	8.37	8.55	8.71	8.86	9.00	9.12
8	.95	3.26	4.04	4.53	4.89	5.17	5.40	5.60	5.77	5.92	6.05	6.18	6.29	6.39	6.48
	.99	4.74	5.63	6.20	6.63	6.96	7.24	7.47	7.68	7.87	8.03	8.18	8.31	8.44	8.55
9	.95	3.20	3.95	4.42	4.76	5.02	5.24	5.43	5.60	5.74	5.87	5.98	6.09	6.19	6.28
	.99	4.60	5.43	5.96	6.35	6.66	6.91	7.13	7.32	7.49	7.65	7.78	7.91	8.03	8.13
10	.95	3.15	3.88	4.33	4.65	4.91	5.12	5.30	5.46	5.60	5.72	5.83	5.93	6.03	6.11
	.99	4.48	5.27	5.77	6.14	6.43	6.67	6.87	7.05	7.21	7.36	7.48	7.60	7.71	7.81
11	.95	3.11	3.82	4.26	4.57	4.82	5.03	5.20	5.35	5.49	5.61	5.71	5.81	5.90	5.99
	.99	4.39	5.14	5.62	5.97	6.25	6.48	6.67	6.84	6.99	7.13	7.26	7.36	7.46	7.56
12	.95	3.08	3.77	4.20	4.51	4.75	4.95	5.12	5.27	5.40	5.51	5.62	5.71	5.80	5.88
	.99	4.32	5.04	5.50	5.84	6.10	6.32	6.51	6.67	6.81	6.94	7.06	7.17	7.26	7.36

Critical Values of the *F* Distribution

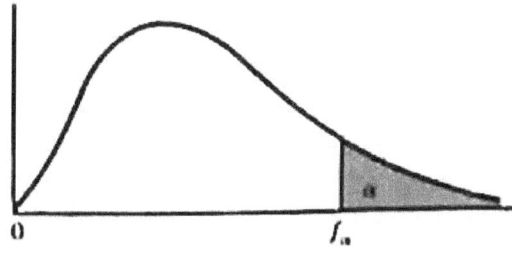

$$f_{0.05}(v_1, v_2)$$

v_2	v_1								
	1	2	3	4	5	6	7	8	9
1	161.4	199.5	215.7	224.6	230.2	234.0	236.8	238.9	240.5
2	18.51	19.00	19.16	19.25	19.30	19.33	19.35	19.37	19.38
3	10.13	9.55	9.28	9.12	9.01	8.94	8.89	8.85	8.81
4	7.71	6.94	6.59	6.39	6.26	6.16	6.09	6.04	6.00
5	6.61	5.79	5.41	5.19	5.05	4.95	4.88	4.82	4.77
6	5.99	5.14	4.76	4.53	4.39	4.28	4.21	4.15	4.10
7	5.59	4.74	4.35	4.12	3.97	3.87	3.79	3.73	3.68
8	5.32	4.46	4.07	3.84	3.69	3.58	3.50	3.44	3.39
9	5.12	4.26	3.86	3.63	3.48	3.37	3.29	3.23	3.18
10	4.96	4.10	3.71	3.48	3.33	3.22	3.14	3.07	3.02
11	4.84	3.98	3.59	3.36	3.20	3.09	3.01	2.95	2.90
12	4.75	3.89	3.49	3.26	3.11	3.00	2.91	2.85	2.80
13	4.67	3.81	3.41	3.18	3.03	2.92	2.83	2.77	2.71
14	4.60	3.74	3.34	3.11	2.96	2.85	2.76	2.70	2.65
15	4.54	3.68	3.29	3.06	2.90	2.79	2.71	2.64	2.59
16	4.49	3.63	3.24	3.01	2.85	2.74	2.66	2.59	2.54
17	4.45	3.59	3.20	2.96	2.81	2.70	2.61	2.55	2.49
18	4.41	3.55	3.16	2.93	2.77	2.66	2.58	2.51	2.46
19	4.38	3.52	3.13	2.90	2.74	2.63	2.54	2.48	2.42
20	4.35	3.49	3.10	2.87	2.71	2.60	2.51	2.45	2.39
21	4.32	3.47	3.07	2.84	2.68	2.57	2.49	2.42	2.37
22	4.30	3.44	3.05	2.82	2.66	2.55	2.46	2.40	2.34
23	4.28	3.42	3.03	2.80	2.64	2.53	2.44	2.37	2.32
24	4.26	3.40	3.01	2.78	2.62	2.51	2.42	2.36	2.30
25	4.24	3.39	2.99	2.76	2.60	2.49	2.40	2.34	2.28
26	4.23	3.37	2.98	2.74	2.59	2.47	2.39	2.32	2.27
27	4.21	3.35	2.96	2.73	2.57	2.46	2.37	2.31	2.25
28	4.20	3.34	2.95	2.71	2.56	2.45	2.36	2.29	2.24
29	4.18	3.33	2.93	2.70	2.55	2.43	2.35	2.28	2.22
30	4.17	3.32	2.92	2.69	2.53	2.42	2.33	2.27	2.21
40	4.08	3.23	2.84	2.61	2.45	2.34	2.25	2.18	2.12
60	4.00	3.15	2.76	2.53	2.37	2.25	2.17	2.10	2.04
120	3.92	3.07	2.68	2.45	2.29	2.17	2.09	2.02	1.96
∞	3.84	3.00	2.60	2.37	2.21	2.10	2.01	1.94	1.88

Critical Values of the F Distribution (*continued*)

$$f_{0.05}(v_1, v_2)$$

v_2	\multicolumn{10}{c}{v_1}
	10	12	15	20	24	30	40	60	120	∞
1	241.9	243.9	245.9	248.0	249.1	250.1	251.1	252.2	253.3	254.3
2	19.40	19.41	19.43	19.45	19.45	19.46	19.47	19.48	19.49	19.50
3	8.79	8.74	8.70	8.66	8.64	8.62	8.59	8.57	8.55	8.53
4	5.96	5.91	5.86	5.80	5.77	5.75	5.72	5.69	5.66	5.63
5	4.74	4.68	4.62	4.56	4.53	4.50	4.46	4.43	4.40	4.36
6	4.06	4.00	3.94	3.87	3.84	3.81	3.77	3.74	3.70	3.67
7	3.64	3.57	3.51	3.44	3.41	3.38	3.34	3.30	3.27	3.23
8	3.35	3.28	3.22	3.15	3.12	3.08	3.04	3.01	2.97	2.93
9	3.14	3.07	3.01	2.94	2.90	2.86	2.83	2.79	2.75	2.71
10	2.98	2.91	2.85	2.77	2.74	2.70	2.66	2.62	2.58	2.54
11	2.85	2.79	2.72	2.65	2.61	2.57	2.53	2.49	2.45	2.40
12	2.75	2.69	2.62	2.54	2.51	2.47	2.43	2.38	2.34	2.30
13	2.67	2.60	2.53	2.46	2.42	2.38	2.34	2.30	2.25	2.21
14	2.60	2.53	2.46	2.39	2.35	2.31	2.27	2.22	2.18	2.13
15	2.54	2.48	2.40	2.33	2.29	2.25	2.20	2.16	2.11	2.07
16	2.49	2.42	2.35	2.28	2.24	2.19	2.15	2.11	2.06	2.01
17	2.45	2.38	2.31	2.23	2.19	2.15	2.10	2.06	2.01	1.96
18	2.41	2.34	2.27	2.19	2.15	2.11	2.06	2.02	1.97	1.92
19	2.38	2.31	2.23	2.16	2.11	2.07	2.03	1.98	1.93	1.88
20	2.35	2.28	2.20	2.12	2.08	2.04	1.99	1.95	1.90	1.84
21	2.32	2.25	2.18	2.10	2.05	2.01	1.96	1.92	1.87	1.81
22	2.30	2.23	2.15	2.07	2.03	1.98	1.94	1.89	1.84	1.78
23	2.27	2.20	2.13	2.05	2.01	1.96	1.91	1.86	1.81	1.76
24	2.25	2.18	2.11	2.03	1.98	1.94	1.89	1.84	1.79	1.73
25	2.24	2.16	2.09	2.01	1.96	1.92	1.87	1.82	1.77	1.71
26	2.22	2.15	2.07	1.99	1.95	1.90	1.85	1.80	1.75	1.69
27	2.20	2.13	2.06	1.97	1.93	1.88	1.84	1.79	1.73	1.67
28	2.19	2.12	2.04	1.96	1.91	1.87	1.82	1.77	1.71	1.65
29	2.18	2.10	2.03	1.94	1.90	1.85	1.81	1.75	1.70	1.64
30	2.16	2.09	2.01	1.93	1.89	1.84	1.79	1.74	1.68	1.62
40	2.08	2.00	1.92	1.84	1.79	1.74	1.69	1.64	1.58	1.51
60	1.99	1.92	1.84	1.75	1.70	1.65	1.59	1.53	1.47	1.39
120	1.91	1.83	1.75	1.66	1.61	1.55	1.50	1.43	1.35	1.25
∞	1.83	1.75	1.67	1.57	1.52	1.46	1.39	1.32	1.22	1.00

Critical Values of the F Distribution (continued)

$$f_{0.01}(v_1, v_2)$$

v_2	v_1								
	1	2	3	4	5	6	7	8	9
1	4052	4999.5	5403	5625	5764	5859	5928	5981	6022
2	98.50	99.00	99.17	99.25	99.30	99.33	99.36	99.37	99.39
3	34.12	30.82	29.46	28.71	28.24	27.91	27.67	27.49	27.35
4	21.20	18.00	16.69	15.98	15.52	15.21	14.98	14.80	14.66
5	16.26	13.27	12.06	11.39	10.97	10.67	10.46	10.29	10.16
6	13.75	10.92	9.78	9.15	8.75	8.47	8.26	8.10	7.98
7	12.25	9.55	8.45	7.85	7.46	7.19	6.99	6.84	6.72
8	11.26	8.65	7.59	7.01	6.63	6.37	6.18	6.03	5.91
9	10.56	8.02	6.99	6.42	6.06	5.80	5.61	5.47	5.35
10	10.04	7.56	6.55	5.99	5.64	5.39	5.20	5.06	4.94
11	9.65	7.21	6.22	5.67	5.32	5.07	4.89	4.74	4.63
12	9.33	6.93	5.95	5.41	5.06	4.82	4.64	4.50	4.39
13	9.07	6.70	5.74	5.21	4.86	4.62	4.44	4.30	4.19
14	8.86	6.51	5.56	5.04	4.69	4.46	4.28	4.14	4.03
15	8.68	6.36	5.42	4.89	4.56	4.32	4.14	4.00	3.89
16	8.53	6.23	5.29	4.77	4.44	4.20	4.03	3.89	3.78
17	8.40	6.11	5.18	4.67	4.34	4.10	3.93	3.79	3.68
18	8.29	6.01	5.09	4.58	4.25	4.01	3.84	3.71	3.60
19	8.18	5.93	5.01	4.50	4.17	3.94	3.77	3.63	3.52
20	8.10	5.85	4.94	4.43	4.10	3.87	3.70	3.56	3.46
21	8.02	5.78	4.87	4.37	4.04	3.81	3.64	3.51	3.40
22	7.95	5.72	4.82	4.31	3.99	3.76	3.59	3.45	3.35
23	7.88	5.66	4.76	4.26	3.94	3.71	3.54	3.41	3.30
24	7.82	5.61	4.72	4.22	3.90	3.67	3.50	3.36	3.26
25	7.77	5.57	4.68	4.18	3.85	3.63	3.46	3.32	3.22
26	7.72	5.53	4.64	4.14	3.82	3.59	3.42	3.29	3.18
27	7.68	5.49	4.60	4.11	3.78	3.56	3.39	3.26	3.15
28	7.64	5.45	4.57	4.07	3.75	3.53	3.36	3.23	3.12
29	7.60	5.42	4.54	4.04	3.73	3.50	3.33	3.20	3.09
30	7.56	5.39	4.51	4.02	3.70	3.47	3.30	3.17	3.07
40	7.31	5.18	4.31	3.83	3.51	3.29	3.12	2.99	2.89
60	7.08	4.98	4.13	3.65	3.34	3.12	2.95	2.82	2.72
120	6.85	4.79	3.95	3.48	3.17	2.96	2.79	2.66	2.56
∞	6.63	4.61	3.78	3.32	3.02	2.80	2.64	2.51	2.41

Critical Values of the F Distribution (continued)

$$f_{0.01}(v_1, v_2)$$

v_2	\multicolumn{10}{c}{v_1}									
	10	12	15	20	24	30	40	60	120	∞
1	6056	6106	6157	6209	6235	6261	6287	6313	6339	6366
2	99.40	99.42	99.43	99.45	99.46	99.47	99.47	99.48	99.49	99.50
3	27.23	27.05	26.87	26.69	26.60	26.50	26.41	26.32	26.22	26.13
4	14.55	14.37	14.20	14.02	13.93	13.84	13.75	13.65	13.56	13.46
5	10.05	9.89	9.72	9.55	9.47	9.38	9.29	9.20	9.11	9.02
6	7.87	7.72	7.56	7.40	7.31	7.23	7.14	7.06	6.97	6.88
7	6.62	6.47	6.31	6.16	6.07	5.99	5.91	5.82	5.74	5.65
8	5.81	5.67	5.52	5.36	5.28	5.20	5.12	5.03	4.95	4.86
9	5.26	5.11	4.96	4.81	4.73	4.65	4.57	4.48	4.40	4.31
10	4.85	4.71	4.56	4.41	4.33	4.25	4.17	4.08	4.00	3.91
11	4.54	4.40	4.25	4.10	4.02	3.94	3.86	3.78	3.69	3.60
12	4.30	4.16	4.01	3.86	3.78	3.70	3.62	3.54	3.45	3.36
13	4.10	3.96	3.82	3.66	3.59	3.51	3.43	3.34	3.25	3.17
14	3.94	3.80	3.66	3.51	3.43	3.35	3.27	3.18	3.09	3.00
15	3.80	3.67	3.52	3.37	3.29	3.21	3.13	3.05	2.96	2.87
16	3.69	3.55	3.41	3.26	3.18	3.10	3.02	2.93	2.84	2.75
17	3.59	3.46	3.31	3.16	3.08	3.00	2.92	2.83	2.75	2.65
18	3.51	3.37	3.23	3.08	3.00	2.92	2.84	2.75	2.66	2.57
19	3.43	3.30	3.15	3.00	2.92	2.84	2.76	2.67	2.58	2.49
20	3.37	3.23	3.09	2.94	2.86	2.78	2.69	2.61	2.52	2.42
21	3.31	3.17	3.03	2.88	2.80	2.72	2.64	2.55	2.46	2.36
22	3.26	3.12	2.98	2.83	2.75	2.67	2.58	2.50	2.40	2.31
23	3.21	3.07	2.93	2.78	2.70	2.62	2.54	2.45	2.35	2.26
24	3.17	3.03	2.89	2.74	2.66	2.58	2.49	2.40	2.31	2.21
25	3.13	2.99	2.85	2.70	2.62	2.54	2.45	2.36	2.27	2.17
26	3.09	2.96	2.81	2.66	2.58	2.50	2.42	2.33	2.23	2.13
27	3.06	2.93	2.78	2.63	2.55	2.47	2.38	2.29	2.20	2.10
28	3.03	2.90	2.75	2.60	2.52	2.44	2.35	2.26	2.17	2.06
29	3.00	2.87	2.73	2.57	2.49	2.41	2.33	2.23	2.14	2.03
30	2.98	2.84	2.70	2.55	2.47	2.39	2.30	2.21	2.11	2.01
40	2.80	2.66	2.52	2.37	2.29	2.20	2.11	2.02	1.92	1.80
60	2.63	2.50	2.35	2.20	2.12	2.03	1.94	1.84	1.73	1.60
120	2.47	2.34	2.19	2.03	1.95	1.86	1.76	1.66	1.53	1.38
∞	2.32	2.18	2.04	1.88	1.79	1.70	1.59	1.47	1.32	1.00

Critical Values of the Chi-square Distribution

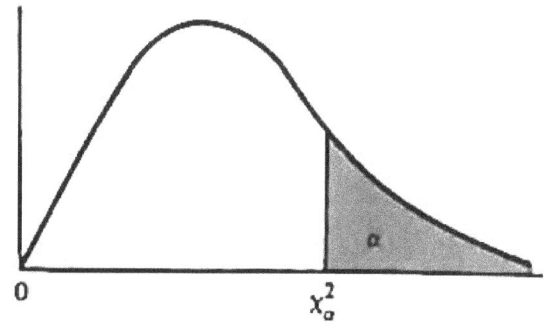

ν	\multicolumn{8}{c}{α}							
	0.995	0.99	0.975	0.95	0.05	0.025	0.01	0.005
1	0.0^4393	0.0^3157	0.0^3982	0.0^2393	3.841	5.024	6.635	7.879
2	0.0100	0.0201	0.0506	0.103	5.991	7.378	9.210	10.597
3	0.0717	0.115	0.216	0.352	7.815	9.348	11.345	12.838
4	0.207	0.297	0.484	0.711	9.488	11.143	13.277	14.860
5	0.412	0.554	0.831	1.145	11.070	12.832	15.086	16.750
6	0.676	0.872	1.237	1.635	12.592	14.449	16.812	18.548
7	0.989	1.239	1.690	2.167	14.067	16.013	18.475	20.278
8	1.344	1.646	2.180	2.733	15.507	17.535	20.090	21.955
9	1.735	2.088	2.700	3.325	16.919	19.023	21.666	23.589
10	2.156	2.558	3.247	3.940	18.307	20.483	23.209	25.188
11	2.603	3.053	3.816	4.575	19.675	21.920	24.725	26.757
12	3.074	3.571	4.404	5.226	21.026	23.337	26.217	28.300
13	3.565	4.107	5.009	5.892	22.362	24.736	27.688	29.819
14	4.075	4.660	5.629	6.571	23.685	26.119	29.141	31.319
15	4.601	5.229	6.262	7.261	24.996	27.488	30.578	32.801
16	5.142	5.812	6.908	7.962	26.296	28.845	32.000	34.267
17	5.697	6.408	7.564	8.672	27.587	30.191	33.409	35.718
18	6.265	7.015	8.231	9.390	28.869	31.526	34.805	37.156
19	6.844	7.633	8.907	10.117	30.144	32.852	36.191	38.582
20	7.434	8.260	9.591	10.851	31.410	34.170	37.566	39.997
21	8.034	8.897	10.283	11.591	32.671	35.479	38.932	41.401
22	8.643	9.542	10.982	12.338	33.924	36.781	40.289	42.796
23	9.260	10.196	11.689	13.091	35.172	38.076	41.638	44.181
24	9.886	10.856	12.401	13.848	36.415	39.364	42.980	45.558
25	10.520	11.524	13.120	14.611	37.652	40.646	44.314	46.928
26	11.160	12.198	13.844	15.379	38.885	41.923	45.642	48.290
27	11.808	12.879	14.573	16.151	40.113	43.194	46.963	49.645
28	12.461	13.565	15.308	16.928	41.337	44.461	48.278	50.993
29	13.121	14.256	16.047	17.708	42.557	45.722	49.588	52.336
30	13.787	14.953	16.791	18.493	43.773	46.979	50.892	53.672

Critical Values of the *t* Distribution

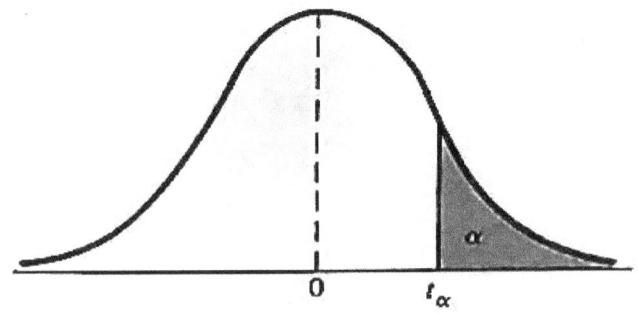

ν	α				
	0.10	0.05	0.025	0.01	0.005
1	3.078	6.314	12.706	31.821	63.657
2	1.886	2.920	4.303	6.965	9.925
3	1.638	2.353	3.182	4.541	5.841
4	1.533	2.132	2.776	3.747	4.604
5	1.476	2.015	2.571	3.365	4.032
6	1.440	1.943	2.447	3.143	3.707
7	1.415	1.895	2.365	2.998	3.499
8	1.397	1.860	2.306	2.896	3.355
9	1.383	1.833	2.262	2.821	3.250
10	1.372	1.812	2.228	2.764	3.169
11	1.363	1.796	2.201	2.718	3.106
12	1.356	1.782	2.179	2.681	3.055
13	1.350	1.771	2.160	2.650	3.012
14	1.345	1.761	2.145	2.624	2.977
15	1.341	1.753	2.131	2.602	2.947
16	1.337	1.746	2.120	2.583	2.921
17	1.333	1.740	2.110	2.567	2.898
18	1.330	1.734	2.101	2.552	2.878
19	1.328	1.729	2.093	2.539	2.861
20	1.325	1.725	2.086	2.528	2.845
21	1.323	1.721	2.080	2.518	2.831
22	1.321	1.717	2.074	2.508	2.819
23	1.319	1.714	2.069	2.500	2.807
24	1.318	1.711	2.064	2.492	2.797
25	1.316	1.708	2.060	2.485	2.787
26	1.315	1.706	2.056	2.479	2.779
27	1.314	1.703	2.052	2.473	2.771
28	1.313	1.701	2.048	2.467	2.763
29	1.311	1.699	2.045	2.462	2.756
inf.	1.282	1.645	1.960	2.326	2.576

* Table A.5 is taken from Table IV of R. A. Fisher: *Statistical Methods for Research Workers*, published by Oliver & Boyd Ltd., Edinburgh, by permission of the author and publishers.

Areas Under the Normal Curve

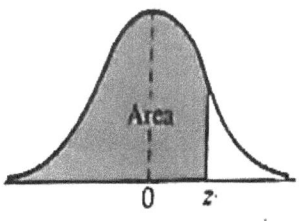

z	0.00	0.01	0.02	0.03	0.04	0.05	0.06	0.07	0.08	0.09
−3.4	0.0003	0.0003	0.0003	0.0003	0.0003	0.0003	0.0003	0.0003	0.0003	0.0002
−3.3	0.0005	0.0005	0.0005	0.0004	0.0004	0.0004	0.0004	0.0004	0.0004	0.0003
−3.2	0.0007	0.0007	0.0006	0.0006	0.0006	0.0006	0.0006	0.0005	0.0005	0.0005
−3.1	0.0010	0.0009	0.0009	0.0009	0.0008	0.0008	0.0008	0.0008	0.0007	0.0007
−3.0	0.0013	0.0013	0.0013	0.0012	0.0012	0.0011	0.0011	0.0011	0.0010	0.0010
−2.9	0.0019	0.0018	0.0017	0.0017	0.0016	0.0016	0.0015	0.0015	0.0014	0.0014
−2.8	0.0026	0.0025	0.0024	0.0023	0.0023	0.0022	0.0021	0.0021	0.0020	0.0019
−2.7	0.0035	0.0034	0.0033	0.0032	0.0031	0.0030	0.0029	0.0028	0.0027	0.0026
−2.6	0.0047	0.0045	0.0044	0.0043	0.0041	0.0040	0.0039	0.0038	0.0037	0.0036
−2.5	0.0062	0.0060	0.0059	0.0057	0.0055	0.0054	0.0052	0.0051	0.0049	0.0048
−2.4	0.0082	0.0080	0.0078	0.0075	0.0073	0.0071	0.0069	0.0068	0.0066	0.0064
−2.3	0.0107	0.0104	0.0102	0.0099	0.0096	0.0094	0.0091	0.0089	0.0087	0.0084
−2.2	0.0139	0.0136	0.0132	0.0129	0.0125	0.0122	0.0119	0.0116	0.0113	0.0110
−2.1	0.0179	0.0174	0.0170	0.0166	0.0162	0.0158	0.0154	0.0150	0.0146	0.0143
−2.0	0.0228	0.0222	0.0217	0.0212	0.0207	0.0202	0.0197	0.0192	0.0188	0.0183
−1.9	0.0287	0.0281	0.0274	0.0268	0.0262	0.0256	0.0250	0.0244	0.0239	0.0233
−1.8	0.0359	0.0352	0.0344	0.0336	0.0329	0.0322	0.0314	0.0307	0.0301	0.0294
−1.7	0.0446	0.0436	0.0427	0.0418	0.0409	0.0401	0.0392	0.0384	0.0375	0.0367
−1.6	0.0548	0.0537	0.0526	0.0516	0.0505	0.0495	0.0485	0.0475	0.0465	0.0455
−1.5	0.0668	0.0655	0.0643	0.0630	0.0618	0.0606	0.0594	0.0582	0.0571	0.0559
−1.4	0.0808	0.0793	0.0778	0.0764	0.0749	0.0735	0.0722	0.0708	0.0694	0.0681
−1.3	0.0968	0.0951	0.0934	0.0918	0.0901	0.0885	0.0869	0.0853	0.0838	0.0823
−1.2	0.1151	0.1131	0.1112	0.1093	0.1075	0.1056	0.1038	0.1020	0.1003	0.0985
−1.1	0.1357	0.1335	0.1314	0.1292	0.1271	0.1251	0.1230	0.1210	0.1190	0.1170
−1.0	0.1587	0.1562	0.1539	0.1515	0.1492	0.1469	0.1446	0.1423	0.1401	0.1379
−0.9	0.1841	0.1814	0.1788	0.1762	0.1736	0.1711	0.1685	0.1660	0.1635	0.1611
−0.8	0.2119	0.2090	0.2061	0.2033	0.2005	0.1977	0.1949	0.1922	0.1894	0.1867
−0.7	0.2420	0.2389	0.2358	0.2327	0.2296	0.2266	0.2236	0.2206	0.2177	0.2148
−0.6	0.2743	0.2709	0.2676	0.2643	0.2611	0.2578	0.2546	0.2514	0.2483	0.2451
−0.5	0.3085	0.3050	0.3015	0.2981	0.2946	0.2912	0.2877	0.2843	0.2810	0.2776
−0.4	0.3446	0.3409	0.3372	0.3336	0.3300	0.3264	0.3228	0.3192	0.3156	0.3121
−0.3	0.3821	0.3783	0.3745	0.3707	0.3669	0.3632	0.3594	0.3557	0.3520	0.3483
−0.2	0.4207	0.4168	0.4129	0.4090	0.4052	0.4013	0.3974	0.3936	0.3897	0.3859
−0.1	0.4602	0.4562	0.4522	0.4483	0.4443	0.4404	0.4364	0.4325	0.4286	0.4247
−0.0	0.5000	0.4960	0.4920	0.4880	0.4840	0.4801	0.4761	0.4721	0.4681	0.4641
0.0	0.5000	0.5040	0.5080	0.5120	0.5160	0.5199	0.5239	0.5279	0.5319	0.5359
0.1	0.5398	0.5438	0.5478	0.5517	0.5557	0.5596	0.5636	0.5675	0.5714	0.5753
0.2	0.5793	0.5832	0.5871	0.5910	0.5948	0.5987	0.6026	0.6064	0.6103	0.6141
0.3	0.6179	0.6217	0.6255	0.6293	0.6331	0.6368	0.6406	0.6443	0.6480	0.6517
0.4	0.6554	0.6591	0.6628	0.6664	0.6700	0.6736	0.6772	0.6808	0.6844	0.6879
0.5	0.6915	0.6950	0.6985	0.7019	0.7054	0.7088	0.7123	0.7157	0.7190	0.7224
0.6	0.7257	0.7291	0.7324	0.7357	0.7389	0.7422	0.7454	0.7486	0.7517	0.7549
0.7	0.7580	0.7611	0.7642	0.7673	0.7704	0.7734	0.7764	0.7794	0.7823	0.7852
0.8	0.7881	0.7910	0.7939	0.7967	0.7995	0.8023	0.8051	0.8078	0.8106	0.8133
0.9	0.8159	0.8186	0.8212	0.8238	0.8264	0.8289	0.8315	0.8340	0.8365	0.8389
1.0	0.8413	0.8438	0.8461	0.8485	0.8508	0.8531	0.8554	0.8577	0.8599	0.8621
1.1	0.8643	0.8665	0.8686	0.8708	0.8729	0.8749	0.8770	0.8790	0.8810	0.8830
1.2	0.8849	0.8869	0.8888	0.8907	0.8925	0.8944	0.8962	0.8980	0.8997	0.9015
1.3	0.9032	0.9049	0.9066	0.9082	0.9099	0.9115	0.9131	0.9147	0.9162	0.9177
1.4	0.9192	0.9207	0.9222	0.9236	0.9251	0.9265	0.9278	0.9292	0.9306	0.9319
1.5	0.9332	0.9345	0.9357	0.9370	0.9382	0.9394	0.9406	0.9418	0.9429	0.9441
1.6	0.9452	0.9463	0.9474	0.9484	0.9495	0.9505	0.9515	0.9525	0.9535	0.9545
1.7	0.9554	0.9564	0.9573	0.9582	0.9591	0.9599	0.9608	0.9616	0.9625	0.9633
1.8	0.9641	0.9649	0.9656	0.9664	0.9671	0.9678	0.9686	0.9693	0.9699	0.9706
1.9	0.9713	0.9719	0.9726	0.9732	0.9738	0.9744	0.9750	0.9756	0.9761	0.9767
2.0	0.9772	0.9778	0.9783	0.9788	0.9793	0.9798	0.9803	0.9808	0.9812	0.9817
2.1	0.9821	0.9826	0.9830	0.9834	0.9838	0.9842	0.9846	0.9850	0.9854	0.9857
2.2	0.9861	0.9864	0.9868	0.9871	0.9875	0.9878	0.9881	0.9884	0.9887	0.9890
2.3	0.9893	0.9896	0.9898	0.9901	0.9904	0.9906	0.9909	0.9911	0.9913	0.9916
2.4	0.9918	0.9920	0.9922	0.9925	0.9927	0.9929	0.9931	0.9932	0.9934	0.9936
2.5	0.9938	0.9940	0.9941	0.9943	0.9945	0.9946	0.9948	0.9949	0.9951	0.9952
2.6	0.9953	0.9955	0.9956	0.9957	0.9959	0.9960	0.9961	0.9962	0.9963	0.9964
2.7	0.9965	0.9966	0.9967	0.9968	0.9969	0.9970	0.9971	0.9972	0.9973	0.9974
2.8	0.9974	0.9975	0.9976	0.9977	0.9977	0.9978	0.9979	0.9979	0.9980	0.9981
2.9	0.9981	0.9982	0.9982	0.9983	0.9984	0.9984	0.9985	0.9985	0.9986	0.9986
3.0	0.9987	0.9987	0.9987	0.9988	0.9988	0.9989	0.9989	0.9989	0.9990	0.9990
3.1	0.9990	0.9991	0.9991	0.9991	0.9992	0.9992	0.9992	0.9992	0.9993	0.9993
3.2	0.9993	0.9993	0.9994	0.9994	0.9994	0.9994	0.9994	0.9995	0.9995	0.9995
3.3	0.9995	0.9995	0.9995	0.9996	0.9996	0.9996	0.9996	0.9996	0.9997	0.9997
3.4	0.9997	0.9997	0.9997	0.9997	0.9997	0.9997	0.9997	0.9997	0.9997	0.9998

Normal curve areas

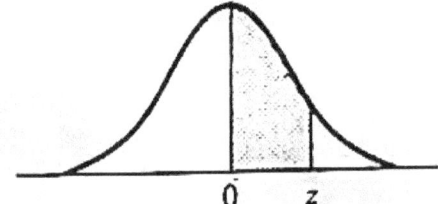

z	.00	.01	.02	.03	.04	.05	.06	.07	.08	.09
0.0	.0000	.0040	.0080	.0120	.0160	.0199	.0239	.0279	.0319	.0359
0.1	.0398	.0438	.0478	.0517	.0557	.0596	.0636	.0675	.0714	.0753
0.2	.0793	.0832	.0871	.0910	.0948	.0987	.1026	.1064	.1103	.1141
0.3	.1179	.1217	.1255	.1293	.1331	.1368	.1406	.1443	.1480	.1517
0.4	.1554	.1591	.1628	.1664	.1700	.1736	.1772	.1808	.1844	.1879
0.5	.1915	.1950	.1985	.2019	.2054	.2088	.2123	.2157	.2190	.2224
0.6	.2257	.2291	.2324	.2357	.2389	.2422	.2454	.2486	.2517	.2549
0.7	.2580	.2611	.2642	.2673	.2704	.2734	.2764	.2794	.2823	.2852
0.8	.2881	.2910	.2939	.2967	.2995	.3023	.3051	.3078	.3106	.3133
0.9	.3159	.3186	.3212	.3238	.3264	.3289	.3315	.3340	.3365	.3389
1.0	.3413	.3438	.3461	.3485	.3508	.3531	.3554	.3577	.3599	.3621
1.1	.3643	.3665	.3686	.3708	.3729	.3749	.3770	.3790	.3810	.3830
1.2	.3849	.3869	.3888	.3907	.3925	.3944	.3962	.3980	.3997	.4015
1.3	.4032	.4049	.4066	.4082	.4099	.4115	.4131	.4147	.4162	.4177
1.4	.4192	.4207	.4222	.4236	.4251	.4265	.4279	.4292	.4306	.4319
1.5	.4332	.4345	.4357	.4370	.4382	.4394	.4406	.4418	.4429	.4441
1.6	.4452	.4463	.4474	.4484	.4495	.4505	.4515	.4525	.4535	.4545
1.7	.4554	.4564	.4573	.4582	.4591	.4599	.4608	.4616	.4625	.4633
1.8	.4641	.4649	.4656	.4664	.4671	.4678	.4686	.4693	.4699	.4706
1.9	.4713	.4719	.4726	.4732	.4738	.4744	.4750	.4756	.4761	.4767
2.0	.4772	.4778	.4783	.4788	.4793	.4798	.4803	.4808	.4812	.4817
2.1	.4821	.4826	.4830	.4834	.4838	.4842	.4846	.4850	.4854	.4857
2.2	.4861	.4864	.4868	.4871	.4875	.4878	.4881	.4884	.4887	.4890
2.3	.4893	.4896	.4898	.4901	.4904	.4906	.4909	.4911	.4913	.4916
2.4	.4918	.4920	.4922	.4925	.4927	.4929	.4931	.4932	.4934	.4936
2.5	.4938	.4940	.4941	.4943	.4945	.4946	.4948	.4949	.4951	.4952
2.6	.4953	.4955	.4956	.4957	.4959	.4960	.4961	.4962	.4963	.4964
2.7	.4965	.4966	.4967	.4968	.4969	.4970	.4971	.4972	.4973	.4974
2.8	.4974	.4975	.4976	.4977	.4977	.4978	.4979	.4979	.4980	.4981
2.9	.4981	.4982	.4982	.4983	.4984	.4984	.4985	.4985	.4986	.4986
3.0	.4987	.4987	.4987	.4988	.4988	.4989	.4989	.4989	.4990	.4990

This table is abridged from Table I of *Statistical Tables and Formulas*, by A. Hald (New York: John Wiley & Sons, Inc., 1952). Reproduced by permission of A. Hald and the publishers, John Wiley & Sons, Inc.

$$f(x;N,n,k) = \frac{\binom{k}{x}\binom{N-k}{n-x}}{\binom{N}{n}}, \qquad F(x;N,n,k) = \sum_{r=0}^{x} \frac{\binom{k}{r}\binom{N-k}{n-r}}{\binom{N}{n}}$$

N	n	k	x	F(x)	f(x)	N	n	k	x	F(x)	f(x)
2	1	1	0	0.500000	0.500000	6	2	2	2	1.000000	0.066667
2	1	1	1	1.000000	0.500000	6	3	1	0	0.500000	0.500000
3	1	1	0	0.666667	0.666667	6	3	1	1	1.000000	0.500000
3	1	1	1	1.000000	0.333333	6	3	2	0	0.200000	0.200000
3	2	1	0	0.333333	0.333333	6	3	2	1	0.800000	0.600000
3	2	1	1	1.000000	0.666667	6	3	2	2	1.000000	0.200000
3	2	2	1	0.666667	0.666667	6	3	3	0	0.050000	0.050000
3	2	2	2	1.000000	0.333333	6	3	3	1	0.500000	0.450000
4	1	1	0	0.750000	0.750000	6	3	3	2	0.950000	0.450000
4	1	1	1	1.000000	0.250000	6	3	3	3	1.000000	0.050000
4	2	1	0	0.500000	0.500000	6	4	1	0	0.333333	0.333333
4	2	1	1	1.000000	0.500000	6	4	1	1	1.000000	0.666667
4	2	2	0	0.166667	0.166667	6	4	2	0	0.066667	0.066667
4	2	2	1	0.833333	0.666667	6	4	2	1	0.600000	0.533333
4	2	2	2	1.000000	0.166667	6	4	2	2	1.000000	0.400000
4	3	1	0	0.250000	0.250000	6	4	3	1	0.200000	0.200000
4	3	1	1	1.000000	0.750000	6	4	3	2	0.800000	0.600000
4	3	2	1	0.500000	0.500000	6	4	3	3	1.000000	0.200000
4	3	2	2	1.000000	0.500000	6	4	4	2	0.400000	0.400000
4	3	3	2	0.750000	0.750000	6	4	4	3	0.933333	0.533333
4	3	3	3	1.000000	0.250000	6	4	4	4	1.000000	0.066667
5	1	1	0	0.800000	0.800000	6	5	1	0	0.166667	0.166667
5	1	1	1	1.000000	0.200000	6	5	1	1	1.000000	0.833333
5	2	1	0	0.600000	0.600000	6	5	2	1	0.333333	0.333333
5	2	1	1	1.000000	0.400000	6	5	2	2	1.000000	0.666667
5	2	2	0	0.300000	0.300000	6	5	3	2	0.500000	0.500000
5	2	2	1	0.900000	0.600000	6	5	3	3	1.000000	0.500000
5	2	2	2	1.000000	0.100000	6	5	4	3	0.666667	0.666667
5	3	1	0	0.400000	0.400000	6	5	4	4	1.000000	0.333333
5	3	1	1	1.000000	0.600000	6	5	5	4	0.833333	0.833333
5	3	2	0	0.100000	0.100000	6	5	5	5	1.000000	0.166667
5	3	2	1	0.700000	0.600000	7	1	1	0	0.857143	0.857143
5	3	2	2	1.000000	0.300000	7	1	1	1	1.000000	0.142857
5	3	3	1	0.300000	0.300000	7	2	1	0	0.714286	0.714286
5	3	3	2	0.900000	0.600000	7	2	1	1	1.000000	0.285714
5	3	3	3	1.000000	0.100000	7	2	2	0	0.476190	0.476190
5	4	1	0	0.200000	0.200000	7	2	2	1	0.952381	0.476190
5	4	1	1	1.000000	0.800000	7	2	2	2	1.000000	0.047619
5	4	2	1	0.400000	0.400000	7	3	1	0	0.571429	0.571429
5	4	2	2	0.000000	0.600000	7	3	1	1	1.000000	0.428571
5	4	3	2	0.600000	0.600000	7	3	2	0	0.285714	0.285714
5	4	3	3	1.000000	0.400000	7	3	2	1	0.857143	0.571429
5	4	4	3	0.800000	0.800000	7	3	2	2	1.000000	0.142857
5	4	4	4	1.000000	0.200000	7	3	3	0	0.114286	0.114286
6	1	1	0	0.833333	0.833333	7	3	3	1	0.628571	0.514286
6	1	1	1	1.000000	0.166667	7	3	3	2	0.971428	0.342857
6	2	1	0	0.666667	0.666667	7	3	3	3	1.000000	0.028571
6	2	1	1	1.000000	0.333333	7	4	1	0	0.428571	0.428571
6	2	2	0	0.400000	0.400000	7	4	1	1	1.000000	0.571429
6	2	2	1	0.933333	0.533333	7	4	2	0	0.142857	0.142857

N	n	k	x	F(x)	f(x)	N	n	k	x	F(x)	f(x)
7	4	2	1	0.714286	0.571429	8	3	3	2	0.982143	0.267857
7	4	2	2	1.000000	0.285714	8	3	3	3	1.000000	0.017857
7	4	3	0	0.028571	0.028571	8	4	1	0	0.500000	0.500000
7	4	3	1	0.371429	0.342857	8	4	1	1	1.000000	0.500000
7	4	3	2	0.885714	0.514286	8	4	2	0	0.214286	0.214286
7	4	3	3	1.000000	0.114286	8	4	2	1.	0.785714	0.571429
7	4	4	1	0.114286	0.114286	8	4	2	2	1.000000	0.214286
7	4	4	2	0.628571	0.514286	8	4	3	0	0.071420	0.071429
7	4	4	3	0.971428	0.342857	8	4	3	1	0.500000	0.428571
7	4	4	4	1.000000	0.028571	8	4	3	2	0.928571	0.428571
											0.071429
7	5	1	0	0.285714	0.285714	8	4	3	3	1.000000	0.014286
7	5	1	1	1.000000	0.714286	8	4	4	0	0.014286	0.228571
7	5	2	0	0.047619	0.047619	8	4	4	1	0.242857	0.514286
7	5	2	1	0.523809	0.476190	8	4	4	2	0.757143	0.228571
7	5	2	2	1.000000	0.476190	8	4	4	3	0.985714	0.014286
7	5	3	1	0.142857	0.142857	8	4	4	4	1.000000	0.375000
7	5	3	2	0.714286	0.571429	8	5	1	0	0.375000	0.625000
7	5	3	3	1.000000	0.285714	8	5	1	1	1.000000	0.107143
7	5	4	2	0.285714	0.285714	8	5	2	0	0.107143	0.535714
7	5	4	3	0.857143	0.571429	8	5	2	1	0.642857	
											0.357143
7	5	4	4	1.000000	0.142857	8	5	2	2	1.000000	0.017857
7	5	5	3	0.476190	0.476190	8	5	3	0	0.017857	0.267857
7	5	5	4	0.952381	0.476190	8	5	3	1	0.285714	0.535714
7	5	5	5	1.000000	0.047619	8	5	3	2	0.821429	0.178571
7	6	1	0	0.142857	0.142857	8	5	3	3	1.000000	0.071429
7	6	1	1	1.000000	0.857143	8	5	4	1	0.071429	0.428571
7	6	2	1	0.285714	0.285714	8	5	4	2	0.500000	0.428571
7	6	2	2	1.000000	0.714286	8	5	4	3	0.928571	0.071429
7	6	3	2	0.428571	0.428571	8	5	4	4	1.000000	0.178571
7	6	3	3	1.000000	0.571429	8	5	5	2	0.178571	
											0.535714
7	6	4	3	0.571429	0.571429	8	5	5	3	0.714286	0.267857
7	6	4	4	1.000000	0.428571	8	5	5	4	0.982143	0.017857
7	6	5	4	0.714286	0.714286	8	5	5	5	1.000000	0.250000
7	6	5	5	1.000000	0.285714	8	6	1	0	0.250000	0.750000
7	6	6	5	0.857143	0.857143	8	6	1	1	1.000000	0.035714
7	6	6	6	1.000000	0.142857	8	6	2	0	0.035714	0.428571
8	1	1	0	0.875000	0.875000	8	6	2	1	0.464286	0.535714
8	1	1	1	1.000000	0.125000	8	6	2	2	1.000000	0.107143
8	2	1	0	0.750000	0.750000	8	6	3	1	0.107143	0.535714
8	2	1	1	1.000000	0.250000	8	6	3	2	0.642857	
											0.357143
8	2	2	0	0.535714	0.535714	8	6	3	3	1.000000	0.214286
8	2	2	1	0.964286	0.428571	8	6	4	2	0.214286	0.571429
8	2	2	2	1.000000	0.035714	8	6	4	3	0.785714	0.214286
8	3	1	0	0.625000	0.625000	8	6	4	4	1.000000	0.357143
8	3	1	1	1.000000	0.375000	8	6	5	3	0.357143	0.535714
8	3	2	0	0.357143	0.357143	8	6	5	4	0.892857	0.107143
8	3	2	1	0.892857	0.535714	8	6	5	5	1.000000	0.535714
8	3	2	2	1.000000	0.107143	8	6	6	4	0.535714	0.428571
8	3	3	0	0.178571	0.178571	8	6	6	5	0.964286	0.035714
8	3	3	1	0.714286	0.535714	8	6	6	6	1.000000	

N	n	k	x	F(x)	f(x)	N	n	k	x	F(x)	f(x)
8	7	1	0	0.125000	0.125000	9	5	3	1	0.404762	0.357143
8	7	1	1	1.000000	0.875000	9	5	3	2	0.880952	0.476190
8	7	2	1	0.250000	0.250000	9	5	3	3	1.000000	0.119048
8	7	2	2	1.000000	0.750000	9	5	4	0	0.007936	0.007936
8	7	3	2	0.375000	0.375000	9	5	4	1	0.166667	0.158730
8	7	3	3	1.000000	0.625000	9	5	4	2	0.642857	0.476190
8	7	4	3	0.500000	0.500000	9	5	4	3	0.960317	0.317460
8	7	4	4	1.000000	0.500000	9	5	4	4	1.000000	0.039683
8	7	5	4	0.625000	0.625000	9	5	5	1	0.039683	0.039683
8	7	5	5	1.000000	0.375000	9	5	5	2	0.357143	0.317460
8	7	6	5	0.750000	0.750000	9	5	5	3	0.833333	0.476190
8	7	6	6	1.000000	0.250000	9	5	5	4	0.992063	0.158730
8	7	7	6	0.875000	0.875000	9	5	5	5	1.000000	0.007936
8	7	7	7	1.000000	0.125000	9	6	1	0	0.333333	0.333333
9	1	1	0	0.888889	0.888889	9	6	1	1	1.000000	0.666667
9	1	1	1	1.000000	0.111111	9	6	2	0	0.083333	0.083333
9	2	1	0	0.777778	0.777778	9	6	2	1	0.583333	0.500000
9	2	1	1	1.000000	0.222222	9	6	2	2	1.000000	0.416667
9	2	2	0	0.583333	0.583333	9	6	3	0	0.011905	0.011905
9	2	2	1	0.972222	0.388889	9	6	3	1	0.226190	0.214286
9	2	2	2	1.000000	0.027778	9	6	3	2	0.761905	0.535714
9	3	1	0	0.666667	0.666667	9	6	3•	3	1.000000	0.238095
9	3	1	1	1.000000	0.333333	9	6	4	1	0.047619	0.047619
9	3	2	0	0.416667	0.416667	9	6	4	2	0.404762	0.357143
9	3	2	1	0.916667	0.500000	9	6	4	3	0.880952	0.476190
9	3	2	2	1.000000	0.083333	9	6	4	4	1.000000	0.119048
9	3	3	0	0.238095	0.238095	9	6	5	2	0.119048	0.119048
9	3	3	1	0.773809	0.535714	9	6	5	3	0.595238	0.476190
9	3	3	2	0.988095	0.214286	9	6	5	4	0.952381	0.357143
9	3	3	3	1.000000	0.011905	9	6	5	5	1.000000	0.047619
9	4	1	0	0.555556	0.555556	9	6	6	3	0.238095	0.238095
9	4	1	1	1.000000	0.444444	9	6	6	4	0.773809	0.535714
9	4	2	0	0.277778	0.277778	9	6	6	5	0.988095	0.214286
9	4	2	1	0.833333	0.555556	9	6	6	6	1.000000	0.011905
9	4	2	2	1.000000	0.166667	9	7	1	0	0.222222	0.222222
9	4	3	0	0.119048	0.119048	9	7	1	1	1.000000	0.777778
9	4	3	1	0.595238	0.476190	9	7	2	0	0.027778	0.027778
9	4	3	2	0.952381	0.357143	9	7	2	1	0.416667	0.388889
9	4	3	3	1.000000	0.047619	9	7	2	2	1.000000	0.583333
9	4	4	0	0.039683	0.039683	9	7	3	1	0.083333	0.083333
9	4	4	1	0.357143	0.317460	9	7	3	2	0.583333	0.500000
9	4	4	2	0.833333	0.476190	9	7	3	3	1.000000	0.416667
9	4	4	3	0.992063	0.158730	9	7	4	2	0.166667	0.166667
9	4	4	4	1.000000	0.007936	9	7	4	3	0.722222	0.555556
9	5	1	0	0.444444	0.444444	9	7	4	4	1.000000	0.277778
9	5	1	1	1.000000	0.555556	9	7	5	3	0.277778	0.277778
9	5	2	0	0.166667	0.166667	9	7	5	4	0.833333	0.555556
9	5	2	1	0.722222	0.555556	9	7	5	5	1.000000	0.166667
9	5	2	2	1.000000	0.277778	9	7	6	4	0.416667	0.416667
9	5	3	0	0.047619	0.047619	9	7	6	5	0.916667	0.500000

N	n	k	x	F(x)	f(x)	N	n	k	x	F(x)	f(x)
9	7	6	6	1.000000	0.083333	10	5	1	0	0.500000	0.500000
9	7	7	5	0.583333	0.583333	10	5	1	1	1.000000	0.500000
9	7	7	6	0.972222	0.388889	10	5	2	0	0.222222	0.222222
9	7	7	7	1.000000	0.027778	10	5	2	1	0.777778	0.555556
9	8	1	0	0.111111	0.111111	10	5	2	2	1.000000	0.222222
9	8	1	1	1.000000	0.888889	10	5	3	0	0.083333	0.083333
9	8	2	1	0.222222	0.222222	10	5	3	1	0.500000	0.416667
9	8	2	2	1.000000	0.777778	10	5	3	2	0.916667	0.416667
9	8	3	2	0.333333	0.333333	10	5	3	3	1.000000	0.083333
9	8	3	3	1.000000	0.666667	10	5	4	0	0.023810	0.023810
9	8	4	3	0.444444	0.444444	10	5	4	1	0.261905	0.238095
9	8	4	4	1.000000	0.555556	10	5	4	2	0.738095	0.476190
9	8	5	4	0.555556	0.555556	10	5	4	3	0.976190	0.238095
9	8	5	5	1.000000	0.444444	10	5	4	4	1.000000	0.023810
9	8	6	5	0.666667	0.666667	10	5	5	0	0.003968	0.003968
9	8	6	6	1.000000	0.333333	10	5	5	1	0.103175	0.099206
9	8	7	6	0.777778	0.777778	10	5	5	2	0.500000	0.396825
9	8	7	7	1.000000	0.222222	10	5	5	3	0.896825	0.396825
9	8	8	7	0.888889	0.888889	10	5	5	4	0.996032	0.099206
9	8	8	8	1.000000	0.111111	10	5	5	5	1.000000	0.003968
10	1	1	0	0.900000	0.900000	10	6	1	0	0.400000	0.400000
10	1	1	1	1.000000	0.100000	10	6	1	1	1.000000	0.600000
10	2	1	0	0.800000	0.800000	10	6	2	0	0.133333	0.133333
10	2	1	1	1.000000	0.200000	10	6	2	1	0.666667	0.533333
10	2	2	0	0.622222	0.622222	10	6	2	2	1.000000	0.333333
10	2	2	1	0.977778	0.355556	10	6	3	0	0.033333	0.033333
10	2	2	2	1.000000	0.022222	10	6	3	1	0.333333	0.300000
10	3	1	0	0.700000	0.700000	10	6	3	2	0.833333	0.500000
10	3	1	1	1.000000	0.300000	10	6	3	3	1.000000	0.166667
10	3	2	0	0.466667	0.466667	10	6	4	0	0.004762	0.004762
10	3	2	1	0.933333	0.466667	10	6	4	1	0.119048	0.114286
10	3	2	2	1.000000	0.066667	10	6	4	2	0.547619	0.428571
10	3	3	0	0.291667	0.291667	10	6	4	3	0.928571	0.380952
10	3	3	1	0.816667	0.525000	10	6	4	4	1.000000	0.071429
10	3	3	2	0.991667	0.175000	10	6	5	1	0.023810	0.023810
10	3	3	3	1.000000	0.008333	10	6	5	2	0.261905	0.238095
10	4	1	0	0.600000	0.600000	10	6	5	3	0.738095	0.476190
10	4	1	1	1.000000	0.400000	10	6	5	4	0.976190	0.238095
10	4	2	0	0.333333	0.333333	10	6	5	5	1.000000	0.023810
10	4	2	1	0.866667	0.533333	10	6	6	2	0.071429	0.071429
10	4	2	2	1.000000	0.133333	10	6	6	3	0.452381	0.380952
10	4	3	0	0.166667	0.166667	10	6	6	4	0.880952	0.428571
10	4	3	1	0.666667	0.500000	10	6	6	5	0.995238	0.114286
10	4	3	2	0.966667	0.300000	10	6	6	6	1.000000	0.004762
10	4	3	3	1.000000	0.033333	10	7	1	0	0.300000	0.300000
10	4	4	0	0.071429	0.071429	10	7	1	1	1.000000	0.700000
10	4	4	1	0.452381	0.380952	10	7	2	0	0.066667	0.066667
10	4	4	2	0.880952	0.428571	10	7	2	1	0.533333	0.466667
10	4	4	3	0.995238	0.114286	10	7	2	2	1.000000	0.466667
10	4	4	4	1.000000	0.004762	10	7	3	0	0.008333	0.008333

393

	00–04	05–09	10–14	15–19	20–24	25–29	30–34	35–39	40–44	45–49
00	54463	22662	65905	70639	79365	67382	29085	69831	47058	08186
01	15389	85205	18850	39226	42249	90669	96325	23248	60933	26927
02	85941	40756	82414	02015	13858	78030	16269	65978	01385	15345
03	61149	69440	11286	88218	58925	03638	52862	62733	33451	77455
04	05219	81619	10651	67079	92511	59888	84502	72095	83463	75577
05	41417	98326	87719	92294	46614	50948	64886	20002	97365	30976
06	28357	94070	20652	35774	16249	75019	21145	05217	47286	76305
07	17783	00015	10806	83091	91530	36466	39981	62481	49177	75779
08	40950	84820	29881	85966	62800	70326	84740	62660	77379	90279
09	82995	64157	66164	41180	10089	41757	78258	96488	88629	37231
10	96754	17676	55659	44105	47361	34833	86679	23930	53249	27083
11	34357	88040	53364	71726	45690	66334	60332	22554	90600	71113
12	06318	37403	49927	57715	50423	67372	63116	48888	21505	80182
13	62111	52820	07243	79931	89292	84767	85693	73947	22278	11551
14	47534	09243	67879	00544	23410	12740	02540	54440	32949	13491
15	98614	75993	84460	62846	59844	14922	48730	73443	48167	34770
16	24856	03648	44898	09351	98795	18644	39765	71058	90368	44104
17	96887	12479	80621	66223	86085	78285	02432	53342	42846	94771
18	90801	21472	42815	77408	37390	76766	52615	32141	30268	18106
19	55165	77312	83666	36028	28420	70219	81369	41943	47366	41067
20	75884	12952	84318	95108	72305	64620	91318	89872	45375	85436
21	16777	37116	58550	42958	21460	43910	01175	87894	81378	10620
22	46230	43877	80207	88877	89380	32992	91380	03164	98656	59337
23	42902	66892	46134	01432	94710	23474	20423	60137	60609	13119
24	81007	00333	39693	28039	10154	95425	39220	19774	31782	49037
25	68089	01122	51111	72373	06902	74373	96199	97017	41273	21546
26	20411	67081	89950	16944	93054	87687	96693	87236	77054	33848
27	58212	13160	06468	15718	82627	76999	05999	58680	96739	63700
28	70577	42866	24969	61210	76046	67699	42054	12696	93758	03283
29	94522	74358	71659	62038	79643	79169	44741	05437	39038	13163
30	42626	86819	85651	88678	17401	03252	99547	32404	17918	62880
31	16051	33763	57194	16752	54450	19031	58580	47629	54132	60631
32	08244	27647	33851	44705	94211	46716	11738	55784	95374	72655
33	59497	04392	09419	89964	51211	04894	72882	17805	21896	83864
34	97155	13428	40293	09985	58434	01412	69124	82171	59058	82859
35	98409	66162	95763	47420	20792	61527	20441	39435	11859	41567
36	45476	84882	65109	96597	25930	66790	65706	61203	53634	22557
37	89300	69700	50741	30329	11658	23166	05400	66669	48708	03887
38	50051	95137	91631	66315	91428	12275	24816	68091	71710	33258
39	31753	85178	31310	89642	98364	02306	24617	09609	83942	22716
40	79152	53829	77250	20190	56535	18760	69942	77448	33278	48805
41	44560	38750	83635	56540	64900	42912	13953	79149	18710	68618
42	68328	83378	63369	71381	39564	05615	42451	64559	97501	65747
43	46939	38689	58625	08342	30459	85863	20781	09284	26333	91777
44	83544	86141	15707	96256	23068	13782	08467	89469	93842	55349
45	91621	00881	04900	54224	46177	55309	17852	27491	89415	23466
46	91896	67126	04151	03795	59077	11848	12630	98375	52068	60142
47	55751	62515	21108	80830	02263	29303	37204	96926	30506	09808
48	85156	87689	95493	88842	00664	55017	55539	17771	69448	87530
49	07521	56898	12236	60277	39102	62315	12239	07105	11844	01117